THE WAR ON SCIENCE

Also by Shawn Otto

Nonfiction

Fool Me Twice:
Fighting the Assault on Science in America

Fiction

Sins of Our Fathers

PRAISE FOR *THE WAR ON SCIENCE*

"Every so often a book comes along that changes the way you view the world.
The War on Science is one of those rare books. If you care about attacks on
climate science and the rise of authoritarianism, if you care about biased media
coverage or shake-your-head political tomfoolery, this book is for you. . . .
A resource for people who ponder what it means to live in a free society and the
important role scientists play in conveying good science to the larger public."
—*Guardian*

"Otto makes a case that can't be refuted. Science is important to all of us,
especially in a democracy. He backs it up with peer-reviewed studies, carefully
researched numbers, and his own extensive experience. He uses the process
of science to prove that we need science in order to remain free. Here's hoping
voters everywhere take him seriously—soon."
—**Bill Nye**

"We're seeing right now a titanic battle between the power of science and the
power of money—and money is winning. This book explains why, and offers
some pointers that might get us back on the right track."
—**Bill McKibben**

"Science is not a body of facts, but rather a structured approach to uncovering
the fundamental laws that govern the natural world. As *The War on Science*
shows, policymakers who choose to ignore those fundamental laws imperil
us all, for the laws of nature will always trump the laws of man."
—**Marcia McNutt, president of the US National Academy of Sciences**

"This insightful, heavily researched book pulls back the curtain to show
exactly where and how the rise of authoritarianism is being accomplished,
via academic, fundamentalist, and public relations attacks on scientists
and the ideas of science that underlie modern democracy. *The War on
Science* is must reading for anyone wanting to understand what's
really going on in today's politics."
—**Michael E. Mann**

"Evidence from science is one of the world's great equalizers, because
it forms an objective basis for public policy. This book illustrates how central
that notion is to the forming of modern democracy, and how current attacks
on science endanger our freedom. Policymakers and voters everywhere
would do well to read *The War on Science*."
—**Walter Mondale, former US Vice President,
senator and ambassador**

"In the struggle of people to be free, there has been one common denominator
on which democracy, like Sherlock Holmes, depends—science, and the
evidence it provides, as a guide to truth, fairness, and justice. This insightful
book explores how science became a necessary prerequisite for democracy,
why it is under attack today, and what we can do to defend truth and freedom."
—**Maria Konnikova, author of *Mastermind:
How to Think Like Sherlock Holmes***

"Before you vote in the next election, read Shawn Otto's *The War on Science*."
—**Ben Bova, award-winning author of the Grand Tour series
and former editorial director of *Omni***

"*The War on Science* dissects the factors creating a perfect storm
of anti-intellectualism, persuading millions to actively vote against their own
interests. This is not a book that will convert Limbaugh dancers. But it just
might help you to draw that smart engineer uncle back toward the light.
It might even encourage such fellows to join a newborn movement,
reviving a science-loving version of conservatism out of the ashes."
—**David Brin, scientist and award-winning author
of *The Transparent Society: Will Technology Force Us
to Choose between Privacy and Freedom?***

"One of the most important books published in the last decade."
—**Don Shelby, Peabody Award–winning news anchor**

THE WAR ON SCIENCE

Who's Waging It
Why It Matters
What We Can Do about It

SHAWN OTTO

milkweed
editions

Published 2016 by Milkweed Editions
Printed in the United States of America
Cover design by Mary Austin Speaker
Author photo by Erika Ludwig
17 18 19 20 5 4 3
First Edition

Milkweed Editions, an independent nonprofit publisher, gratefully acknowledges sustaining support from the Jerome Foundation; the Lindquist & Vennum Foundation; the McKnight Foundation; the National Endowment for the Arts; the Target Foundation; and other generous contributions from foundations, corporations, and individuals. Also, this activity is made possible by the voters of Minnesota through a Minnesota State Arts Board Operating Support grant, thanks to a legislative appropriation from the arts and cultural heritage fund, and a grant from the Wells Fargo Foundation Minnesota. For a full listing of Milkweed Editions supporters, please visit www.milkweed.org.

Library of Congress Cataloging-in-Publication Data

Names: Otto, Shawn Lawrence.
Title: The war on science : who's waging it, why it matters, what we can do
    about it / Shawn Otto.
Description: Minneapolis, Minnesota : Milkweed Editions, [2016] | Includes
    bibliographical references and index.
Identifiers: LCCN 2016002797 (print) | LCCN 2016009938 (ebook) | ISBN
    9781571313539 (pbk. : alk. paper) | ISBN 9781571319524 (e-book)
Subjects: LCSH: Science--Political aspects. | Science--Political
    aspects--History. | Science--Social aspects. | Science--Social
    aspects--History.
Classification: LCC Q175.5 .O88 2016 (print) | LCC Q175.5 (ebook) | DDC
    303.48/3--dc23
LC record available at http://lccn.loc.gov/2016002797

To the underdogs

Contents

PART IV

WINNING THE WAR

FOREWORD

"The mind once enlightened cannot again become dark."
—Thomas Paine

Thomas Paine's remark above is most certainly true for individuals. This is, after all, the central purpose of education: to lift the veil of darkness for young people, with the hope that once lifted, the enlightenment that results will help build wiser adults, with a brighter communal future.

But what may be true for individuals is not necessarily true for societies. The scientific wisdom of the Greeks was largely abandoned in the Middle Ages. The Arab countries, once the heart of mathematics and scholarship, did not partake in many of the benefits of the European Enlightenment, in part because of the emerging influence of fundamentalism in the tenth and eleventh centuries.

The United States itself was founded by people, like Thomas Paine, Thomas Jefferson, and Benjamin Franklin, for whom science and enlightenment were paramount. As a nation the United States has benefited more than any other because it became a center for technological innovation and progress.

When I was a child, born in the 1950s, emerging as a young man in the 1960s, the allure of science was unmistakable, and it seemed impossible to imagine a time when ideology, myth, superstition, and ignorance might gain the upper hand in determining the future in the developing world.

Yet, as my friend Shawn Otto describes in meticulous detail in this book, there are forces at work coming from many directions that serve to undermine the simple proposition that public policy should be based on rational reflections on sound empirical evidence. From efforts to obstruct the teaching of evolution in schools, the removal of fluoridation in water, restrictions on vaccination, restrictions on other scientific research and related attacks on the efficacy of the scientific enterprise itself, and most recently wholesale and broadly organized efforts to deny the science associated with human-induced climate change, science, as a basis for helping address real-world challenges, has been under attack. Not just from local school boards in rural America, but from a significant fraction of presidential candidates in 2016.

Shawn is not a professional scientist, but he is the epitome of a responsible citizen scientist. We met and came together with several other odd bedfellows in 2007 to form an organization called ScienceDebate 2008 because we felt that the key issues that would really determine the success of the next presidential administration were being ignored in public debate, and we thought it would be worthwhile trying to create an opportunity for the presidential candidates to discuss these issues in a national public forum. The idea caught on among a broad and diverse segment of the population and we came close in our goal. While a televised debate didn't occur, we did get the candidates from both major parties in the United States to answer a series of questions about science policy that were later posted online, and made nearly a billion media impressions during the course of the campaign.

We have continued our efforts with each presidential campaign, and we remain hopeful, in spite of the obstacles. Shawn has continued to spearhead the program, with unflagging energy and enthusiasm. But he has done more than this. The effort to restore science as an important basis of public policy, and to ensure that the benefits of scientific and technological research continue to promote the health and welfare of the public, have focused his writing.

This book details the long and checkered set of battles that have been won and lost over the years to try and ensure the Enlightenment is not lost in a new dark age. Meticulously documenting case by case, and area by area, this book provides a valuable resource to put our present conundrum in proper perspective.

But in a characteristic manner Shawn does not merely catalogue here a series of problems and challenges. He also outlines a broad set of specific proposals to address these problems. One might not agree with all of his proposals, but the process of exploring them can only help to enlighten readers as to what they might do, and what they might demand their legislatures do, to ensure a reality-based system of government that addresses the real challenges of the twenty-first century. And as Paine emphasized, such enlightenment cannot be darkened. There is thus reason for hope.

Lawrence M. Krauss, 2016

FOREWORD

"The mind once enlightened cannot again become dark."

—Thomas Paine

Thomas Paine's remark above is most certainly true for individuals. This is, after all, the central purpose of education: to lift the veil of darkness for young people, with the hope that once lifted, the enlightenment that results will help build wiser adults, with a brighter communal future.

But what may be true for individuals is not necessarily true for societies. The scientific wisdom of the Greeks was largely abandoned in the Middle Ages. The Arab countries, once the heart of mathematics and scholarship, did not partake in many of the benefits of the European Enlightenment, in part because of the emerging influence of fundamentalism in the tenth and eleventh centuries.

The United States itself was founded by people, like Thomas Paine, Thomas Jefferson, and Benjamin Franklin, for whom science and enlightenment were paramount. As a nation the United States has benefited more than any other because it became a center for technological innovation and progress.

When I was a child, born in the 1950s, emerging as a young man in the 1960s, the allure of science was unmistakable, and it seemed impossible to imagine a time when ideology, myth, superstition, and ignorance might gain the upper hand in determining the future in the developing world.

Yet, as my friend Shawn Otto describes in meticulous detail in this book, there are forces at work coming from many directions that serve to undermine the simple proposition that public policy should be based on rational reflections on sound empirical evidence. From efforts to obstruct the teaching of evolution in schools, the removal of fluoridation in water, restrictions on vaccination, restrictions on other scientific research and related attacks on the efficacy of the scientific enterprise itself, and most recently wholesale and broadly organized efforts to deny the science associated with human-induced climate change, science, as a basis for helping address real-world challenges, has been under attack. Not just from local school boards in rural America, but from a significant fraction of presidential candidates in 2016.

Shawn is not a professional scientist, but he is the epitome of a responsible citizen scientist. We met and came together with several other odd bedfellows in 2007 to form an organization called ScienceDebate 2008 because we felt that the key issues that would really determine the success of the next presidential administration were being ignored in public debate, and we thought it would be worthwhile trying to create an opportunity for the presidential candidates to discuss these issues in a national public forum. The idea caught on among a broad and diverse segment of the population and we came close in our goal. While a televised debate didn't occur, we did get the candidates from both major parties in the United States to answer a series of questions about science policy that were later posted online, and made nearly a billion media impressions during the course of the campaign.

We have continued our efforts with each presidential campaign, and we remain hopeful, in spite of the obstacles. Shawn has continued to spearhead the program, with unflagging energy and enthusiasm. But he has done more than this. The effort to restore science as an important basis of public policy, and to ensure that the benefits of scientific and technological research continue to promote the health and welfare of the public, have focused his writing.

This book details the long and checkered set of battles that have been won and lost over the years to try and ensure the Enlightenment is not lost in a new dark age. Meticulously documenting case by case, and area by area, this book provides a valuable resource to put our present conundrum in proper perspective.

But in a characteristic manner Shawn does not merely catalogue here a series of problems and challenges. He also outlines a broad set of specific proposals to address these problems. One might not agree with all of his proposals, but the process of exploring them can only help to enlighten readers as to what they might do, and what they might demand their legislatures do, to ensure a reality-based system of government that addresses the real challenges of the twenty-first century. And as Paine emphasized, such enlightenment cannot be darkened. There is thus reason for hope.

Lawrence M. Krauss, 2016

THE WAR ON SCIENCE

This war is not an ordinary war. It is not a conflict for markets or territories. It is a desperate struggle for the possession of the souls of men.
—Harold Ickes, May 18, 1941

PART I

Democracy's Science Problem

Chapter 1

THE WAR ON SCIENCE

> Wherever the people are well informed they can be trusted with their
> own government; that whenever things get so far wrong as to attract
> their notice, they may be relied on to set them to rights.
>
> —Thomas Jefferson, January 8, 1789

Houston, We Have a Problem

Thomas Jefferson's trust in the well-informed voter lies at the heart of the modern democracy that has, over the course of two centuries, come to guide the world. Much like the "invisible hand" that guides Adam Smith's economic marketplace, so too does the invisible hand of the people's will guide the democratic process. Faith in this idea is so central to democracy that George Washington emphasized it in the nation's first inaugural address. "No people can be bound to acknowledge and adore the invisible hand, which conducts the Affairs of men more than the People of the United States," he told a joint session of Congress gathered in Federal Hall, which stood kitty-corner to today's New York Stock Exchange.

But lately the invisible hand seems confused and indecisive. Democratic governments the world over are increasingly paralyzed, unable to act on many key issues that threaten the economic and environmental stability of their countries and the world. They often enact policies that seem to run against their own interests, quashing or directly contradicting well-known evidence. Ideology and rhetoric guide policy discussions, often with a brazenly willful denial of facts. Even elected officials seem willing to defy laws, often paying negligible prices. And the civil society we once knew now seems divided and angry, defiantly embracing unreason. Everyone, we are told, has his or her own experience of reality, and history is written by the victors. What could be happening?

At the same time, science and technology have come to affect every aspect of life on the planet. There is a phase change going on in the scientific revolution: a shifting from one state to another, as from a solid to a liquid. There is a sudden, *quantitative* expansion of the number of scientists and engineers around the globe, coupled with a sudden *qualitative* expansion of their ability to collaborate with each other over the Internet.

These two changes are dramatically speeding up the process of discovery and the convergence of knowledge across once-separate fields, a process Harvard entomologist Edward O. Wilson named consilience. We now have fields where economics merges with environmental science, electrical engineering with neuroscience and physics, computer science with biology and genetics, astronomy with biology, and many more. This consilience is shedding new light on long-held assumptions about the world we live in and the nature of life.

Over the course of the next forty years, science is poised to create more knowledge than humans have created in all of recorded history, completely redefining our concepts about—and power over—life and the physical and mental worlds as we assume editing control over the genetic code and mastery in our understanding of the brain. One only has to recall the political battles fought over past scientific advances to see that we are in for a rocky ride. How that rush of new knowledge will impact life, how it will be applied through technology and law, and whether our societies and governments will be able to withstand the immense social and economic upheavals it will bring depends upon whether we can update our political process to accommodate it. Can we manage the next phase of the scientific revolution to our advantage, or will we become its unwilling victims?

If that were not enough, the explosion of information technology is creating a power struggle between individual privacy and the public good, and between the organizations—businesses, criminal enterprises, terrorist groups, and governments—who seek to use this new technology for influence and control. Sensing technology and robotics are threatening to replace millions of truck drivers and taxi drivers over the next decade, and to mechanize warfare with tiny autonomous robots that carry enough charge and intelligence to hunt and kill humans. These advancements have prompted many of the world's leading scientists and engineers to warn that we must get ahead of artificial intelligence before it gets ahead of us.

As we are being overwhelmed by new scientific and technological developments, we also are facing a host of legacy challenges caused by commercialization of the incomplete scientific knowledge of the past. Thanks to early science, humans have prospered, but at a cost: significant climate disruption, unprecedented environmental degradation, massive extinction of other species, vast economic and power inequities, and a world armed to the teeth with the products of a military-industrial complex, including weapons that could destroy nearly all life on the planet.

Without a better way of incorporating science into our policymaking, democracy may ultimately fail its promise. We now have a population that we cannot support without destroying our environment—and the developing world is advancing by using the same model of unsustainable development. We are 100 percent dependent on science and technology to find a solution.

The Whipsaw of Science

Between these two areas—the wild future that is rapidly emerging and the unsustainable present whose repercussions we can no longer ignore—science and technology are poised to whipsaw us in the coming decades like never before. This has the potential to produce even more intense social upheaval and political paralysis at the very time we can least afford them.

Perhaps the most troubling aspect of the problem is the dearth of conversation about the issues in the policymaking process. Imagine for a moment the potential science-themed questions one could ask a candidate for president, for example, or Parliament, or Congress, in a debate, forum, or news interview. There are multitudes of them, each with profound relevance to both today's problems and those of the near future. Because of this, they are political, but they are rarely asked or answered in the political process. A small sampling could include:

What is your vision for maintaining a competitive edge as other countries work toward becoming global forces in science and technology? Will you support tripling our investment in mental-health research? Will you support using science to study the underlying causes of gun violence? What are your thoughts on balancing energy and the environment? How should we manage biosecurity in an age of rapid international travel while preserving civil liberties? What should we do about the world's aging nuclear weapons? How will you tackle climate disruption? Do you support embryonic stem-cell research? What steps will you take to stop the collapse of pollinator colonies and promote pollinator health? What can we do to stimulate and incentivize the transition to a low-carbon economy? How should we handle immigration of highly skilled workers? In an era of intense droughts, what steps will you take to better manage our freshwater resources? What should we do to prevent ocean fisheries collapses? Will you support federal funding to make public broadband Internet universally available? Is Internet access a universal human right? How can we better protect the health of the world's oceans? How can we improve science education? What steps can we take to better incorporate science information into our policymaking process? What will you do to slow the sixth mass extinction? Should we require children to be vaccinated against human papillomavirus, the leading cause of cervical cancer? Should only evolution be taught in science classes, or should intelligent design also be taught? When is it acceptable for a president or prime minister to implement policies that are contradicted by science? Should pharmaceutical companies be allowed to advertise on public airwaves? What will you do to incentivize the production of generic pharmaceuticals to prevent shortages and extreme price increases? Should foods made from genetically modified crops be labeled? Should we regulate the use of nanoparticles in the environment? Do you support federal renewable energy tax credits? What would you do to end the war on drugs and transition to treating drugs as a public-health problem? Will you support increased funding for curiosity-driven basic research? What steps would you take to repair the postdoctoral employment pipeline so that highly trained workers can get jobs in their fields? Will you support federal funding to study science denial and the threat it poses to democracy? Do you support banning the use of antibiotics in animal feed? What other steps should we take to stop the rise of antibiotic-resistant bacteria? Should pharmacists be allowed to deny prescriptions on the basis of their religion? Should public officials be allowed to deny services on the basis of their religion? Should the federal government regulate hydraulic fracturing? Should parents be required to vaccinate their children? Under what circumstances should there be an exemption? Do vaccines cause autism? Will you support adoption of new, cleaner nuclear reactors for power generation? Do you support water fluoridation? Will you prioritize an Apollo Project for clean energy innovation to stimulate the economy? Should we initiate a manned mission to Mars? What steps would you take to transition to a sustainable or circular economy? Do you oppose or support plans to mine copper and other nonferrous minerals in or near water-rich areas? How should we balance privacy with freedom and security on the Internet? Do you support reinstating the Fairness Doctrine in broadcast journalism? What steps would you take to control the global population? Do you support or oppose efforts to prosecute energy companies for funding denial of climate science? How can we stop anti-science disinformation campaigns from stalling public policy while protecting freedom of expression? Would you use foreign and economic policy to demand trading partners adopt uniform environmental standards? What will you do about anticipated economic disruptions posed by driverless vehicles and other robotic outsourcing of jobs? What is your position on deploying autonomous, artificially intelligent killer robots in the battlefield? Will you support restoring funding for the US Congress's nonpartisan science advisory body, the Office of Technology Assessment? Should the morning-after pill be available off the shelf in pharmacies?

The length of the above sample is part of the point—the list is of course much longer—and it is growing as science advances. Yet almost none of these issues are discussed on the campaign trail. All of them evoke strong reactions, and, in each of these cases, policy has become stuck because of our broken way of incorporating evidence from science into the policymaking and political processes. Something's got to change.

The Battle for the Future

Science and engineering are providing us with increasingly clear pictures of how to solve many of our challenges, but policymakers are increasingly unwilling to pursue the remedies that scientific evidence suggests. Instead, they take one of two routes: deny the science, or pretend the problems don't exist. Vast areas of scientific knowledge and the people who work in them are under daily attack in a fierce worldwide war on science. Scientific advances in public health, biology and the environment are being resisted or rolled back. Political and religious institutions are pushing back against science and reason in a way that is threatening social and economic stability.

This pullback is affecting leading and emerging economies alike. The name of the radical pan-national Islamist group *Boko Haram* roughly translates as "Western education is forbidden." The Islamic drive for *al-asala*, or authenticity, leads some fundamentalist Muslims to reject Western science in favor of Quranic instructions, says Islamic scholar Bassam Tibi. But radical Islam is not alone in this rejection. The vanguard of the retreat is in the Western democracies, where Christian fundamentalists; postmodernist academics, teachers, and journalists; liberal new age purists; and industry front groups all attack science for their own reasons.

Politically, the war on science is coming from both left and right. But the antiscience of those on the right—a coalition of fundamentalist churches and corporations largely in the resource extraction, petrochemical and agrochemical industries—has far more dangerous public-policy implications because it's about forestalling policy based on evidence to protect destructive business models. As well, the right generally has far more money with which to spread disinformation and attack science on a host of issues.

Those on the political left often unwittingly abet the right's antiscience efforts by arguing that truth is relative, harboring suspicions about hidden dangers to health and the environment that are not supported by evidence, and selectively rejecting science that doesn't affirm their health-food and back-to-Eden value system. While they are right that there are serious environmental and health threats afoot from poorly regulated industries, they undermine their credibility when they extend these suspicions to scientifically unsupported ideas like vaccines cause autism, cell phones cause brain cancer, or genetically modified crops are unsafe to eat. By seeking arguments that support preexisting beliefs (however laudable the concerns that motivate them) instead of looking to scientific evidence, these progressives give up the very critical-thinking and argumentation tools liberals once used to defend modern society against its authoritarian attackers.

The split is happening not just in science, but across the engineering world as well. Unlike a generation ago, when a radio could be made sitting at one's kitchen table, a smartphone cannot be made in the same way. This lack of plain accessibility is making complex science and technology less a matter of knowhow and more magical. Smartphones and flying brooms are both made by people cloistered away wearing long robes and uttering strange incantations. This inaccessibility makes science and technology more into a matter of belief than know-how, making people more vulnerable to disinformation campaigns. It is also increasingly difficult for the non-science-literate to accurately perceive the threats, challenges, and opportunities of this complex new world so dominated by inaccessible and magical science and technology (something that, for the reader, will hopefully change by the end of this book).

This is having effects across society from education to law enforcement. Consider the case of Xiaoxing Xi, the chair of Temple University's physics department. Xi was arrested by the FBI in 2015 for leaking top-secret technology information to China. The FBI had intercepted schematics of a sophisticated device known as a pocket heater, used in classified superconductor research, that Xi had sent to scientists in China.

The only problem was that Xi had done nothing of the kind. Independent experts in superconductor research looked at Xi's schematics and one after another told lawyers that the design wasn't for a pocket heater at all. Xi was simply doing what scientists and engineers the world over do every day: he was collaborating with colleagues over the Internet. It was an embarrassing acknowledgment that prosecutors and FBI agents did not understand the science involved in their case—and did not make enough of an attempt to learn it—before bringing charges that jeopardized Xi's career and left the impression that he was a spy for China. "I don't expect them to understand everything I do," Xi told the *New York Times* after the Justice Department dropped all charges. "But the fact that they don't consult with experts and then charge me? Put my family through all this? Damage my reputation? They shouldn't do this. This is not a joke. This is not a game."

Something similar happened in the fall of 2015. Ahmed Mohamed, a ninth grader from Irving, Texas, brought a clock to school that he had designed and built himself using some integrated circuit chips and a circuit board. He mounted it in an aluminum project case with a big LCD readout and took it in to show his new engineering teacher. He had been part of his middle school robotics team and now, a few days into high school, he was anxious to impress his new teacher with what he could do. The teacher told him that it was nice but advised him not to show anyone else. When the clock beeped during English class, the English teacher asked to see what was in his backpack. Ahmed, who was wearing a NASA T-shirt, brought it forward. The English teacher examined it and said "that looks like a bomb." The police were called and Ahmed was arrested. When he was brought into the principal's office, one of the police officers said, "That's who I thought it was." He was interrogated by five police officers and the principal before being handcuffed and taken to the police station, where he was fingerprinted.

"They interrogated me and searched through my stuff and took my tablet and my invention," Ahmed told MSNBC. "They were like, 'So you tried to make a bomb?' I told them no, I was trying to make a clock." But the officer said, "It looks like a movie bomb to me."

With the help of the sci-tech community, the story went viral on social media. The school and police officers were caught by their ignorance of electronic engineering. In the face of that ignorance, fear and racism took over, as our worst qualities often do when we are ignorant and afraid. In the end, Ahmed received invitations from Facebook CEO Mark Zuckerberg and President Barack Obama, who tweeted "Cool clock, Ahmed. Want to bring it to the White

House? We should inspire more kids like you to like science. It's what makes America great."

The clash between the science-literate and a science-illiterate society creates unique problems not just for hapless individuals who run afoul of ignorant or racist authorities, but for the mainstream media as well. Budget-strapped and increasingly unable to discern between knowledge and opinion, science-illiterate journalists too often aid the slide into unreason. Many journalists believe there is no such thing as objectivity, rendering otherwise brilliant minds unable to discern between objective knowledge developed from years of scientific investigation, on the one hand, and a well-argued opinion made by an impassioned and charismatic advocate on the other. This problem extends beyond journalists. Cumulatively, newspaper editors have allowed themselves to be heavily manipulated by antiscience public-relations campaigns. One cannot be certain exactly why an opinion editor chooses to run one piece and not another, for example, but in December, 2015, the nonprofit *Media Matters* did an analysis of opinion pieces that mentioned the recently concluded Paris climate talks and ran in the ten largest-circulation newspapers in the United States: *USA Today*, the *Wall Street Journal*, the *New York Times*, the *Orange County Register*, the *Los Angeles Times*, the *San Jose Mercury News*, the *New York Post*, the *New York Daily News*, *Newsday*, and the *Washington Post*. Nine of the pieces, or 17 percent, included climate-science denial. Just 3 percent of climate scientists in any way dispute human-caused disruption of the Earth's climate system. This means that the major US papers expressed views that were more than five times as doubtful about climate change as the actual climate scientists publishing in the field. By engaging in this sort of misrepresentation, the media deprives the public of the reliable information necessary for self-governance.

A vast war on science is underway, and the winners will chart the future of power, democracy, and freedom itself. This book is an account of that war, and what we—concerned citizens of all political persuasions, in all countries—can do to win it.

The Silence of the Invisible Hand

The idea behind democracy was pretty simple. The invisible hand of the people's will was supposed to guide us in our own interests. That was the American Founding Fathers' thinking, and it worked, more or less, for about two hundred years, slowly marching forward toward the stated goals of liberty and justice for all. Not with perfection, not without injustice, but with undeniable progress. But something changed in the fundamental formula in the last four decades as science has advanced. The times we live in have in some ways become absurd: a century that could rightly be called the century of science whose voters are increasingly willing to reject science and to elect ardently antiscientific politicians.

Can it be that science has simply advanced too far? That the problems are too big or too complex, or that knowledge is now too inaccessible to normal citizens to make good decisions—decisions in their own best interest? In a world dominated by science that requires extensive education to fully grasp, can democracy still prosper, or will the invisible hand finally fall idle? Are the people still sufficiently well informed to be trusted with their own government?

Judging from the US Congress, or recent Canadian or Australian parliaments, or a number of other governments particularly in the developed world, the answer seems to be no. In an age when most major public-policy challenges revolve around science, fewer than 1 percent of US congresspersons have professional backgrounds in it. The membership of the 114th Congress, which ran from January 2015 to January 2017, included just three scientists: one physicist, one chemist, and one microbiologist. If one counts the eight engineers, it's a total of eleven out of 535 members, or 2 percent. Similarly low ratios are present in Canada's parliament, where the combined number is about 4 percent; Australia's, where it's 4 percent; and in many of the world's other major governments.

In contrast, how many representatives and senators might one suppose have law degrees—and often avoided college science classes in favor one of the top four prelaw majors: business, English, history, and political science? In the United States, it's 213, or roughly 40 percent. So it's little wonder we see more rhetoric than facts in global policymaking. In an age when most major policy issues have large inputs from science, this disparity can be a problem. Scientists and lawyers approach arguments very differently. Lawyers are trained to start with a conclusion, discover evidence to support that conclusion, and craft it into a compelling narrative to win the argument. They rely on the opposing counsel to do the same, and on an impartial third party—the judge, jury, or in government, Congress or Parliament as a whole—to determine who has made the more compelling case. But as any trial lawyer will tell people, such an approach uses facts selectively and only for the purposes of winning the argument, not for establishing the truth. That is the opposite of the approach of science, which starts with observation, accumulates evidence from studying nature, and forms a conclusion based on what the preponderance of the evidence as a whole suggests.

This disconnect creates opportunities for our policies to be led away from evidence by compelling propagandists. The problem is even more pronounced in presidential politics and among the journalists who cover it. Consider the disruption of Earth's climate system, arguably the greatest public-policy challenge facing the planet. In late 2007, the League of Conservation Voters analyzed the questions asked of the candidates for US president by the five top prime-time TV journalists: CNN's Wolf Blitzer, ABC's George Stephanopoulos, MSNBC's Tim Russert, Fox News's Chris Wallace, and CBS's Bob Schieffer. By January 25, 2008, these journalists had conducted 171 interviews with the candidates. Of the 2,975 questions they asked, how many might one reasonably suppose mentioned the words "climate change" or "global warming"?

In fact it was six. To put that in perspective, three questions mentioned UFOs.

By 2015, political journalists had shown little improvement. In December, 195 countries had reached an historic and unanimous accord in Paris to begin to find ways to limit greenhouse gases. The non-binding agreement involved re-envisioning the global economy and paying hundreds of billions of dollars to poorer countries. Just a few days later, both the Republicans and Democrats running for US president held primary debates. Despite the profound potential implications of both action and inaction and the strong differences between the parties on the topic of climate change, the journalists moderating the two debates didn't ask a single question about it.

Similar things could be said of any one of several major topics surrounding science, each of them with vast policy implications. Not a single candidate for president spoke about them, and humanities-trained political journalists did not ask about them. It was as if they didn't exist. But in a world increasingly dominated by complex science, the answers to such questions will determine the future. Certainly they should be contemplated by voters when making electoral decisions. What could have happened to the media, to make it so derelict in its duties in this regard?

Let's Have a Science Debate

In the fall of 2007, this divergence was noticed by a British expat: Charles Darwin's great-great-grandson Matthew Chapman, who wondered what could be going on. A science writer, film director, and the screenwriter of films including 2003's *Runaway Jury*, Chapman picked up the phone and began calling friends to see if they, too, had noticed this. He reached physicist Lawrence Krauss, science journalist Chris Mooney, energy scientist and science blogger Sheril Kirshenbaum, science philosopher Austin Dacey, and me. We all agreed that the silence on science issues was astounding. As a group, we founded ScienceDebate 2008 (later ScienceDebate.org), an effort to get the candidates for president to debate the major science policy issues.

We put up a website, placed op-eds in national publications, and reached out to contacts and leading science bloggers. The effort went viral. One of those bloggers, Darlene Cavalier of ScienceCheerleader.com, connected us with the US National Academies and became part of our core team, as did Michael Halpern, a senior staffer in the Scientific Integrity program at the Union of Concerned Scientists. Within weeks, thirty-nine thousand people from across the political spectrum had signed on, including Nobel laureates, prominent scientists, the presidents of most major American universities, the CEOs of several large corporations, and political movers ranging from John Podesta, President Bill Clinton's former chief of staff, on the left, to former house speaker Newt Gingrich on the right. Feeling affirmed, we reached out to the campaigns.

They ignored us. This is, of course, a classic campaign tactic. You never give energy to anything that you wish would go away. You simply do not engage, because the moment you do there is a story, the thing gets legs, and if you don't have your message already developed, you can lose control of your narrative. The question was why they wouldn't *want* to engage.

I went on Ira Flatow's US National Public Radio program *Talk of the Nation: Science Friday*. The American Association for the Advancement of Science (AAAS); the US National Academies (of Sciences, Engineering, and Medicine); and the nonprofit Council on Competitiveness signed on as cosponsors. Soon we represented more than 125 million people through our signatory organizations. It was the largest political initiative in the history of science.

Presidential Candidates Debate Religion, Not Science

Still, the candidates refused to even return phone calls and e-mails. So we decided to organize a presidential debate ourselves, and turned to the national media outlets for help. We brought on PBS's flagship science series *Nova* and its then-news program *Now on PBS* as broadcast partners. David Brancaccio, *Now*'s host, would moderate. We set a date shortly before the crucial Pennsylvania primaries and teamed up with the venerable Franklin Institute in Center City Philadelphia to host.

But despite the urging of advisors like EMILY's List founder Ellen Malcolm, who was involved with Senator Hillary Clinton's (D-NY) campaign, and Nobel laureate Harold Varmus, who was supporting Senator Barack Obama (D-IL), both of those candidates refused invitations to a debate that would center on the US economy and science and technology issues. Senator John McCain (R-AZ) ignored the invitation entirely. Instead, Clinton and Obama chose to debate religion at Messiah College in Harrisburg, Pennsylvania—where, ironically, they answered questions about science.

An old joke tells of the three things one never discusses in polite company: sex, politics, and religion. How has political culture come to a point where science is more taboo to discuss than religion? What little news coverage there was of this stunning development didn't seem to affect the campaigns at all. The candidates continued their policies of non-engagement. It wasn't because they felt inhibited about opining on issues outside their expertise. They waxed on about foreign policy and military affairs even though none were diplomats or generals. They offered economic plans even though they had little knowledge of economics. They talked about morality and religion even though they were not rabbis or priests. But they refused to debate the many crucial issues presented by science.

Marveling at this odd situation, I began speaking to news directors and editors, asking them why they weren't covering this remarkable situation. Here we had virtually the entire US science and technology enterprise—which, by the way, is the main engine of the economy—calling for the presidential candidates to debate enormous science policy issues, and the candidates were dodging us. That sounded a lot like news, and yet it was getting very little coverage.

The people I spoke to said they thought it was a niche topic, and the public wasn't interested. So ScienceDebate and the nonprofit Research!America teamed up to do a little science to test that assumption. We commissioned a national poll and found that 85 percent of the American public thought that the candidates should debate the major science issues. Support was virtually identical among Democrats and Republicans. Religious people clearly were not put off by the idea either. Only the candidates and the press, it seemed, were reticent.

Fear and Loathing on the Campaign Trail

With the window closing for a debate before the endorsing conventions, I recruited Jane Lubchenco, a marine scientist and former AAAS president, to help organize a science debate in Oregon in August. But Obama and McCain refused this one too, opting instead to hold yet another faith forum, this time at Saddleback Church in Lake Forest, California. The scientific, academic, and high-tech communities were stunned. Science was responsible for more than half of all US economic growth since World War II. It lies at the core of most major unsolved policy challenges the world over. How could people who wanted to lead America avoid talking about science? Intel chairman Craig Barrett reached out to former Hewlett-Packard CEO Carly Fiorina on the McCain side to encourage his participation, and Varmus redoubled his efforts to convince Obama.

Meanwhile, our supporters had submitted more than 3,400 questions that they wanted to ask the candidates. Political staffers at the campaigns told me they were concerned about candidates appearing foolish. One Republican said they wanted to avoid a "Dan Quayle moment." I explained that we weren't interested in asking them technical questions about the third digit of pi or the details of cell mitosis. We were interested in big science policy questions. Still, they were skeptical. So, working with several leading science organizations, I culled the crowdsourced questions into "The Top 14 Science Questions Facing America," and released them publicly as a sort of open-book test.

The Original Top 14 Science Questions Facing America

These are the original fourteen final questions we arrived at. We stated them very broadly—some might say too broadly—in an effort to show how policy-oriented they were. But the candidates still ignored us.

1. Innovation. Science and technology have been responsible for half of the growth of the American economy since World War II. But several recent reports question America's continued leadership in these vital areas. What policies will you support to ensure that America remains the world leader in innovation?
2. Climate Change. Earth's climate is changing and there is concern about the potentially adverse effects of these changes on life on the planet. Please set out what your positions are on the following measures

that have been proposed to address global climate change: a cap-and-trade system, a carbon tax, increased fuel-economy standards, firm carbon-emissions targets, and/or research. What other policies would you support?

3. Energy. Many policymakers and scientists say energy security and sustainability are major problems facing the United States during this century. What policies would you support to meet demand for energy while ensuring an economically and environmentally sustainable future?

4. Education. A comparison of fifteen-year-olds in thirty wealthy nations found that average science scores among US students ranked seventeenth, while average US math scores ranked twenty-fourth. What role do you think the federal government should play in preparing K–12 students for the science- and technology-driven twenty-first century?

5. National Security. Science and technology are at the core of national security like never before. What is your view of how science and technology can best be used to ensure national security, and where should we put our focus?

6. Pandemics and Biosecurity. Some estimates suggest that an emerging pandemic could kill more than three hundred million people. In an era of constant and rapid international travel, what steps should the United States take to protect our population from global pandemics and deliberate biological attacks?

7. Genetics Research. The field of genetics has the potential to improve human health and nutrition, but many people are concerned about the effects of genetic modification both in humans and in agriculture. What is the right policy balance between the benefits of genetic advances and their potential risks?

8. Stem Cells. Stem-cell-research advocates say it may successfully lead to treatments for many chronic diseases and injuries, saving lives, but opponents argue that using embryos as a source for stem cells destroys human life. What are your positions on government regulation and funding of stem-cell research?

9. Ocean Health. Scientists estimate that some 75 percent of the world's fisheries are in serious decline and habitats around the world like coral reefs are seriously threatened. What steps, if any, should the United States take during your term to protect ocean health?

10. Water. Thirty-nine states expect some level of water shortage over the next decade, and scientific studies suggest that a majority of our water resources are at risk. What policies would you support to meet demand for water resources?

11. Space. The study of Earth from space can yield important information about climate change; focus on the cosmos can advance our understanding of the universe; and manned space travel can help us inspire new generations of youth to go into science. Can we afford all of them? How would you prioritize space in your administration?

12. Scientific Integrity. Many government scientists have reported political interference in their work. Is it acceptable for elected officials to hold back or alter scientific reports if they conflict with their own views, and how will you balance scientific information with politics and personal beliefs in your decision making?

13. Research. For many years, Congress has recognized the importance of science and engineering research to realizing our national goals. Given that the next Congress will likely face spending constraints, what priority would you give to investment in basic research in upcoming budgets?

14. Health. Americans are increasingly concerned about the cost, quality, and availability of health care. How do you see science, research, and technology contributing to improved health and quality of life, and what do you believe is the solution to America's "health-care crisis"?

Presidential Antiscience

The war on science wasn't limited to candidates and the media. The George W. Bush White House had become notoriously antiscience, which legitimized science denial in a way the world is still dealing with. Bush appointed ideologues to key agency posts throughout the federal government and empowered them to hold back or alter scientific reports with which they disagreed. This represented a marked change from the Republican Party of just ten years prior. Consider the following quote by President George H. W. Bush, George W.'s father:

Science, like any field of endeavor, relies on freedom of inquiry; and one of the hallmarks of that freedom is objectivity. Now more than ever, on issues ranging from climate change to AIDS research to genetic engineering to food additives, government relies on the impartial perspective of science for guidance.

Then consider this one by his son's White House spokesman, Scott McClellan, thirteen years later:

This administration looks at the facts, and reviews the best available science based on what's right for the American people.

The first approach uses knowledge as a basis for public policy. The second looks first to a predetermined political agenda ("what's right for the American people") and seeks only those facts that support it. It is antiscience.

After Bush's 2004 reelection, scientists noticed that the problem was becoming worse. One example was Bush's appointment of George Deutsch, a twenty-four-year-old Texas A&M University dropout and Bush campaign intern, to a key position in NASA's public-relations department. Deutsch set to work muzzling NASA's top climate scientist, James Hansen, refusing to allow him to interview with National Public Radio because it was "the most liberal" media outlet in the country and telling a contractor that the word "theory" had to be inserted after every mention of the big bang on NASA's website presentations being prepared for middle-school students. The big bang is "not proven fact; it is opinion," Deutsch told the contractor. "It is not NASA's place, nor should it be to make a declaration such as this about the existence of the universe that discounts intelligent design by a creator. . . . This is more than a science issue, it is a religious issue. And I would hate to think that young people would only be getting one-half of this debate from NASA." Deutsch later resigned after it was revealed that he had fabricated his academic credentials, and did not graduate from college.

Other Bush public-relations appointees were muzzling scientists at other agencies, or altering scientific information in official agency reports to fit a preconceived ideological agenda, angering many scientists. (The same tactics would be employed by Canada's Conservative Harper government just a few years later.)

The problem became so widespread that, in early 2007, the House Oversight Committee held hearings investigating the distortions. The Centers for Disease Control and Prevention was forced to discontinue a project called Programs-That-Work, which identified sex education programs found to be effective in scientific studies, none of which were abstinence-based. On the National Cancer Institute's website, breast cancer was falsely linked to abortion. The morning-after pill, an emergency contraceptive that prevents ovulation after unprotected sex and may in rare circumstances prevent an already-fertilized egg from attaching to the uterus, was held back from Food and Drug Administration (FDA) approval for over-the-counter sale even though scientists and physicians had judged it to be safe and determined that it was actually likely to *reduce* the number of abortions. (Later, it would also be partially held back by the Obama administration, again contrary to the recommendations of panels of scientists.) "Faith-based" initiatives like abstinence-only sex education, by contrast, were federally funded at high levels, even when they were contradicted or shown ineffective by science. And business-friendly FDA administrators failed to remove the arthritis drug rofecoxib (Vioxx) from the market even after it became apparent that it was causing heart attacks, resulting in more than fifty thousand American deaths—nearly as many as the number of American soldiers lost in Vietnam. FDA administrators made calls to a whistleblower-protection attorney and a leading medical journal in an attempt to discredit the scientist who brought the problem to light.

The Watershed

By the 2008 election, antiscience views had become entrenched as mainstream political planks of the Republican Party. The focus was on three main areas: denying the science of reproductive medicine, denying the science of evolution, and denying the science of climate change. Its messaging followed a public-relations playbook that had been developed in part by Southern US tobacco companies to fight the emerging science-based conclusion that smoking causes cancer, and by US agribusiness in fighting the revelations that pesticides were disrupting the environment and hazardous to health. Like climate disruption, these had been facts that, if widely accepted, could undermine entire industries. "Doubt is our product," a tobacco executive wrote in a 1969 memo to fellow tobacco executives, "since it is the best means of competing with the 'body of fact' that exists in the minds of the general public. It is also the means of establishing a controversy."

Controversy is the most common aspect of modern antiscientific attacks, because it takes advantage of the reasonable-sounding but incorrect idea that a "healthy debate" reveals the truth. When such a debate pits knowledge against a passionately articulated opinion, the opinion often wins. "For what a man had rather were true," as the father of modern science, Francis Bacon, noted, "he more readily believes."

Today, this is called motivated reasoning, or, more simply, confirmation bias. During a 2006 Alaska gubernatorial debate, Republican Sarah Palin provided a good example of the problem when she came out in favor of teaching creationism in science class. "Teach both," she said. "You know, don't be afraid of education. Healthy debate is so important, and it's so valuable in our schools."

By 2008, it had become doubtful whether a Republican candidate for president could get the party's endorsement without taking a stridently antiscience position. Democrats, in turn, seemed terrified of offending evangelical swing voters, preferring instead to either out-conservative the conservatives or avoid discussing science and technology altogether. Antiscience advocates on the left could be just as vicious as the right's climate deniers when scientists pointed out that their ideas that cell phones cause brain cancer, vaccines cause autism, genetically modified crops are unhealthy to eat, and similar notions were not supported by the evidence. Scientists hoped that John McCain would somehow rebuff this trend. McCain had long crafted a reputation as a "maverick" and a "straight shooter." If anyone could stem the tide, they thought, he could. But they couldn't even get Obama to engage in a debate, much less McCain.

Finally, on the eve of the Democratic National Convention, I was hiking in Rocky National Park with my son Jake. He was thirteen, and as the two if us stood on the continental divide, leaning into a fifty-mile-per-hour wind, I was struck by the irony the location symbolized: left and right, past and future, proscience and antiscience. So many divides, and I was trying to straddle them all as the leader of the ScienceDebate effort, while the political wind was trying to blow me away. I had been very public in the attempt to get the candidates to engage, and felt I was letting down my cofounders as well as the entire scientific enterprise that had gotten behind us. But I was there with Jake in the backcountry and it was beautiful on top of the Rockies. We hiked back down and made our way into Estes Park, Colorado, and as we got back into cell-phone range my phone started buzzing with voicemails from the Obama campaign. While Obama wouldn't participate in a televised forum, he would participate in an online "debate." Scientists were jubilant. Finally, someone was listening.

Days later, McCain agreed as well, and the press, given a classic conflict frame, was finally interested. The ScienceDebate story, and the candidates' answers to "The Top 14 Science Questions Facing America" made nearly one billion media impressions—an enormous opening of the floodgates on stories that had previously been ignored. The public finally started seeing discussions of the candidates' positions on climate change, energy, health care, the environment, economic competitiveness, and a host of other science policy topics. Obama used our mission statement—to "restore science to its rightful place"—in his inauguration speech. And once in office, the candidate who had started out not particularly friendly toward science seemed to embrace it as a central part of his strategic approach. He appointed several of our early supporters to cabinet-level posts. Steven Chu became energy secretary. John Holdren became presidential science advisor. Jane Lubchenco became undersecretary of commerce and director of the National Oceanographic and Atmospheric Administration (NOAA). Harold Varmus led the National Cancer Institute. Marcia McNutt became director of the US Geological Survey. John Podesta led Obama's transition team. The administration had more scientists than any in memory. Perhaps, scientists dared to hope, the dark days of unreason had finally passed. They couldn't have been more wrong.

Are There Really Two Sides to Every Story?

The problem, as we've already seen, wasn't limited to the candidates. Many reporters (and editors, who often direct reporters' lines of questioning) are—like many politicians—humanities majors who were required to take few or no science classes in college. The classes were hard, and they ducked them, and now few seem to understand science's unique importance to the democracies they report on. Most seem to think, incorrectly, that the public shares their disinterest. In an age when so many major policy problems are dominated by science, this is a concern.

Another part of the problem may be that journalists, scientists, and politicians each approach questions of fact from differing perspectives. Journalists look for conflict to find an angle, so there are always two sides to every story. Bob says 2 + 2 = 4. Mary says it is 6. It sounds surprising, but Mary may have legitimate reasons for her perspective. The media outlet gets a good headline and an interesting story, the controversy rages, and newspapers or web clicks are sold. A scientist would say that, based on the knowledge built up from observation, one of these claims can be shown to be objectively false and it's poor reporting to paint this as a controversy, because it's not. Using four apples, the scientist can quickly and objectively demonstrate that Bob is right. Not so fast, a politician might answer. How about a compromise? Soon we see a new law affirming that 2 + 2 = 5. This is democracy's problem, in a nutshell, in the age of science.

The modern journalistic approach does not work when applied to scientific questions, and it tends to skew public policy in counterfactual directions, as the above example shows. This is a bit ironic because journalistic techniques were originally developed as a means of fact-checking, akin to replication and peer review in scientific research. For example, reporters would get multiple sources to corroborate a story (which is an account of events in our shared, objective reality), establishing a relative confidence in its veracity, or they wouldn't run the story. But today, journalism schools teach a mantra that scientists will say is completely false: "there is no such thing as objectivity"—a phrase frequently repeated by some of the profession's leading figures, and contained in many newspaper reporters' guidelines.

This conceit may be true when reporting on politics or interviewing the witnesses to a crime, but it is decidedly not true when it comes to reporting on events or issues that have large inputs of objective knowledge from science, even when those issues or events are political. For such stories, we have developed a unique, reproducible, peer-reviewed method of scientific research whose very purpose is to create the objective knowledge reporters seem to think cannot be had. The process of science is designed to cull out reliable knowledge—no matter who does the investigating or reports on the outcome—from our gender identities, our political identities, our religious identities, our sexual identities, our cultural identities, and so on, trimming away all those subjective forms of bias reporters think we can never escape until we are left with knowledge that is provisionally objective in the stories we tell about reality. While it may not be possible to attain total objectivity, approaching it is what science is all about, and the reliable knowledge it produces is responsible for every advance in the modern world. The proof of the pudding is in the eating.

Peabody-winning news anchor Don Shelby lectures journalists about this misconception often.

"Some journalists don't even attempt to establish the reality or truth of a story. Instead, they go out of their way to present 'both sides,' as if this were admirable," he says.

> And what I tell them is that "balance" doesn't mean you present
> stories evenhandedly. It means you present them like a set of scales,
> and if the vast weight of the evidence is on one side of the argument,

that's the side that should get the vast weight of your reporting. You don't push on the other side to falsely balance the scales. You tell the truth. That's the "balance" we used to talk about in journalism. Today what we too often see is called "false balance," because it presents both sides as if they have equal weight of the evidence, when that is objectively not true.

The first casualty of this "false balance" is journalism's own credibility, and journalists' ability to speak truth to or about power, which is one of the field's main functions. (It is, incidentally, also one of the functions of the journalistic aspect of a scientist's recounting of an experiment.) If one side's account is based on the accumulated knowledge gained from tens of thousands of painstaking experiments done by thousands of scientists working over fifty years taking and reporting on billions of measurements reproducible by others, as in the case of climate science, and the other side is a persuasive opinion articulated by a passionate advocate who is intent on convincing viewers of the rightness of his or her perspective, by presenting them as a debate, journalism becomes an implicit advocate for extreme views, weighting them and presenting them to the public as if they had equal merit with tested knowledge. Journalism thus fuels the extreme partisanship we see in public dialogue today, and feeds into the hands of the very power journalists exist to challenge—vested interests who seek to circumvent evidence and undermine the democratic process to achieve a desired outcome.

It should be noted that many journalists argue that their job is not to establish truth, but simply to relay information fairly. This laissez-faire, hands-off view has come to dominate mainstream political journalism. David Gregory, NBC News's chief White House correspondent during the George W. Bush administration, put it quite clearly in his defense of the White House press corps for not pushing President Bush on the lack of credible evidence of Saddam Hussein's "weapons of mass destruction" and the inconsistencies in Bush's rationale for invasion before the United States entered Iraq. "I think there are a lot of critics who think that . . . if we did not stand up and say this is bogus, and you're a liar, and why are you doing this, that we didn't do our job," said Gregory. "I respectfully disagree. It's not our role."

But if it is not the press's role, whose role is it? How are the people to make well-informed decisions about momentous policies without accurate, reasonably objective information and a questioning of the powerful, asking for evidence?

Similarly, the tendency of politicians to look for compromises on disputed questions of fact instead of basing decisions on an objective standard of knowledge is eroding the country's ability to solve its problems, leaving it mired in policies that don't work and political battles that go on forever. And by allowing the teaching of "alternative theories" on politically contentious topics like evolution or climate change or birth control in science classes, those same politicians damage children's ability to learn critical thinking, to compete in a science-driven global economy, and to live in a world increasingly impacted by climate disruption.

This dumbing down of the people for ideological reasons is, of course, not new. It is an age-old authoritarian tactic. It happened in China during the Cultural Revolution. It happened in the Roman Empire, the Ottoman Empire, Renaissance Italy, twentieth-century Russia, and Nazi Germany—all of them societies whose leaders turned their backs on science, making it subordinate to an authoritarian ideology, and the societies collapsed.

The Great Dumbing Down

In trying to understand why mainstream journalists weren't fairly covering the important science issues of the day, I continued probing editors and news directors, and I learned something else. There is a long-standing tradition in newsrooms for editors and news directors to forbid political reporters from covering science issues and to rarely place science stories in the politics pages. Science has been relegated to its own specialized section, and those sections are being eliminated.

This is a problem in a time when science is so central to our policy challenges. No other major human endeavor is so ghettoized. The religion and ethics beat has long since crossed over into the politics pages, as has the business and economics beat. Military affairs and foreign policy have been there all along.

Partly, this growing ghettoization is due to economics. Facing increased competition from cable TV and a largely free model of news on the Internet, commercial news media have been trimming costs. Among the first things to go were the most expensive: investigative and science reporters. A Joan Shorenstein Center on the Press, Politics and Public Policy report from 2005—early in the science-news crisis—showed that from 1989 to 2005, the number of major US newspapers with weekly science sections fell from ninety-five to thirty-four. By 2005, just 7 percent of the approximately 2,400 members of the

US National Association of Science Writers had full-time positions at media outlets that reached the general public.

In May of 2008, the *Washington Post* killed its famed science section. In November, NBCUniversal fired the Weather Channel's entire *Forecast Earth* staff—during the NBC network's Green Week promotion—ending the station's only environmental series that focused on global warming. In December, CNN fired its entire science, technology, and environment news unit. In March of 2009, the *Boston Globe*, located in a worldwide capital of scientific research, closed its renowned science and health section. Later that year, Columbia University announced that it was closing its Earth and Environmental Science Journalism program because of "current weakness in the job market for environmental journalists." US newspapers had ninety-five weekly science sections in 1989, but just nineteen were left in 2012. The bloodshed continued. In early January 2013, the *New York Times* closed its environmental desk just two months after Hurricane Sandy (which scientists say was made worse by climate disruption) decimated the city. That same year, Johns Hopkins University retired its thirty-year-old science-writing program. The massacre left only about a dozen environmental reporters still standing throughout the five largest US newspapers—with the *Los Angeles Times* the only one to still have an environmental desk—at a time when the US and the world face the gravest environmental issues in human history.

As a result, we now live in a dangerous situation, when many major challenges revolve around science, and few reporters are covering them. That means they tend to get far less attention than they deserve. Some efforts have emerged to combat this. England's Science Media Centre seeks to provide general-assignment reporters with the science angle on major policy stories, and similar centers have been set up in Canada, Australia, New Zealand, and Japan. Nonprofit news outlets have begun to spring up, some focusing on the expensive professions of investigative or science reporting. And blogging is creating a new, intimate relationship between educated readers and scientists.

In continental Europe, science coverage has actually increased in the mainstream media. A 2008 analysis of prime-time news on selected European TV stations, for example, showed that there were 218 science-related stories (including science and technology, environment, and health) among the 2,676 news stories aired during the same week in the years 2003 and 2004, an elevenfold *increase* since 1989.

The European Union of Science Journalists' Associations, under the leadership of Hanns-J. Neubert and Wolfgang Goede, also promoted a 2009 German parliamentary science debate patterned loosely on the US effort, and similar efforts have begun in Estonia, as well as Sweden, Belgium, Switzerland, and the EU as a whole. Goede considers it an essential component of modern democracy, an argument he and I have made together to journalists around the world. The group is also pushing a new initiative called NUCLEUS, along with some two dozen universities across Europe and in Beijing, to devise other ways that scientific institutions can work to counter the communication gap between science and the public.

In the developing world, science reporting is described, for the time being, as "flourishing," but journalists there have reported that many of the same problems are beginning to emerge.

Congressional Antiscience

Of course, the war on science isn't limited to presidential campaigns and the media. It is present in city councils, state legislatures, and congressional and parliamentary delegations the world over, particularly in several of the leading democracies with strong corporate economies and liberal interpretations of the right to freedom of expression.

In the US Congress, the war first began to widely emerge in the religious and patriotic fervor following the September 11, 2001, attacks on the World Trade Center by terrorists from the fundamentalist Islamist group al-Qaeda. In April of 2002, then–house majority whip Tom DeLay (R-TX) quoted the evangelical Christian authors of a 1999 book when he told a Texas church group, "Only Christianity offers a comprehensive worldview that covers all areas of life and thought, every aspect of creation." DeLay, who would soon become House majority leader, said he wanted to promote "a biblical worldview" in American politics. "Our entire system is built on the Judeo-Christian ethic, but it fell apart when we started denying God," he had said in 2001. After the 1999 Columbine school shootings, DeLay had given a speech on the House floor in which, his voice dripping with sarcasm, he suggested the tragedy "couldn't have been because our school systems teach our children that they are nothing but glorified apes who have evolutionized out of some primordial soup of mud by teaching evolution as fact." Ironically, DeLay's Bachelor of Science degree, from the University of Houston, was in biology.

DeLay's views ran throughout his caucus, particularly among the increasingly powerful baby boomers in the House. In March of 2002, eventual house speaker John Boehner (R-OH) wrote to the Ohio State Board of Education to urge that the state's science curriculum content standards require teaching creationism, saying,

> It's important that the implementation of these science standards not be used to censor debate on controversial issues in science, including Darwin's theory of evolution. . . . Students should be allowed to hear the scientific arguments on more than one side of a controversial topic. Censorship of opposing points of view retards true scholarship and prevents students from developing their critical thinking skills.

This language was coded to boost creationists, who were promoting their "scientific" argument for "intelligent design" in the latest attack on the teaching of evolution. There is no scientific controversy about the theory of evolution. Boehner's letter was antiscience doublespeak.

Antiscience grew politically stronger as evangelicals were swept into public office in the years immediately following 9/11. Their early battles were over the teaching of evolution in public schools and the anticipated arrival of gay marriage, but they were also upset about scientific characterizations of origins, from the big bang to the scientific definition of when a woman can be said to be pregnant, all of which they saw as an assault on Christian values. The willingness to reject science by these candidates and the vast numbers of motivated foot soldiers they had drawn into the ranks of the GOP provided a unique opportunity for vested corporate interests increasingly vulnerable to political action on climate change.

Outvoting Galileo

The amount of money those vested interests—particularly those aligned with the energy and extraction industries (oil, gas, coal, and minerals)—were willing to spend to battle science, and the power of their public-relations efforts to do so, became the overriding force in American politics for the first two decades of the twenty-first century.

In February 2009, President Obama asked Congress to send him legislation that placed a market-based cap on carbon emissions. The House passed the bill later that summer, with eight Republican votes. Throughout 2009 and 2010, raging battles were fought in GOP primaries throughout the country as energy-industry-funded groups recruited and promoted Tea Party candidates to run against Republicans who had voted for the cap-and-trade bill, utilizing evangelical Republican foot soldiers, and knocking the offenders out with relatively small investments. Climate science became equated with Obama and socialism in Republican talking points, and the technique of bashing science or promoting brazenly antiscientific positions became a political identity statement. By late 2010, fully ninety-four of one hundred newly elected Republican members of Congress either denied that global warming was happening (it was all a vast hoax by scientists, they said) or signed pledges to oppose mitigation.

By the 2012 elections, when Republican presidential hopefuls hit the campaign trail, they were propelled by a strong antiscientific wind. It became a predictable pattern: when a conservative candidate was sinking in the polls, he or she would make an antiscience statement in an effort to get a bounce. Texas Governor Rick Perry compared himself to Galileo when denying in a Florida primary debate that climate science is settled. "The idea that we would put Americans' economy at—at—at jeopardy based on scientific theory that's not settled yet, to me, is just—is nonsense. I mean, it—I mean—and I tell somebody, I said, just because you have a group of scientists that have stood up and said here is the fact, Galileo got outvoted for a spell."

It seemed lost on Perry that the people who "outvoted" Galileo were the members of the Roman Catholic Inquisition, who, like Perry, chose ideology over science. The US National Academy of Sciences had in 2010 stated that man-made climate change was supported by so many independent lines of data that its existence and causes should be "regarded as settled facts."

Other candidates made similar antiscientific assertions. Herman Cain, who was previously well-respected in business circles, said that "man-made global warming is poppycock." Congresswoman Michele Bachmann, a former IRS tax attorney and an ardent evangelical campaigner against gay marriage, talked of how the human papillomavirus vaccine had caused "mental retardation," while Congressman Ron Paul, a medical doctor, agreed that it was "not good medicine." Former Pennsylvania Senator Rick Santorum, also an attorney, said "absolutely not, I don't believe in" evolution. Even Newt Gingrich, an academic historian who once told me, "I'm very interested in doing anything I can to support science" and signed on as a supporter of ScienceDebate.org, felt compelled to announce that he was "opposed to killing children in order to get research material." He was talking about stem-cell research, and characterizing a frozen blastocyst of about 150 cells that would otherwise be discarded by law as "killing children in order to get research material." He experienced a slight bump in the polls.

Former Massachusetts governor Mitt Romney and former Utah governor Jon Hunstman Jr., both successful Mormon businessmen, were alone in not kowtowing to the new antiscience fervor spreading among evangelical Republican Party activists. "Listen," Huntsman said in a September GOP primary debate in California, "when you make comments that fly in the face of what ninety-eight out of one hundred climate scientists have said, when you call into question evolution, all I'm saying is that in order for the Republican Party to win, we can't run from science." Huntsman plummeted in the polls.

Appeasing Republican primary voters while not sounding so absurdly antiscience that one alienated mainstream voters was clearly a delicate balancing act. The activists were being propelled into a sort of anticlimate frenzy by right-wing media organizations and radio that amplified ideas from organizations like the Heartland Institute, Americans for Prosperity, and the George C. Marshall Institute. Romney was forced to retreat from the affirmative stance regarding anthropogenic global warming that he had previously taken, eventually saying that "we don't know what's causing climate change." At the Republican National Convention, he turned President Obama's efforts to address climate change into a laugh line.

Lies Straight from the Pit of Hell

By late 2012, antiscientific rhetoric had become normalized in US politics. Public statements that once would have been considered ludicrous and career-ending were accepted by media and voters without challenge, mostly on the Republican side of the aisle, and mostly on issues surrounding climate change, contraception, and evolution. That's not to say that Democrats didn't have their own issues with accepting science they didn't agree with politically—they did—but they weren't running loudly against science the way Republicans were.

Congressman John Shimkus (R-IL), chairman of the House Subcommittee on Environment and the Economy, waved his gilded Bible in a congressional hearing on climate change, declaring that "the earth will end only when God says it's time to be over. Man will not destroy this Earth, this Earth will not end in a flood." He added that "there is a theological argument that this is a carbon-starved planet."

Congressman Todd Akin (R-MO), who sat on the House Science, Space, and Technology Committee, was asked whether he opposed abortion even in the case of rape. He replied that "if it's a legitimate rape, the female body has ways to try to shut that whole thing down." He could not explain what a "legitimate rape" was, and what little science there is shows that pregnancies from rape seem to run at around 8 percent, about twice the pregnancy rate from consensual sex.

Congressman Paul Broun (R-GA), a medical doctor who served on the House Science, Space, and Technology Committee, told a luncheon crowd that

> All that stuff that I was taught about evolution and embryology, big bang theory, all that, is lies straight from the pit of hell. And it's lies to try to keep me and all the folks who were taught that from understanding that they need a savior. You see, there are a lot of scientific data that I've found out as a scientist that show that this is really a young Earth. I don't believe that the earth's but about 9,000 years old. I believe it was created in six days as we know them. That's what the Bible says. And what I've come to learn is that it's the manufacturer's handbook, is what I call it. It teaches us how to run our lives, individual. How to run our families. How to run our churches. But it teaches us how to run all of public policy and everything in society. And that's the reason as your congressman, I hold the Holy Bible as being the major directions to me of how I vote in Washington, D.C., and I'll continue to do that.

That is the opposite of the idea Thomas Jefferson originally had in mind for the United States.

The Death of Evidence

"We are sliding back into a dark era, and there seems little we can do about it," AAAS president Nina Fedoroff lamented on a cool, cloudy February day in 2012. Fedoroff was attending the world's most prestigious scientific organization's annual meeting in Vancouver, British Columbia, and she confessed that she was "scared to death" by the vast war on science that was spreading through the Western world.

As she spoke, the revolution was in full swing across the rest of Canada. Beginning shortly after taking power in 2006, Prime Minister Stephen Harper had imitated George W. Bush in his efforts to muzzle scientists. He and several other conservative members of Parliament had ties with top US Republican activists and elected officials, ranging from climate denier Senator James Inhofe to anti-tax activist Grover Norquist. In 2007, the Harper government established rules that required Environment Canada scientists to obtain permission before speaking with reporters, reducing their engagement on climate change by 80 percent. In 2008, Harper abolished the position of the National Science Advisor, and his administration soon began closing research libraries. The public didn't seem to notice at first. Canadian scientists were the only ones really feeling the thumb or watching the destruction of knowledge, and eventually they decided they needed to do more to raise public awareness.

On July 10, 2012, five months after Fedoroff's lament, thousands of scientists marched on Parliament Hill in downtown Ottawa to demonstrate against policies that cut science funding and prevented government scientists from speaking to the press, attending conferences, or even speaking to groups of high-school students without permission.

Wearing white lab coats or dressed all in black, they marched through Ottawa, chanting, "No science, no evidence, no truth, no democracy." On Parliament Hill, they held a mock funeral for evidence, with speakers delivering eulogies and describing how science was being threatened by the conservative, pro-extraction-industry government, whose members seemed only interested in research that served business.

"If you are fed up with the closure of federal scientific programs and muzzling of scientists, if you think that decisions should be based on evidence and facts instead of ideology, then please come out and show your support," the scientists' announcement said. Katie Gibbs, then a PhD student in biology, organized the rally and would subsequently set her career in science aside to lead the advocacy group Evidence for Democracy. "You can't have a functioning democracy if you don't have informed citizens, if you don't have the facts," Gibbs told me.

But the Harper government's attack on science continued unabated. Fully 90 percent of federal Canadian scientists said they could no longer speak freely. Eighty-six percent said that when faced with a departmental decision or action that could harm public health, safety, or the environment, they did not believe they could share their concerns with the public or media without censure or retaliation. Research libraries were closed, and much of their contents— thousand and thousands of volumes—were discarded in Dumpsters. In 2013, there were eleven Department of Fisheries and Oceans research libraries across the country. By 2015, there were four, and scientists complained that the information they contained—critical historical data records—was no longer available, making it impossible to track how ecological measurements such as ocean temperatures and fish populations were changing over time.

But the antiscience wasn't limited to conservative politicians seeking to quash opposition over environmental issues. Canadian scientists were also banding together to battle left-leaning antiscience from alternative-medicine providers who were giving doses of "Influenzinum," a homeopathic alternative to the flu vaccine that has no scientific basis. Other community activists across Canada were worried about microwave radiation and so-called "electromagnetic hypersensitivity," which has no scientific basis, and had begun actively lobbying local governments to ban Wi-Fi. The beleaguered scientists formed an organization called Bad Science Watch to monitor Influenzinum, Wi-Fi bans, and other issues, saying that "the media has been all too willing to fan the flames of controversy and has contributed to a growing false uncertainty over the safety of Wi-Fi. As a result many school boards, libraries, and town councils across Canada have been called on by concerned citizens to limit or remove Wi-Fi networks."

Worldwide Antiscience

By November of 2013 the antiscience cancer had spread to several other countries, most of them places like the United States and Canada that had heavy deposits of fossil fuels or minerals, often driven by PR efforts against policies regulating mining and emissions, sometimes with direct involvement from US Republican Party operatives. Within days of taking office in September of 2013, for example, Australian prime minister Tony Abbott—who had, in 2010, said, "The climate-change argument is absolute crap. However, the politics are tough for us because 80 percent of people believe climate change is a real and present danger"—followed Harper's lead and abolished the position of minister for science, a post that had existed since 1931.

In 2010, his predecessor, Labor (political left) prime minister Julia Gillard, had said, "There will be no carbon tax under the government I lead." Australia is a major producer of coal, worth about $60 billion annually, and Australians emit more carbon dioxide (CO_2) per capita—at the time, about 18 tons annually—than anyone except Saudi Arabians.

To form a minority government, however, Gillard cut a deal with the Greens in Parliament and agreed to a "carbon price." She immediately found herself the focus of a public-relations attack that coordinated right-wing talk-radio personalities, Rupert Murdoch–owned newspapers (whose articles ran 82 percent negative), the conservative coalition parties, and mining companies.

The tax-by-another-name went into effect in 2011 and was held up by the International Atomic Energy Agency as a model for other nations. But it drew more climate-denial public-relations efforts as the global environmental battle briefly shifted to Australia, ultimately costing Gillard her position as prime minister. Abbott, who at the time was the Liberal (political right) opposition leader and was running for prime minister, made a "pledge in blood" to repeal the tax. After being elected, he succeeded.

Send in the Clowns

In 2012, in an effort to retain power and court Christian conservative swing voters, Gillard had also pushed through a program funding chaplains in secular public schools nationally, something that had already been in practice in the state of Queensland. Australia's high court struck the program down as unconstitutional, but in 2014 Abbott used a legal technicality to provide $244 million to restore the program, sending 2,339 religious workers into schools across Australia. Trained Humanist chaplains and secular mental-health workers were excluded. Prominent astronomer and atheist Lawrence Krauss suggested that the idea that the chaplains would not be preaching in school was ridiculous: "It's like sending in clowns and telling them not to be funny."

Antiscience, now legitimized and empowered by top Australian government officials, began spreading. In early 2013, Merilyn Haines, an activist for the innocuous-sounding Queenslanders for Safe Water, began traveling across the state, urging city councils to stop water fluoridation. Fluoride, a naturally occurring nutrient, helps prevent tooth decay and promotes healthy bone growth. It has been used in public water supplies for seventy years, and health officials have identified it as "one of 10 great public health achievements of the 20th century." In fact, since fluoride occurs naturally in well water to varying degrees, the correct term for what cities do is really "fluoride regulation," i.e., adding it or removing it as necessary to achieve the optimum health benefit at .7 to 1.2 mg per litre.

In Queensland, however, the government made its use optional. "I look forward to listening to all views this evening," said a Charters Towers city councilor, and Haines told the council of how fluoride is "used as a schedule six poison" and "as an insecticide, particularly for roaches and ants," as well as for "electroplating."

"Please take it out," another woman told the Fraser Coast Regional Council. "It's killing Earth and the rest of us." A man testified, "It's a toxic by-product that cost manufactures a lot of money to dispose of so instead we want to have it our water for to us drink it."

Within weeks, nine cities encompassing almost a half a million residents abandoned the public-health measure. Moreton Bay Regional Councilor James Houghton, who believes fluoridation is mass medication without consent, explained why he was pushing for removal:

> I'm not wrong. Galileo was proven right even though they said he was wrong. Columbus was proven right when others said he was wrong. When I was younger they used to spray us with DDT. Spray us! That's been proven wrong. So science is—makes themselves, provided there's proper research done—they will come up and prove previous reports wrong, so I've adopted an old adage, when in doubt, anyway, leave out. When in doubt, leave out.

Lawrence Springborg, health minister for Queensland, defended the new optional policy with classic Tea Party–like rhetoric. "We understand people want to make different decisions to central government agencies and not be dictated to and we respect that as well," Springborg told Matt Wordsworth of *ABC News*. "We just ask them to have a debate on all the information, if upon that they feel they're uncomfortable to proceed with fluoridation, we do respect that."

Astounded, Wordsworth asked him, "Even if it's to their own detriment?"

"That's a very subjective thing," Springborg replied.

With repeal of the carbon price, religious chaplains in secular public schools, fluoridation elimination, and other issues, Australian antiscience was in full swing. In 2015, on the day after Pope Francis urged world leaders to cut back on fossil-fuel emissions, the Abbott government appointed a "wind czar" to crack down on wind farms, and to research whether they might damage people's health. Australians have "concerns over the localized impacts of wind energy and they deserve a right to be heard," said Environment Minister Greg Hunt.

Banning fluoride wasn't just an Aussie problem—it was happening worldwide. On the other side of the Pacific, the left-leaning US city of Portland, Oregon, voted in May of that year to ban fluoridation. Ban supporters called the mineral a "toxin" and pointed to a discredited propaganda piece that linked fluoride to IQ problems, despite the fact that no such problems have been observed in seventy years of use in the United States. The piece was most likely the spur that drove a worldwide wave of antifluoridation campaigns. Other North American cities banned fluoride as well, including the conservative cities of Calgary, Alberta, and Wichita, Kansas, where the Kansas Republican Assembly campaigned for the ban, and dozens of other, smaller cities. Across the Atlantic, the city council in Dublin, Ireland, voted to oppose it, while the entire country of Israel banned fluoride after Health Minister Yael German, a history major and former mayor, ruled it must be removed from public water supplies over the criticisms of medical associations. Previously she had raised health concerns over cell-phone towers.

Eugenics

In Western Europe, many more countries decided against fluoridation, including Belgium, Denmark, Finland, Greece, Iceland, Italy, Luxembourg, Netherlands, Norway, Portugal, Scotland, and Sweden. In the EU, however, the antiscience purity quest was mostly focused not on what we drink but on what we eat, with conservatives complaining that liberals were panicking over genetically modified food.

"There is a danger, almost unintentionally, that we become antiscience," British prime minister Tony Blair had warned in 2000, speaking of increasing attacks by Green Party members and progressives against government scientists who worked with genetically modified crops. Genetic engineering is, in Europe, still politically tied to the Nazi practice of eugenics, and therefore still causes strong political reactions. Additionally, in Northern Europe especially, the left-wing focus on alternative medicine, holistic health, and bodily purity are major concerns that, when taken to an extreme, drive widespread opposition to fluoride, vaccinations, and genetically modified foods, all antiscience problems that are common in the EU.

While GM crops pose some legitimate economic-justice concerns, science does not support the contention of many anti-GM activists that they are unhealthy to eat. To the contrary, genetic engineering is safer than previous plant-breeding methods because it is more precise, altering just one or a few genes under controlled circumstances instead of blindly altering many through hybridization, or exposing plants to radiation, carcinogenic chemicals, or both in order to cause mutations, a process called mutagenesis (plants modified using mutagenesis can, however, be labeled as "organic"). "Our conviction about what is natural or right should not inhibit the role of science in discovering the truth," Blair said. "Rather, it should inform our judgment about the implications and consequences of the truth science uncovers."

By 2014, the food fight had become even more pronounced, with the United Kingdom threatening to split from European Union authority over science funding, claiming that policies promulgated in Brussels risked "condemning Europe to a new Dark Age."

Tainted

GM opposition wasn't limited to Europe. In China—a far-left country whose politburo is now dominated by engineers but that was once famous for anti-science under Chairman Mao Zedong—anti-GMO sentiment is becoming widespread, with activists describing their cause as "patriotic." The government seems unsure just how to manage the situation. At one 2013 protest, a group chanted slogans calling for the eradication of "traitors" who support GM food, recalling images of scientists and intellectuals being "struggled against" during the Cultural Revolution. Even though the government has prioritized GM research, it is treading cautiously when it comes to actual implementation. Only two GM crops—cotton and papaya—have been approved for production in China. "We must be bold in studying it, [but] be cautious in promoting it," Chinese president Xi Jinping said in a December 2013 speech. Rumors abound that American seed companies are seeking to control the people of China, and that Americans want to weaken or poison China with "dodgy GM food." These factors lead many people—including government officials—to oppose GM crops that could lead to higher yields and, in the case of golden rice, prevent blindness due to vitamin A deficiency.

A 2014 study of public attitudes on environmental issues in twenty leading countries found China at the top, with 91 percent of Chinese agreeing that "we are heading for an environmental disaster unless we change our ways quickly." The United States, in contrast, was at the bottom, with only 57 percent agreeing. Reacting to decades of lax or nonexistent health and environmental regulations, unbreathable air, and undrinkable water, China is seeing the birth of a growing, and potentially massive, environmental movement. When it comes to trusting scientists, 75 percent agreed that "even the scientists don't really know what they are talking about on environmental issues." Ninety-three percent felt that "companies don't pay enough attention to the environment," but individuals felt that they were trying to help the situation, with 88 percent saying, "I try to recycle as much as I can." Chinese were also at the top of international rankings when it came to climate change, with 93 percent saying that "the climate change we are currently seeing is largely the result of human activity." Here, too, the United States brought up the rear of all countries measured, with only 54 percent agreeing.

The distrust of scientists when it comes to environmental issues appears to be a factor in other areas as well. One of the most common worries in today's China is over food safety, a concern that dates back to the great famine under Mao's agricultural policies, when corrupt local officials inflated food production figures to curry political favor, giving the party more than its share as millions of people starved. Even today, local officials are frequently exposed as corrupt, and are often found to be in league with shoddy businesses that cut corners in food safety and production.

The scandals have touched shores around the world with stories of poisoned food. In 2008, it was milk powder tainted with the industrial chemical melamine, which is used in producing plastics and can make milk appear to have more protein, thus increasing its value. The milk powder killed six infants and sickened more than three hundred thousand. Chinese officials tried to cover up the story, fearing public unrest. Similar melamine poisoning occurred in pet food shipped from China to the United States. Then there was the lead paint on wooden baby toys, and the frequent toxic-bean-sprout scandals, the result of growers treating bean sprouts with sodium nitrite, urea, antibiotics, and a plant hormone called 6-benzyladenine in order to make the sprouts grow faster and look "shinier." Then there was the admission by officials in Guangdong Province that 44 percent of local rice tested was laced with dangerous levels of cadmium. After those reports, Shenzhen authorities tested foods made with flour, including dumplings and steamed buns, and found that 28 percent had levels of aluminum above national standards. The contamination was blamed on excessive use of baking powder that contained the metal.

So the Chinese public's reticence about accepting genetically modified foods, given the track record of the agricultural and food sectors in China, is perhaps a reasonable reaction. The difference is that genetic modification is not an ingredient or chemical that can be added to food to fool consumers or slip one by regulators in order to make more money. It is a method of plant breeding—one that could help China feed its burgeoning population at a time when yields are no longer increasing from the use of nitrogen and pesticides, and when climate disruption is placing new stresses on agricultural crops.

Supreme Antiscience

By March 2014, the antiscience cancer had spread full circle, arriving back in America and the thinking of US Supreme Court justices in *Burwell v. Hobby Lobby Stores*. The owners of the Oklahoma-based Hobby Lobby retail chain argued they should not be forced by the government to provide employees with insurance that covers forms of contraception such as the morning-after pill and three similar pills. They believed those types of birth control cause abortions, and therefore had religious objections to them. The future of the Affordable Care Act (the US health care law) hung in the balance.

Several scientific studies have shown that the pills work by keeping a woman's ovary from releasing an egg, not by causing an implanted egg to abort. In rare cases there is also a possibility that a fertilized egg may be prevented from implanting. Some religious conservatives define pregnancy as occurring when an egg is fertilized, but US federal law follows the scientific definition that says a woman can only be considered pregnant when a fertilized egg implants in the uterine wall. Many fertilized eggs are flushed naturally. And yet the justices repeatedly referred to the pills as "abortifacients" during oral arguments—a term adopted from fundamentalists and often used by journalists to describe the pills—even though that is scientifically false, seemingly tilting the scales of justice toward the plaintiff's argument.

Writing for the majority, Justice Samuel Alito said that the pills "may have the effect of preventing an already fertilized egg from developing any further by inhibiting its attachment to the uterus." In a footnote, he conceded that Hobby Lobby's religious-based assertions were in fact contradicted by the science-based federal regulations of the Food and Drug Administration, which had studied the issue for over a decade: "The owners of the companies involved in these cases and others who believe that life begins at conception regard these four methods as causing abortions, but federal regulations, which define pregnancy as beginning at implantation, do not so classify them." It is only at implantation that a woman can be considered pregnant, because it is only at this point that her body begins to undergo the chemical and biological changes of hosting a fertilized egg. However, the court accepted the argument for a religious objection even though it was contraindicated by the scientific definition of pregnancy and federal regulations elsewhere. The administration of the Affordable Care Act, the legal definition of pregnancy in the United States (at least in some cases), and the right of corporations to be considered "religious persons" is now based on political views rather than evidence.

That this is a problem becomes clear when we strip away politics. Consider a science issue in which ideology was not a significant factor in a Supreme Court decision. In June 2013, the court ruled on gene patents in *Association for Molecular Pathology v. Myriad Genetics.* At issue was whether Myriad Genetics, which had isolated two naturally occurring genes implicated in breast cancer, could patent the genes to protect their diagnostic procedures. US patent law limits patents to "any new and useful process, machine, manufacture, or composition of matter, or any new and useful improvement thereof." Clearly, the genes are naturally occurring and so should not have been granted patents, and the High Court got this right. But its opinion incorrectly defined key scientific terms that are material to the case and showed little grasp of the underlying scientific issues being debated. "It's troubling that the highest court in the land can't get even the basic facts of molecular biology right when writing a decision that has such fundamental importance to genetic testing, the biotechnology industry, and health care," wrote Johns Hopkins biomedical engineering professor Steven Salzberg in *Forbes* magazine. "I cannot pretend to know who they got to do their biology background research, but any genetics graduate student could have done far better."

In a world in which advanced molecular biology will increasingly present legal challenges as we parse out what it means to have the power to analyze, edit and design life, this raises serious questions about whether our judicial system is up to the task. The High Court's willingness to redefine medical or scientific terms to accommodate ideological concerns, and its poor grasp of the science underlying major decisions, raises doubts about its ability to deliver justice in an age of advanced science where exact definitions matter even more than they do in the law.

Climate of Denial

By the 2016 US presidential election, the trend had grown worse, with neurosurgeon Ben Carson telling audiences he didn't believe in evolution or the big bang, and Donald Trump telling audiences that vaccines can cause autism and saying he didn't believe in climate change. Marco Rubio and Ted Cruz denied climate change, while Chris Christie said the climate has always been changing and was not a crisis. Bernie Sanders tweeted, regarding climate change, "For those of us who believe in science, you simply cannot ignore what the scientific community is saying almost unanimously"—but he kicked an even greater scientific consensus aside (that GM foods are safe to eat) in favor of requiring labels warning consumers of "what's in the food that they eat." (GM is, again, a process for plant breeding, not an ingredient. GM crops have the same nutritional profile as their non-GM parents.)

Meanwhile, Neil Chatterjee, a top aide to US Republican senate majority leader Mitch McConnell, spent time visiting representatives from foreign embassies to make it clear that Republicans intended to fight any international agreement on climate change. McConnell himself worked to spread the anti-climate agenda internationally, warning foreign governments that "our international partners should proceed with caution before entering into a binding, unattainable deal." *New York* magazine writer Jonathan Chait registered the disbelief and frustration of many: "The speed at which Republicans have changed from insisting other countries would never reduce their greenhouse-gas emissions to warning other countries not to do so—without a peep of protest from within the party or the conservative movement—says everything you need to know about the party's stance on climate change."

Leading up to the 2015 climate summit in Paris, Texas Republican Lamar Smith, chair of the House Science Committee, conducted a months-long probe of National Oceanic and Atmospheric Administration researchers. Beginning in July, Smith sent NOAA letters and subpoenas asking the agency to provide "all documents and communications" related to a study published in *Science* that refuted the so-called global-warming "hiatus"—a favored theory of climate deniers that the planet had not warmed since 1998. NOAA refused to comply with Smith's requests for e-mails, citing the importance of confidentiality among scientists. The American Association for the Advancement of Science and several other scientific societies publicly deplored the science committee's move. "In one fell swoop, you have accused a host of different individuals of wrongdoing," fellow Texas representative Eddie Bernice Johnson, the ranking Democrat, wrote Smith.

> You have accused NOAA's top research scientists of scientific misconduct. By extension, you have also accused the peer-reviewers at one of our nation's most prestigious academic journals, *Science*, of participating in this misconduct (or at least being too incompetent to notice what was going on). If that weren't enough, you

are intimating a grand conspiracy between NOAA and the White House to doctor climate science to advance administration policy. Presumably this accusation extends to [NOAA] Administrator [Kathryn] Sullivan herself. And all of these indictments are conjured out of thin air, without you presenting any factual basis for these sweeping accusations—exposing this so-called "investigation" for what it truly is: a witch hunt designed to smear the reputations of eminent scientists for partisan gain.

The NOAA investigation wasn't the first ideologically motivated attack on science and individual scientists that Smith's committee had conducted. It had previously investigated the National Science Foundation and the Environmental Protection Agency over conclusions Smith didn't agree with. In October 2015, Smith used the committee to begin investigating Jagadish Shukla, a climate scientist at George Mason University who led a group of scientists calling for a Department of Justice investigation into whether the fossil-fuel industry orchestrated a cover-up of dangers from climate change.

In December 2015, during the Paris climate talks, Texas Republican and presidential candidate Ted Cruz, chairman of the Senate Subcommittee on Space, Science, and Competitiveness, held a similarly combative hearing "on the ongoing debate over climate science, the impact of federal funding on the objectivity of climate research, and the ways in which political pressure can suppress opposing viewpoints in the field of climate science." The hearing featured prominent climate deniers repeating discredited talking points designed to cast doubt on the science.

In the end, the Paris climate accord was non-binding, as McConnell had hoped.

Through the Looking Glass, Darkly

How could we have gotten here? How could science, our greatest global source of health, wealth, and power, have somehow become a partisan political football? How did it come to be dismissed out of hand, denied, debated, even reviled by politicians, by large swaths of the voting public, and by judges—even by Supreme Court justices—with no consequences from the media, the law, the government, or the public?

And what will it mean for democracy, in an age dominated by increasingly complex science, that our political and governance structures seem to have so little regard for the role of scientific evidence in democratic decision making?

In short, are the people still sufficiently well-informed to be trusted with their own government?

To understand what is happening—and we must, if we are to have hope of winning the war—we must first understand science's complex and fraught relationship with political power.

Chapter 2

The Politics of Science

There is nothing which can better deserve your patronage than the promotion of Science and Literature. Knowledge is in every country the surest basis of public happiness. In one in which the measures of Government receive their impression so immediately from the sense of the Community as in ours it is proportionably [*sic*] essential.

—George Washington, January 8, 1790

How to Ruffle a Scientist's Feathers

When speaking to scientists, there is one thing that will almost always raise their indignation, and that is the suggestion that science is political. Science, they will respond, has nothing to do with politics.

But is that true?

Let's consider the relationship between knowledge and power. "Knowledge and power go hand in hand," said Francis Bacon, "so that the way to increase in power is to increase in knowledge."

At its core, science is a reliable method for creating knowledge, and thus power. To the extent that I have knowledge about the world, I can affect it, and that exercise of power is political. Because science pushes the boundaries of knowledge, it pushes us to constantly refine our ethics and morality to incorporate new knowledge, and that, too, is political. In these two realms—the socioeconomic and the moral-ethical-legal—science disrupts hierarchical power structures and vested interests (including those based on previous science) in a long drive to grant knowledge, and thus power, to the individual. That process is always and inherently political.

The politics of science is nothing new. Galileo, for example, committed a political act in 1610 when he wrote about his observations through a telescope. Jupiter had moons and Venus had phases, he wrote, which proved that

Copernicus had been right in 1543: the celestial bodies did not all revolve around Earth. In fact, Earth revolved around the sun, not the other way around, as contemporary opinion—and the Roman Catholic Church—held. These were simple observations, immediately obvious to anyone who wanted to look through Galileo's telescope.

But the statement of an observable fact is a political act that either supports or challenges the current power structure. Every time a scientist makes a factual assertion—Earth goes around the sun, there is such a thing as evolution, humans are causing climate change—it either supports or challenges somebody's vested interests.

Consider Galileo's 1633 indictment by the Roman Catholic Church, which was at the time the seat of global political and economic power:

> The proposition that the sun is in the center of the world and immovable from its place is absurd, philosophically false, and formally heretical; because it is expressly contrary to Holy Scriptures.
>
> The proposition that the earth is not the center of the world, nor immovable, but that it moves, and also with a diurnal action, is also absurd, philosophically false, and, theologically considered, at least erroneous in faith.
>
> Therefore . . . invoking the most holy name of our Lord Jesus Christ and of His Most Glorious Mother Mary, We pronounce this Our final sentence: We pronounce, judge, and declare, that you, the said Galileo . . . have rendered yourself vehemently suspected by this Holy Office of heresy, that is, of having believed and held the doctrine (which is false and contrary to the Holy and Divine Scriptures) that the sun is the center of the world, and that it does not move from east to west, and that the earth does move, and is not the center of the world; also, that an opinion can be held and supported as probable, after it has been declared and finally decreed contrary to the Holy Scripture.

Why did the church go to such lengths to deal with Galileo? For the same reasons we fight political battles over issues like climate disruption today: facts and observations are inherently powerful, and that power means they are political. Failing to acknowledge this leaves both science and citizens vulnerable to attack by antiscience propaganda—propaganda that has come to infiltrate politics and much news media coverage and educational curricula in the early twenty-first century. The war on science has steered modern democracy away from the vision held by its founders, and is threatening its survival.

Wishing to sidestep the painful moral and ethical parsing that their discoveries sometimes compel, scientists for the last two generations saw their role as the creators of knowledge and believed they should leave the moral, ethical, and political implications to others to sort out. But the practice of science itself cannot possibly be apolitical, because it takes nothing on faith. The very essence of the scientific process is to question long-held assumptions about the nature of the universe, to dream up experiments that test those questions, and, based on the resulting observations, to incrementally build knowledge that is independent of our beliefs, assumptions, and identities, and independently verifiable no matter who does the measuring—in other words, that is objective. A scientifically testable claim is transparent and can be shown to be either most probably true, or to be false, whether the claim is made by a king, a president, a prime minister, a pope, or a common citizen. Because it takes nothing on faith, science is inherently antiauthoritarian, and a great equalizer of political power. That is why it is under attack.

Who Defines *Your* Reality?

The scientific revolution has proven to be more beneficial to humanity than anything previously developed. By painstakingly building objective knowledge about the way things really are in nature instead of how we would wish them to be, we have been able to double our life spans and boost the productivity of our farms by thirty-five times. With careful observation, recording, testing, and replication, we have been able to give children to those who were "barren" and the fertile the freedom to decide when—and whether—to reproduce, freeing women by providing choice. Science has released us from a life that was, according to Thomas Hobbes, "a war . . . of every man against every man . . . solitary, poor, nasty, brutish, and short."

In Hobbes's era, economics was a zero-sum game: "Without a common power to keep them all in awe," he wrote, men fell into war. There was finite wealth and opportunity, and to get ahead I had to take some of it away from you. In its capacity to create knowledge, science broke that zero-sum economic model and generated wealth, health, freedom, and power beyond Hobbes's wildest dreams. It produced tremendous insights into our place in the cosmos, into the inner workings of our own bodies, and into our capacity as human beings to exercise our highest aspirations of love, hope, creativity, courage, and charity.

Each step forward has come at the price of a political battle. Also, too often, they have come at the cost of the environment. As we continue to refine our knowledge of the way nature really is, independent of our beliefs, perceptions, identities, and wishes for it, we must also refine our ethics and morality, assuming more responsibility for our choices. Inevitably, this is uncomfortable, because the process throws many reassuring notions into conflict with our new knowledge—notions that are often our most deeply rooted, ancient, and awe-struck explanations about the primacy of our clan, the wonders of creation, the specialness of our identities, and the possibility of life after death.

The Power of the Scientific Method

How do we create knowledge? There is no one "scientific method"; rather, there is a collection of strategies that have proven effective in answering our questions about how things in nature really work, as opposed to how they at first appeared to work to our common senses, or to scientists or theologians with less precise measuring tools than the ones we now have. How do plants grow? What is stuff made of? How do viruses work? Why are montane voles promiscuous sex fiends while prairie voles are loyal lifelong mates? The process usually begins with a question about something, and that suggests a strategy for making and recording observations and measurements. If we want to learn how plants grow, for example, we begin by looking at plants, not rocks.

These initial recorded observations suggest a hypothesis: a possible explanation for the observations that partially or fully answers the initial question. This hypothesis must make a risky prediction, one that, if true, might confirm our conclusion or, if false, will destroy it. If there's no possible way to prove the hypothesis is false, then we aren't really doing science. Saying plants grow because God wills it is a statement of faith rather than a statement of science because (a) it's not limited to the natural world, and (b) it cannot be disproved. Therefore it can't be tested and so it can't produce any real knowledge. An article of faith is an assertion. A statement of science can be tested by observing nature to see if it's likely true or not. Nature is the judge.

After we set out our hypothesis, we design and conduct experiments that test the hypothesis and try to disprove it. If we can't disprove it, we conclude that it may be true and write a paper detailing what we did and concluded, and outlining ways the conclusion could be tested further. We send it to a professional journal, which sends it to others who have knowledge of the field (peer reviewers) to see if they can tear any holes in it. Was our method sound? Did we make any mistakes? Did we control all the possible influences on the outcome? Are there other explanations we didn't think of? Was our math right?

If these peer reviewers discover any holes in our logic or methodology, they send it back for more work. But if they conclude that it is solid and transparent enough to stake their reputations on, they recommend our paper for publication. Once it is published the process is not over. Others who read it may then set out to disprove it. If they can, their stars rise and ours fall proportionately. But if they confirm what we found, the conclusion becomes a little more reliable. In this way, we slowly, meticulously create knowledge that is objective: it is independent of our identities, and replicable by anyone.

The method is fallible, since our senses and our logical processes are easily influenced by our assumptions and wishes, and so they often mislead us. But over time the method tends to catch those errors and correct them via peer review and replication. Thus, bit by careful, painstaking bit, we build a literature of what we know, as distinct from our beliefs and our opinions—and as we do, we gain power.

How Old is Earth?

One example of knowledge, as opposed to belief or opinion, is the age of Earth. Geological measurements show, over and over, no matter who does the measuring, that Earth is about 4.54 billion years old. This is something one can learn to measure for oneself. It's called radiometric dating, and it's pretty simple. Radioactive uranium isotopes decay at known, measurable rates into stable (nonradioactive) lead isotopes, and radioactive potassium isotopes decay at known, measurable rates into stable argon isotopes. By using a mass spectrometer one can buy on eBay for about $2,000, one can count how many atoms of a particular uranium isotope are left in a rock and how many atoms of its "daughter," or decayed isotope of lead, there are. One can do the same thing with potassium and argon. Doing some simple math lets one then figure out how old that rock is.

Because the whole solar system is thought to have formed at the same time, we can also look at the ages of meteorites and of rocks astronauts brought back from the moon to get a pretty complete picture of Earth's history, and its age consistently comes up to be about 4.54 billion years old. This isn't something we believe, it's something we measure, like the distance between Minneapolis and Dallas. Claiming that Earth is just six thousand years old is mathematically akin to saying, "No, the distance between Minneapolis and Dallas is not around 942 miles. It's about six and a half feet."

An Old Book or Your Own Eyes

The measurements we've learned how to do using science sometimes conflict with translations of ancient statements in the Bible that, if taken literally, date Earth's age at around six thousand years. These statements were made before we knew how to measure such things as the age of rocks, or what an atom was, much less how to count them with a mass spectrometer. A reasonable person would note this and ask: Is it better to base our knowledge on reading an old book, or on observations of nature before our very eyes? It's a question as old as the scientific revolution. Galileo ran into it often after he began lecturing about what he had seen through his telescope. In 1610, he wrote his friend, the German mathematician Johannes Kepler, "My dear Kepler, what would you say of the learned here, who, replete with the pertinacity of the asp, have steadfastly refused to cast a glance through the telescope? What shall we make of this? Shall we laugh, or shall we cry?"

In 1632, in his book *Dialogue Considering the Two Chief World Systems*, Galileo recounted a tale of a learned man, a natural philosopher, visiting the home of a Venetian anatomist. The anatomist had invited a group in to watch and learn as he dissected a human cadaver. He knew the philosopher believed, as he had read in an old book, that the nerves originated in the heart, and that this was wrong. Here's how Galileo told the story:

> The anatomist showed that the great trunk of nerves, leaving the brain and passing through the nape, extended on down the spine and then branched out through the whole body, and that only a single strand as fine as a thread arrived at the heart. Turning to a gentleman whom he knew to be a Peripatetic philosopher, and on whose account he had been exhibiting and demonstrating everything with unusual care, he asked this man whether he was at last satisfied and convinced that the nerves originated in the brain and not in the heart. The philosopher, after considering for a while, answered: "You have made me see this matter so plainly and palpably that if Aristotle's text were not contrary to it, stating clearly that the nerves originate in the heart, I should be forced to admit it to be true.

If we choose the careful, repeatable science of observation and measurement tied back to nature over the estimates more roughly crafted from the creation stories in the Bible or other old and translated texts, must we reject the rest of religion? Or does it still have value in leading a moral life? But then others ask: Is religion even required for morality? If it is inaccurate, can we take any of it seriously? One can see how easily new knowledge can throw worldviews into conflict.

When Does Life Begin?

Another example of the thorny intersection of science with traditional ideas, law, and politics comes from the biosciences. Careful, reproducible observations and measurements have forced us to repeatedly refine our ideas about what life is and when it begins. Is a human being first a life when it emerges from the birth canal? Does it have any legal rights as a person before then? Or is it a life when it is able to survive independently outside of the womb even if it is removed early, as can happen naturally with premature birth or with a Caesarean section? Or is it perhaps a life at quickening (the moment a mother first feels a fetus move, at about four months), as was the legal standard for a life when America was formed? But wait! Perhaps it is really a life when a fertilized egg first implants in the uterine lining, which, based on observations, is the medical definition of when a pregnancy begins. A woman cannot be said to be pregnant until her body begins the chemical and biological changes that accompany a symbiotic hosting of the embryo, can she? If it does not implant, the egg, even if fertilized, is simply flushed. Here we get into a tricky area, because many religious conservatives say, "No, it is a life when egg and sperm meet," whether or not the fertilized egg ever implants.

But then, a scientist would ask the fundamentalist, is it *still* a life at the moment of fertilization, even if we know from careful observation that one-third to one-half of fertilized eggs never implant, and as many as three-quarters fail to lead to an ongoing pregnancy? And, of course, that brings up more questions: What are fertilized eggs that never implant? How should we define them, if life occurs at fertilization? As miscarriages? Abortions? Nonpregnancies? Suicides? Murders? Something else? What implications might that definition have—legally, ethically, morally—for the use of birth-control pills that inhibit implantation? Is that abortion, murder, or pregnancy prevention?

As our careful observations of life continue, so does our power both to assist and prevent pregnancy. But as our skills improve, new, more troubling questions form. What if we remove the uterus from the process entirely? Is it a life when sperm and egg are joined in a test tube at a fertility clinic and allowed to divide into a group of, say, sixteen cells that are then frozen for future implantation in a woman desperate to have children? Can the woman be said to be "pregnant" as long as this microscopic clump of frozen cells exists? What does *Burwell v. Hobby Lobby Stores* say about that? What, if any, rights should these frozen cells possess? And is a child conceived in this way—a "test-tube baby," as we once called them—without a soul, as was suggested by some religious conservatives in the 1970s? Once born, are the joy they bring and the contributions they make less valuable? If we make a special exception for them, by agreeing that in vitro fertilization is not interfering with God's plan, or by acknowledging that they do appear to have souls, why? On what basis? And what does that make the dozens of frozen cells we discard after a successful pregnancy?

While we're pondering these linguistic, legal, and ethical quandaries, our observations lead us to yet another new understanding. We don't need sperm to fertilize an egg; we can do it with the nucleus of another cell from the same being. We try this, and sure enough, we find we can create many identical genetic copies of a sheep or mouse. We call them clones. But then we have to ask: Is it a life if it is just an ovum that has had its nucleus removed and replaced by the nucleus of another cell, and has then been chemically or electrically shocked to induce the natural process of cell division, without fertilization by sperm? If egg and sperm have never met, is it a life? Or is that creature— possibly, one day, a human—damned or soulless as it was once argued "test tube babies" would be?

Observations tell us that beings produced in nontraditional ways seem to be the same as any other creatures. We have to ask, then, is *every* one of the roughly 1.5 million eggs a woman has in her ovaries at birth a life with rights? When, exactly, does life begin? Is it true, as the comedy troupe Monty Python sang in *The Meaning of Life*, that "every sperm is sacred"?

What happens if we transform adult skin cells into stem cells, and those into sperm and egg, and then fertilize one with the other? Is that a clone or something else? What if we take the troublesome term "fertilization" out of the picture? Is it a life if we design its genome on a computer (as scientists at the J. Craig Venter Institute have done), buy a high-quality DNA synthesizer on eBay for $8,000 or so, use it to make fragments of the genome we designed, chemically stitch the fragments together, inject the complete genome into a cell with an empty nucleus, and shock it into replicating? Here, we have made a living, reproducing thing starting with a computer design and a few common chemicals. What does that mean for our ideas about life and our definition about conception? Is it wrong to be doing this? To be asking these questions? Applying these observations? Gaining these powers?

What is life? Is life an unbroken chain of genetic code, running down through the generations, endlessly recombining in new forms? Is it *software*? Does the software beget the hardware? When does it become an individual with rights? Where do we draw the legal line? The moral line? Can we draw a line at all? Is that the right way to be thinking about it? And if we do, how do we define the terms conception, fertilization, implantation, and pregnancy?

In each of the above cases, new knowledge was gained by applying the scientific method of making careful observations and measurements of nature and recording the data, then testing and drawing conclusions based on the results instead of on assumptions or beliefs, and then publishing those conclusions about how things really appear to be in nature for others to review and attempt to disprove if they can. The knowledge gained through this incredible process has given us new power over the physical world, but it also forces us to reevaluate our intuitive assumptions, and to refine—and, in some cases, redefine—the meanings of words and values we thought we understood when we didn't know what was really going on.

This power and these new definitions have moral, ethical, and legal implications for how we conduct our lives, and this is where science, democracy, and our legal system can come into conflict. As our knowledge becomes more refined and precise, so too must our social contract, and this process is disruptive to moral, ethical, economic, and political authority based on prior definitions and understandings. Science itself is inherently political, and inherently antiauthoritarian.

Antiauthoritarian Politics

Because this is the case, it's reasonable to ask how science fits into political thought. As science writer Timothy Ferris pointed out, in politics there are not just two forces, the progressive left (encouraging change) and the conservative right (encouraging retention). In fact, there are four. Imagined on a vertical axis, there are also the authoritarian (totalitarian, closed, and controlling, at the bottom of the axis) and the antiauthoritarian (liberal, open, and freedom-loving, at the top), which one can argue have actually played much more fundamental roles in human history. Politics, then, can be more accurately thought of as a box with four quadrants rather than as a linear continuum from left to right.

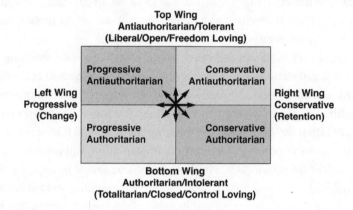

Any one of the infinite gradations of political thought can be placed on the plane around these axes.

When looked at in historical perspective, it's clear that while science and republican democracy are antiauthoritarian systems of knowledge and of governance, respectively, they are neither progressive nor conservative, but are both. Both communism on the left and fascism on the right are authoritarian and opposed to the freedom of inquiry and expression that characterize science and democracy, just as fundamentalist and authoritarian religions are.

Alternatively, left-leaning progressives and right-leaning conservatives can find common cause in the antiauthoritarian principles of freedom of inquiry and expression, universal education, and individual human rights that go hand in hand with the liberal (meaning "free") thinking that informs science and democracy. The life of conservative writer David Horowitz, who was a part of the radical US new-left movement in the late 1960s but is now on the radical right, provides an example of how one can move 180 degrees ideologically from left to right, but still maintain the same general level of liberal antiauthoritarianism vertically.

Democracy: An Endangered Species?

The challenge to authority that science presents is one of many reasons why it has flourished in free, democratic societies, and why those same societies have fallen when they have turned their backs on the freedom science requires in favor of authoritarianism. Nazi Germany is an excellent example. In the

1930s, Berlin was the world pinnacle of science, art, and culture. As we will see later, it was the power of new technology created by that liberal culture that allowed Hitler to come to power, and it was a cultural turn away from freedom that then led to the fleeing of Germany's scientists and artists, and its eventual downfall. It is not a coincidence that the ongoing scientific revolution has been led in significant part by the United States and other free, democratic societies. But it is also partly why, since the late twentieth century, the political climate has increasingly hampered US policymakers and those in other leading democracies in dealing with so many critical science policy issues, and why, by turning away from it, the United States may soon cede both its leadership in scientific research and development and the economic, social, and cultural influence that leadership provides.

Without a well-informed voter, the very exercise of democracy becomes removed from the problems it is charged with solving. The more complex the world becomes, the more challenging it is for democracy to function, because it places an increased burden of education and information upon the people—and in the twenty-first century, that includes science education and science reporting. Without the mooring provided by the well-informed opinion of the people, governments may become paralyzed or, worse, corrupted by powerful interests seeking to oppress and enslave.

For this reason and others, Jefferson was a staunch advocate of free public education and freedom of the press, the primary purposes of which were to ensure an educated and well-informed people. In 1787, he wrote to James Madison,

> And say, finally, whether peace is best preserved by giving energy to the government, or information to the people. This last is the most certain, and the most legitimate engine of government. Educate and inform the whole mass of the people. Enable them to see that it is their interest to preserve peace and order, and they will preserve them. And it requires no very high degree of education to convince them of this. They are the only sure reliance for the preservation of our liberty.

But what do we do when the level of complexity actually *does* require a "very high degree of education"? Can democracy still function effectively?

PART II

The History of Modern Science Politics

Chapter 3
RELIGION, MEET SCIENCE

The value of science to a republican people, the security it gives to liberty by enlightening the minds of its citizens, the protection it affords against foreign power, the virtue it inculcates, the just emulation of the distinction it confers on nations foremost in it; in short, its identification with power, morals, order and happiness (which merits to it premiums of encouragement rather than repressive taxes), are topics, which your petitioners do not permit themselves to urge on the wisdom of Congress, before whose minds these considerations are always present, and bearing with their just weight.

—Thomas Jefferson, 1821

In the Beginning

To understand the ironic situation in which modern democracies find themselves—their major challenges largely revolving around science, yet few elected leaders understanding these issues and few political reporters reporting on them—we have to understand what makes the relationship what it is, and the best place to start is to look at how science shaped or didn't shape the forming of the world's oldest democracy: the United States of America, on which so many other democracies are patterned.

Was America always a nation of science? Or was it founded as a Christian nation? Has there always been a conflict between religion and science? What, exactly, are the relationships among science, politics, freedom, and religion in America and, by extension, in other democracies? Why did science get so advanced there? Most important, why do we keep having conflicts over it? Can democracy—and Earth's ecosystem—survive them?

Contrary to what many fundamentalist politicians and televangelists have claimed, America was not founded as a Christian nation. The land was initially settled by Puritans, but the country was *founded* on the principles of science, something the Puritans valued greatly. In fact, 150 years after the Pilgrims' arrival, the Founding Fathers took great pains to expunge religious thinking from the writings that laid the legal and philosophical foundations for the country they wished to form, beginning with the Declaration of Independence.

They carefully carved out a new, secular form of government based on limited powers for the authorities and reservation of most freedoms for the people, including freedom of inquiry and expression, and freedom of and from religion. The founding documents guaranteed protection of these freedoms, and of the people's right to experiment with and modify their government, by using an eighteenth-century version of crowdsourcing they called democracy. This was something entirely new. It was not coincidental with the scientific revolution, but, rather, a natural outgrowth of it. The liberties these founding principles afforded have in turn produced the highest standard of living, the greatest scientific and technological advances, and the greatest power in the history of the planet.

God's Natural Law Is Reason

Among the first to seize upon the discovery of the New World were the English merchants of Jamestown, Virginia. When their party of 104 men and boys landed in 1607, they were looking for gold, but instead they wound up growing and selling something even more profitable: tobacco.

They were followed some thirteen years later by those seeking not fortune, but freedom of religion. The Puritans began to settle Massachusetts as early as 1620, forming fundamentalist enclaves in niches carved out of the "savage" new continent. They disliked the Catholic-leaning authority of the Church of England and they believed in progress and innovation, that the Bible was God's true law, and that it provided a plan for living.

Puritanism wasn't just a theology, it was a whole set of ideas that included taking an antiauthoritarian, experimental, empirical approach to discovering the *natural laws* by which God's creation abided. In exercising his will, God did not contradict reason. Rather, he revealed himself to humans through two books: the Book of Revelation, made accessible by faith, and the Book of Nature, made accessible by observation and reason. Science was the "handmaiden" to theology, assisting in the study of "the vast library of creation" as a vehicle to religious understanding.

This thinking can be traced to that of the Islamic Mu'tazilites several hundred years before, at a time when Islam was the keeper of scientific knowledge during Europe's Dark Ages. The Mu'tazilites' primary ethos had been to celebrate the power of reason and the human intellect. God spoke not only through the Quran, but through his creations, so we could discern his will by studying nature. It is our intellect, not a literalist reading of an old book, they argued, that guides us toward a true knowledge of God and the basis of morality. The idea resurfaced among Puritans of the late sixteenth century, whose more prominent members had read the old Mu'tazilite books on science.

This idea that God does not contradict reason and that his laws are implicit in nature also lies at the foundation of English common law, as first set forth in Christopher St. Germain's 1518 treatise *The Doctor and Student*, which relates a hypothetical conversation between a doctor of divinity and a student of the laws of England and established common law's moral basis.

St. Germain was a Protestant polemicist during the reign of King Henry VIII, a time when a great battle was raging between the Catholic Church, which was the highest authority in all matters, and the antiauthoritarian Protestants, who promoted do-it-yourself study of the Bible and nature. In fact, *The Doctor and Student* was published just a year after Martin Luther posted his Ninety-Five Theses on a church door. As the reform movement swept through Europe, monks were thrown out of monasteries and told to marry nuns, as Luther did in 1525 when he married an ex-nun. Adherents to Luther's philosophy destroyed and looted the Catholic churches of the bones of their saints and other relics and jewels, condemning the objects as false idols. The truth was to be found in the Bible and in direct experience, they believed, not in the pronouncements of the pope in Rome.

This thinking required a reexamination of the world and the devising of a new order based not upon the authority of the church, but upon reason. The question at hand for St. Germain amid this upheaval was "what be the very grounds of the law of England." He offered first that the law of God underlies reason:

> The law of God is a certain law given by revelation to a reasonable creature, shewing him the will of God, willing that creatures reasonable be bound to do a thing, or not to do it, for obtaining of the felicity eternal.

He then declared that reason and natural law are synonymous:

> As when any thing is grounded upon the law of nature, they say, that reason will that such a thing be done; and if it be prohibited by the law of nature, they say it is against reason, or that reason will not suffer that to be done.

Therefore, nature was knowable, and God's will could be understood by studying nature to discern its laws. This was a powerful idea that put man into an immediate relationship with God and nature, without an intermediary authority figure. Evidence from the study of nature was to be the basis of the laws of England. To the Puritans, then, there were no conflicts in the ideas of religion, law, reason, and science. All were varying examinations of natural law.

The idea of natural law developed over the next ninety years as Protestantism flourished in England. In 1608, the great English jurist Edward Coke sought to more clearly define it. Coke was a Puritan sympathizer who spent his career working to protect individual liberty and make sure the monarchy's arbitrary authority was circumscribed by the rule of law, an idea the Puritans very much favored.

His report, *Calvin's Case*, became a foundational document in English law. In it, Coke wrote,

> The law of nature is that which God at the time of creation of the nature of man infused into his heart, for his preservation and direction.

And in an oft-quoted section of his 1628 *Institutes of the Lawes of England*, Coke broadened this idea in an important way that for the first time turned to the crowdsourcing model the American founders would eventually adopt. Seeking to limit the caprice of the "royal prerogative" by which the king claimed the authority to do whatever he wanted, Coke argued that, while natural law motivated individual men, "an infinite number of grave and learned men" working over "successions of ages" could refine and perfect the laws derived from this initial natural moral basis. In other words, this was an early form of scientific literature and peer review, but also of the idea of the rule of law. Coke called this aggregation "artificiall [*sic*] reason," which he defined as "perfect reason, which commands those things that are proper and necessary and which prohibits contrary things."

Thus, law had a basis in physical reality through the hard-wired biological instincts of humans that God had infused into their hearts at the time of their creation, but its full force and power came from socially aggregating those insights. No one man's authority, even the king's, stood above it.

This science-friendly Protestant perspective—that one could establish law and understand God's will by studying nature and, over time, aggregating and refining a body of knowledge that bound even the king—stood in stark contrast to the position taken by the Roman Catholic Church when, in 1633, church authorities denied the validity of astronomical science and indicted Galileo for heresy for simply describing what he found by observing nature.

The poet John Milton, author of *Paradise Lost*, visited Galileo at Arcetri, the hilly area to the South of Florence where he was under house arrest. Milton told the British Parliament of his visit in 1644, when protesting an order by England to make authors submit their writings first to the government for approval, and he warned them of the dangers of censorship.

> I could recount what I have seen and heard in other countries, where this kind of inquisition tyrannizes; when I have sat among their learned men, for that honor I had, and been counted happy to be born in such a place of philosophic freedom as they supposed England was, while themselves did nothing but bemoan the servile condition into which learning amongst them was brought; that this was it which had damped the glory of Italian wits; that nothing had been there written now these many years but flattery and fustian. There it was that I found and visited the famous Galileo, grown old, a prisoner to the Inquisition, for thinking in astronomy otherwise than the Franciscan and Dominican licensers thought. And though I knew that England then was groaning loudest under the prelatical [church] yoke, nevertheless I took it as a pledge of future happiness that other nations were so persuaded of her liberty.

By the end of the seventeenth century, as Anglican clergy in London were preaching Newton's science, Italian scientists were standing trial in Naples for stating "that there had been men before Adam composed of atoms equal to those of other animals."

The Islamic Keepers of Science

This second fall of Italy as a world leader of science (the first being the fall of the Roman Empire) was not unlike what had happened to the Islamic Empire at the very dawn of the Renaissance. Through the long centuries of the Dark Ages, it was not Christianity but Islam that had kept the flame of science alive. Turkish Ottoman muskets and superior military technology had conquered the Balkans, Ukraine, Crimea, Palestine, Lebanon, Syria, Arabia, and much of North Africa, creating a vast Ottoman empire. Scholars in this golden age of Islam laid the foundation for much of modern Western thinking in ways few people realize today, down to the language and the numerical and mathematical systems we use. The word algebra, for instance, comes from *al-gabr*, Arabic for "completion," one of two ways of solving quadratic equations developed by "the father of algebra," Muhammad ibn Musa al-Khwārizmī, whose last name (al-Khwārizmī), when translated into Latin, is *Algoritmi*, the root of the word algorithm.

Baghdad's House of Wisdom madrasa, where al-Khwārizmī taught, was the largest university and the greatest repository of books in the medieval world. Its scholars studied the ancient Greek, Persian, and Sanskrit texts and, based on what they learned there, developed their own science in astronomy, cartography, chemistry, geography, mathematics, medicine, and zoology. In fact, the approach of using the empirical observation of nature to discover the objective truth of things was first used not by Francis Bacon but by an eminent Islamic scientist, Ibn al-Haytham.

It was al-Haytham's *Optics* and other Arabic texts, translated into Latin, that informed and inspired the early European thinkers of the Renaissance. The first English astronomer, Walcher of Malvern, noted for using an astrolabe to measure the time of several solar and lunar eclipses, was also the first English scholar of Arabic and one of the first translators of Arabic treatises into Latin (from which he likely learned much of his science), in the late eleventh century. Roger Bacon, the thirteenth-century scientist and Franciscan friar, described a cycle of observation, hypothesis, experimentation, and independent verification, which sounds an awful lot like the modern scientific method, and which he got from studying *Optics*. Al-Haytham is the first scientist we know of, as the term is used today, to describe someone guided by empirical observation of nature. *Optics*, written between 1028 and 1038, was translated into Latin and printed in Europe in 1572, and was read by the most influential scientists of the day, including Kepler, Galileo, and Descartes. The book describes the scientific method Francis Bacon would soon champion—to start with observation and induction, being cautious about conclusions and wary of the swaying power of opinion. As al-Haytham put it,

> [We should begin] our investigation with an inspection of the things that exist and a survey of the conditions of visible objects. We should distinguish the properties of particulars, and gather by induction what pertains to the eye when vision occurs and what is found in the manner of sensation to be uniform, unchanging, manifest and not subject to doubt. After which we should ascend in our enquiry and reasoning, gradually and orderly, criticizing premises and exercising caution in regard to conclusions—our aim in all that we make subject to inspection and review being to employ justice, not to follow prejudice, and to take care in all that we judge and criticize that we seek the truth and not be swayed by opinion.

But by the time al-Haytham was read by the great minds of Western science, Muslim freedom of inquiry had long since been sacrificed, and Islamic science was no more. British nuclear physicist and Iraqi scholar Jim al-Khalili, who has written extensively about early Islamic science, notes,

> There were very few . . . Christian scholars whose achievements could rival their Muslim counterparts until the end of the fifteenth century and the arrival of Renaissance geniuses such as Leonardo da Vinci. By that time, European universities would have contained the Latin translations of the works of all the giants of Islam, such as Ibn Sīna, Ibn al-Haytham, Ibn Rushd, al-Rāzi, al-Khwārizmi and many others. In medicine in particular, translations of Arabic books continued to be studied and printed well into the eighteenth century.
> Among the European scholars influenced by their Islamic counterparts before them were Roger Bacon, whose work on lenses relied heavily on his study of Ibn al-Haytham's Optics, and Leonardo of Pisa (Fibonacci), who introduced algebra and the Arabic numeral characters after being strongly influenced by the work of al-Khwārizmi. Some historians have even argued that the great German astronomer Johannes Kepler may have been inspired to develop his groundbreaking work on elliptical orbits after studying the work of the

twelfth-century Andalusian astronomer al-Bitrūji (Alpetragius), who had tried and failed to modify the Ptolemaic model.

But at the very moment Protestantism and the practitioners of the new science were blossoming in Italy, Germany, France, and England, and beginning to draw on the works of their Muslim counterparts, science was shutting down in the Islamic world.

The first reason was politics. A conservative, literalist scientist-theologian named al-Ghazāli, who is influential in Muslim thinking to this day, wrote a critique of Muslim scientists, or Mu'tazilites, called *The Incoherence of the Philosophers*, in which he attacked their assimilation of the ideas of Aristotle and the concept of a natural causality of things. Writing of fire burning cotton, he said,

> The one who enacts the burning by creating blackness in the cotton, [causing] separation in its parts, and making it cinder or ashes is God, either through the mediation of His angels or without mediation. As for fire, which is inanimate, it has no action. For what proof is there that it is the agent? They have no proof other than observing the occurrence of the burning at the [juncture of] only contact with the fire. Observation, however, [only] shows the occurrence [of burning] at [the time of the contact with the fire], but does not show the occurrence [of burning] by [the fire] and that there is no other cause for it.

They should stick closer to the text of the Quran, he argued. Be more "authentic." The cause of things was not nature, but God.

Those who followed al-Ghazāli adopted the same literalist view as Christian fundamentalists do today: the only cause of anything was God, and the texts of the old books, in this case not the Bible but the Quran and the Hadīth (the recorded conversations of the Prophet Muhammad), gave Muslims everything they would ever need to know about their faith, and so the sort of philosophical debate and reasoning practiced by the Mu'tazilites was not only unnecessary—it was un-Islamic. As for science, what was the point? If the cause of everything was God, God was the only answer.

This fundamentalist interpretation led to many of the antiscience, anti-Western beliefs that have held back progress in more fundamentalist Muslim countries to this day. "The innate religious conservatism of the school of thought that grew around [al-Ghazāli's] work inflicted lasting damage on the spirit of rationalism and marked a turning point in Islamic philosophy," argues Al-Khalili.

But the second reason was perhaps even more powerful: the Islamic world's failure to do what the Europeans, and particularly the followers of Martin Luther, were doing: adopt the printing press, a new technology that was making knowledge much more widely available. While devout Muslim scholars were painstakingly hand-copying holy books with artistic fealty, Lutherans were printing Bibles by the thousands and putting knowledge in the hands of the people to judge for themselves.

The DNA of Western Thought

Each arm of the double helix of Western Christianity—Roman Catholicism and the emerging Protestantism—embodied the two distinct worldviews of the authoritarian and the antiauthoritarian: that rules, methods, and laws were either proscribed from on high or built up by individuals in consensus.

These two views had always been present, but they were amplified in 1517, when Martin Luther posted his Ninety-Five Theses challenging church authorities to debate principles that seemed defensible only by virtue of the church's authority over its subjects. In Luther's view, the church had become corrupt, telling people they could buy their way into heaven by purchasing "indulgences," the proceeds of which the church used to finance building St. Peter's Basilica in Rome. "Why does not the pope, whose wealth is today greater than the wealth of the richest Crassus [a legendarily greedy first-century Roman businessman]," Luther asked, "build this one basilica of St. Peter with his own money rather than the money of poor believers?" Luther's theses split the church between those who clung to authority and tradition and those who believed in man's individual connection to God. Protestantism, with its streak of populist antiauthoritarianism, was born.

Luther's grand movement, and the very idea that knowledge could be accessible by individuals without an intervening authority, had been made possible by the 1451 invention of the printing press. For the first time, books could be mass-produced, permitting knowledge and its attendant power to be spread widely. Luther used this new technology to distribute power among the people with his 1534 translation of the Bible from Greek, Hebrew, and Latin into common German. More than one hundred thousand copies of the Luther Bible were sold within forty years of its publication (an unfathomable number for the time), and millions heard its message. People could suddenly study the Bible and come to their own conclusions without the intercession of a pope or priest. The printing press laid the intellectual foundation for the scientific revolution that was to come.

This marked an important moment in human history, when Western thought was split into twin, competing paths: the authoritarian and the anti-authoritarian. The other three major sources of human power—government, economics, and science—developed similar authoritarian, top-down and anti-authoritarian, bottom-up strains of thought over the ensuing centuries as power was demystified.

As in religion, in government there are authoritarian, totalitarian models such as monarchy, dictatorship, and fascism on the one hand and antiauthoritarian models like democracy and anarchy on the other. In economics, communism and capitalism are the opposing theories, as are (less extremely) the ideas of John Maynard Keynes about the need for government stewardship of the economy on the one hand and Milton Friedman's laissez-faire, free-market focus on the other. And in renaissance science, the split fell between the two competing paths of knowledge that were first proposed by the Catholic René Descartes and the Protestant Francis Bacon.

Descartes versus Bacon

Descartes stressed the importance of deduction, a method of reasoning from the top down using "first principles," beginning with his famous line "I think, therefore I am." In his 1637 *Discourse on the Method of Rightly Conducting One's Reason and of Seeking Truth in the Sciences,* he began by embracing skepticism and approaching the entire world with doubt, free of preconceived notions. There was nothing, he said, he could count on as real if all things were subject to skepticism. Ah, but wait—*he* was here, thinking these thoughts. Therefore, he must be real.

Beginning with his mind as the only reliable foundation, Descartes regarded the senses as unreliable and the sources of untruth and illusion. He concluded that reliable truth about reality could be determined only by a mind that was separate and distinct from the physical body, thereby originating the concept of the mind-body split, or Cartesian (from "Descartes") dualism. To Descartes, a conclusion is valid if *and only if* it follows logically from the premise, as do the syllogisms Aristotle defined in his classic book on logic, *Organon*. For example:

All men are mortal. Socrates is a man. Therefore, Socrates is mortal.

Bacon, in contrast, thought Aristotle had gotten it all wrong. He liked the ideas of Ibn al-Haytham and Roger Bacon before him, and he stressed nearly the opposite approach. Bacon was a lawyer who worked under Edward Coke, the attorney general, a position he would eventually assume himself. Toward the end of his legal career, he turned more of his attention to science and published what would become a foundational volume, *Novum Organum Scientiarum*, or "New Organon"—a "new instrument" of science. It was a devastating attack on Aristotle's book and the logic of the Greeks with its emphasis on top-down reasoning and disdain for experimentation. In it he argued instead for using the *inductive* method of reasoning, which underlies much of the scientific method we use today. Inductive reasoning proceeds from the bottom up by observing with the senses and then building in logical steps to reach a general conclusion about reality. An example would be:

All observed swans are white; therefore, all swans are white.

This method clearly has a limitation: its conclusions are provisional and always subject to disproof. All it takes is the discovery of a single nonwhite swan to invalidate the statement. This is why one hears scientists talking about the "theory" of evolution. It is not an observed fact; rather, it is a conclusion that is *supported by all the facts observed so far*, but one can never be absolutely sure because one can never see the whole universe at once, and because of the provisional nature of inductive reasoning, scientists hold out the possibility, no matter how small, that it could be invalidated. Science thus demands intellectual honesty, and a scientific conclusion will always contain a provisional statement:

All observed swans are white; therefore, all swans are *probably* white.

In practice, Bacon's method doesn't bother scientists, or most reasonable people, because the chances of being wrong, while present, are usually in a practical sense very small. It is, for example, theoretically *possible* that chemical processes taking place in your body could cause you to spontaneously combust, but we don't live our lives worrying about it because the *probability* is extremely small. That is why math and statistics have become such important parts of science: they quantify the relative probability that a conclusion is true or false.

Puritan Science

Since Protestantism was rooted in a protest against Catholic authority, Puritans did not take kindly to the Catholic Church's indictment of Galileo, or to the idea that opinions that were supported by observation of nature, and thus were evidence of God's law, could be decreed contrary to holy scripture. In fact, the growing conflict between the Puritans and the Church of England—established in 1534 because the Roman Catholic Church would not annul the marriage of King Henry VIII to Catherine of Aragon, preventing him from marrying again—arose because the Puritans thought the Church of England was not anti-Catholic enough. Thus their name: Puritans.

In 1604, their frustration led King James to authorize a new translation of the Bible, the King James version, to address their concerns. Nonetheless, many Puritans viewed having a monarch as the spiritual leader of the church (as is the case with the Church of England) as an irreconcilable compromise, a substitution of king for pope that had been made solely for the matrimonial benefit of Henry VIII. The monarchy's royal prerogative seemed to them yet another hypocritical corruption of authority, akin to the Catholic Church's "indulgences."

Puritans became even more upset when James's successor, King Charles I, hurriedly married the French Catholic princess Henrietta Maria within two months of his coronation, before Parliament could meet to forbid it. Soon after, he began appointing Catholic Lords to his court. He appointed William Laud the new Archbishop of Canterbury in 1633. Laud replaced the wooden communion tables with stone altars, installed railings around them, and ordered the use of candles, causing Puritans to complain that he was altering the Anglican churches to be more Catholic. Laud responded by closing Puritan churches and firing nonconformist clergy.

Separatist Congregationalist Puritans began emigrating to America again, fearing a return of Catholic absolutism and partisan political reprisals. Other Puritans organized as dissenters to King Charles's use of arcane laws to levy personal taxes and his aggressive authoritarian power grab. Edward Coke, now in Parliament, sought to limit the bottom-wing king's powers, for example by serving as chief author of the Petition of Right, passed in 1628, which laid out several basic rights that the United States would later adopt. Among these were that taxes could be levied only by Parliament, not by the king; that martial law could not be imposed in peacetime; that prisoners had to be able to challenge the legitimacy of their detentions through a writ of habeas corpus; and that soldiers could not be billeted in private residences. But in 1629, Charles rebuffed this attempt, dissolved Parliament, and asserted personal rule by extended royal prerogative. This state of affairs lasted until the English Civil Wars, from 1642 to 1646, in 1648, and from 1650 to 1651, which pitted Royalists (authoritarians) against the largely Puritan Parliamentarians (antiauthoritarians) and eventually led to both Laud's and Charles's beheadings.

After the Civil Wars ended and the Church of England was restored, many Puritans broke away. At first these "nonconformists" were again persecuted, but by the 1660s they were tolerated. Because of their emphasis on individual liberty over external authority, their adherents included some of the greatest minds of the age, including Isaac Newton.

Newton provides an example of how the idea of "science" had not yet fully emerged as something separate from religion in early Enlightenment thinking. In fact, during the seventeenth century, the word "scientist" was not commonly used to describe experimenters at all; they were called "natural philosophers," an extension of the Puritan idea of the study of the Book of Nature. Science had also not fully emerged as a separate concept, but was sometimes thought of as a method or style of study in the arts, rather than a discretely defined set of disciplines. This was true even into Thomas Jefferson's day. Jefferson himself usually used the word to mean what today we call the hard sciences, but sometimes he used it to refer simply to the rigorous study of other fields, such as the "sciences" of language, mathematics, and philosophy.

By 1663, a time when Puritans were a decided minority in England, 62 percent of the natural philosophers of the famed Royal Society of London were Puritans, including Newton, who had studied Ibn al-Haytham's work on light and refraction, and who wrote far more on religion and alchemy than he did on science. Newton believed in the inerrancy of scripture, biblical prophecy, and that the apocalypse would come in 2060. He was "not the first of the age of reason. He was the last of the magicians," said economist John Maynard Keynes, who purchased a collection of Newton's papers in 1936 and was astounded to find more than one million words on alchemy and four million on theology, dwarfing his scientific work. Newton went on to create calculus and to publish *Philosophiae Naturalis Principia Mathematica*, or *Mathematical Principles of*

Natural Philosophy (today, *Mathematical Principles of Science*), upon which modern physics was founded.

Eighty-nine years later, *Principia* was one of the main sources Thomas Jefferson drew upon for inspiration as he sat in the two second-story rooms he had rented from Jacob Graff in Philadelphia, writing the Declaration of Independence.

The Scientist-Politician

Racked by the threat of war and with its political power resting on uncertain ground, in June 1776 the Continental Congress appointed Jefferson, along with Benjamin Franklin, John Adams, Roger Sherman, and Robert Livingston to secretly draft the document. The committee delegated the writing of the first draft to Jefferson.

Like Bacon, who had died of pneumonia after conducting an experiment on preserving meat with snow, Jefferson was both an accomplished attorney and a passionate scientist. On July 4, 1776, the day the Continental Congress eventually adopted the Declaration of Independence, Jefferson took the time to record the local temperature on four separate occasions as part of a broader research project he was conducting. His measurements typically also included barometric pressure and wind speed. His goal was to improve meteorological science to refine farmer's almanacs and improve weather forecasting throughout the colonies, both of which were of personal importance to Jefferson as a farmer.

Jefferson also had knowledge of physics, mechanics, anatomy, architecture, botany, archeology, paleontology, and civil engineering. He was an avid astronomer. He carried a small telescope with him wherever he went and recorded the eclipse of 1778 with great precision, although he was frustrated by the cloudy conditions. As president, he commissioned the Lewis and Clark expedition. He sold it to Congress as an economic initiative, but he sent his presidential secretary, Meriwether Lewis, for training with the top scientists of the day and instructed him to conduct it as a scientific expedition.

Jefferson's love of science is well known among students of science policy. "Science is my passion, politics my duty," he said. In writing to a friend just prior to the end of his term as president of the United States, he said,

> Never did a prisoner, released from his chains, feel such relief as I shall on shaking off the shackles of power. Nature intended me for the tranquil pursuits of science, by rendering them my supreme

delight, but the enormity of the times in which I have lived, has forced me to take part in resisting them, and to commit myself to the boisterous ocean of political passions.

Jefferson was also very familiar with Coke, whose *Institutes* he had studied as a law student. This heady mix of science, law, and politics, and the idea of circumscribing the power of the monarch, would lead Jefferson to carve out a founding document for the United States that was based not on religion or God, but on knowledge and reason. Whereas religious authority and proximity to God could be endlessly argued between different faiths or countries, Jefferson reasoned that a country based on the more narrowly defined rule of men—a democracy—was removed from this, freeing both religion and the government. This being the Enlightenment, Jefferson needed to convince the world's nations that American independence should be respected as rational and correct, and that they should not intercede in the revolution, so he had to build the most inspiring and unassailable Enlightenment argument possible. As his friend and advisor Benjamin Franklin later noted dryly after signing the declaration Jefferson would craft, "We must all hang together or most assuredly we will all hang separately." Their very lives would depend on the quality of Jefferson's argument.

How Do We *Know* Things?

Holed up in his rented rooms, faced with this awesome responsibility, the thirty-three-year-old took up his quill pen. He considered Francis Bacon, Isaac Newton, and John Locke, whom he had studied at the College of William and Mary, to be the three most important thinkers of all time. He called them "my trinity of the three greatest men the world had ever produced." Writing on a portable "lap desk" of his own design, he labored to create a document that reflected the clear, axiomatic logic of John Locke, who instructed that "in all sorts of reasoning, every single Argument should be managed as a mathematical demonstration; where the connexion of ideas must be followed till the mind is brought to the source on which it bottoms." Bottoming his argument out on an irrefutable foundation was what Jefferson needed to do to avoid hanging.

Like Newton and Bacon, Locke was an Englishman and a Protestant, and he is credited with creating the philosophy of empiricism, on which much of modern science is based. He divided human thought into two categories: knowledge and belief. Locke was aware of the many divisions within Christianity, with each faith arguing that it was the one true religion. This was true not only of the great divide between Protestantism and Catholicism, but also of lesser divides between German Lutheranism and English Protestantism, as well as between the Church of England and the dissenters: the Puritans, and within them the sects of Presbyterians, separatist Congregationalists (from whose congregations came the American Pilgrims), and Baptists. Each could not be the one true religion, so some method of ascertaining truth or falsehood had to be developed, or the conflicting claims were likely to go on forever. This led him to ask some fundamental questions: How do we know something to be true? What is the basis of knowledge?

Locke's *An Essay Concerning Human Understanding*, published in 1689, just two years after Newton's *Principia*, strove to answer that question, by laying out what can be known empirically, how it is that we know it, and the inherent limits of knowledge. He began, building on Bacon and Ibn al-Haytham, whom he, too, had studied, with observation of the natural world. He then divided knowledge into three types: intuitive, demonstrative, and sensitive.

Intuitive knowledge is "self-evident" to anyone looking at it, and it carries the least doubt of the three types of knowledge. Three is more than two, black is not white, and the presence or absence of a thing are examples of intuitive knowledge.

Demonstrative knowledge, the second type of knowledge, is slightly less certain than intuitive knowledge. Agreement or disagreement is not immediately clear, but instead depends on the use of reason to demonstrate "by necessary consequences, as incontestable as those in mathematics," that something is so. Each step in a reasoning process—which, as St. Germain had described, was the process of discovering the natural law of things—must in and of itself be intuitively evident. For example:

> I can show you using these two apples in my left hands and these two apples in my right hand that two plus two equals four.

or:

> A feather falls more slowly in air than a penny does. When we remove the air with a vacuum pump, the feather and the penny fall

at the same rate. Therefore, air changes how gravity acts on different objects.

These intermediate steps in reasoning are called "proofs." Each conclusion is reliable because it is ultimately traceable, step by demonstrable step, back to a self-evident foundation in the natural world. It is this mathematical reasoning that would distinguish knowledge as reliable and separate from opinion.

Locke called the third kind of knowledge "sensitive knowledge," meaning that we get it directly from our senses. For example, we may become aware of a rose by its scent, then look for its presence.

But, in a nod to Descartes, he acknowledged that our senses are often wrong. Sometimes it's not a rose we smell, but perfume; sometimes we see not a pond, but a mirage. Locke argued that sensitive knowledge is thus much less certain than intuitive or demonstrative knowledge.

Finally, he said,

> Whatever comes short of one of these, with what assurance soever embraced, is but faith, or opinion, but not knowledge, at least in all general truths.

Seventeen Days in June

This approach was critical to Jefferson because it laid the foundational argument for democracy, which was implicit in a different form in Coke's argument for the primacy of English common law: If we can discover the truth by using reason and observation—i.e. by using science—then anyone can discover the truth, and therefore no one is naturally better able or more entitled to discover the truth than anyone else. Because of this, political leaders and others in positions of authority do not have the right to impose their beliefs on other people. By natural law, the people themselves retain this inalienable right. Based on Locke's ideas of knowledge, and Coke's ideas of law, the antiauthoritarian equality of all men in their ability to use reason to discern the truth for themselves is logically self-evident. It is intuitive knowledge. And that's the heart of—and the most powerful argument for—democracy.

Jefferson worked for seventeen days to craft a document that was grand and yet achieved the unassailable quality of a logical proof. The axiomatic beauty of the argument he was reaching for would indelibly tie science, knowledge, law, freedom, and democracy together in a single common cause of human advancement, and it would proclaim the inalienable right of the people to reject authoritarian tyranny as illegitimate.

But despite his best intentions, in his rough draft, Jefferson foundered on the shoals of authoritarian religious assumptions left over from Hobbes's bleak and brutal era—a fact that illustrates how deeply rooted these assumptions are, even for a scientist like Jefferson, and how slow and careful a process is required to tease out what is knowledge from what, to quote Locke, is "but faith, or opinion." The misstep occurred in the opening of the second paragraph, when Jefferson wrote:

> We hold these truths to be sacred and undeniable; that all men are created equal.

The Edit That Changed the World

When Jefferson showed his draft to Franklin, Franklin made several firm, bold deletions, striking the words "sacred and undeniable." Drawing on Locke, whom he too admired, he replaced Jefferson's reference to divine authority with the antiauthoritarian words "self-evident," which Locke had used in his *Essay*:

> The idea of a supreme being, infinite in power, goodness, and wisdom, whose workmanship we are, and on whom we depend; and the idea of ourselves, as understanding rational beings, being such as are clear in us, would, I suppose, if duly considered and pursued, afford such foundations of our duty and rules of action, as might place morality amongst the sciences capable of demonstration: wherein I doubt not but from self-evident propositions, by necessary consequences, as incontestable as those in mathematics, the measures of right and wrong might be made out to any one that will apply himself with the same indifference [sic] and attention to the one, as he does to the other of these sciences.

Franklin's edit, it may be argued, helped to make the United States into the scientific and technological powerhouse it became, and helped to define democracy as a secular form of government instead of a theocratic one. At the time America's most renowned scientist, Franklin was also an admirer of Newton's *Principia* and a friend of the Scottish economist David Hume. Hume had written extensively on natural law and liberty, which Jefferson had drawn on in the sentence, and he defined liberty as freedom of choice:

> By liberty, then, we can only mean a power of acting or not acting, according to the determinations of the will; this is, if we choose to remain at rest, we may; if we choose to move, we also may.

And even though Newton did not see a conflict between science and religion, neither did he insist upon applying religious thinking to the realm of science, which is the realm of "understanding," as he put it. "A man may imagine things that are false," Newton said, "but he can only understand things that are true."

Newton and Hume both instead rested their arguments on empiricism, "bottoming them out" in the natural world with evidence, and so it had to be with Jefferson's argument for liberty. Hume argued,

> Whatever definition we may give of liberty, we should be careful to observe two requisite circumstances; first, that it be consistent with plain matter of fact; secondly, that it be consistent with itself. If we observe these circumstances, and render our definition intelligible, I am persuaded that all mankind will be found of one opinion with regard to it.

This is the persuasive power that Jefferson was reaching for by tying his arguments back to the plain matter of fact laid bare by his venerated "trinity" of three great men, together with the aggregated authority of grave and learned men in English common law as established by Coke, so that "all mankind [would] be found of one opinion with regard to" the right of the United States to declare its independence.

Franklin understood that Jefferson's words had inadvertently confused the realms of knowledge and faith, resting the principle being argued—that all men are created equal and are endowed by their creator with certain inalienable rights—on an authoritarian, religious assertion, which, as Locke himself had shown, could be argued indefinitely. It was therefore weak as a political argument, a matter of mere *belief*, and anyone with a slightly different interpretation of faith could simply disregard it.

Franklin knew Jefferson was reaching for something more powerful, and he knew how to take it there. He instead rested the principle on reason and Locke's intuitive knowledge, moving the founding argument for the United States firmly out of the realm of religious authority (as in "sacred and undeniable," i.e., "God is on our side," or "God save the monarchy," always arguable assertions) and into the realm of man, reason, and the laws of nature that flowed from empiricism, antiauthoritarianism, and nature itself.

It was *self-evident*.

In the process they created something entirely new: a nation that respected and tolerated religion in every sense, but did not base its authority on religion. A nation whose authority was instead based on the underlying principles of liberty, reason, and science.

Chapter 4

SCIENCE, MEET FREEDOM

When, in the course of human events, it becomes necessary for one people to dissolve the political bonds which have connected them with another, and to assume among the powers of the earth, the separate and equal station to which the laws of Nature and of Nature's God entitle them, a decent respect to the opinions of mankind requires that they should declare the causes which impel them to the separation.

We hold these truths to be self-evident, that all men are created equal, that they are endowed by their Creator with certain unalienable rights, that among these are life, liberty and the pursuit of happiness.

—US Declaration of Independence

The Biology of Democracy

The implications of empiricism for government were profound. Suddenly, kings and popes logically—*empirically*—had no greater claim to authority than anyone else. This was a self-evident truth that proceeded from careful observation of nature through our knowledge-building process—the same process that we use in science. And if it was the case that we were equal and free in the senses these foundational thinkers had described the concepts, then the only form of government that made sense was government of, by, and for the people. The question was how best to implement it.

It can be argued that the method the framers hit upon, democracy, is based to some extent on biology, via its roots in natural law. This argument is supported by recent research in the field of opinion dynamics by Princeton University biologist Simon Levin. His observations of animal herds show that, like human social groups, they follow certain innate rules of organization. A vote is an expression of opinion, and herd animals, quite literally, vote with their feet when determining the overall herd's grazing patterns.

But the voting is not entirely egalitarian. There are opinion leaders in these herds, just as there are in human social groups. The idea of having equality of opportunity—in practical terms, having the right to vote—exists, but, just as in human society, all individuals do not have equal influence. In America, the framers of the Declaration of Independence had more influence in shaping our democracy than the average Virginia tobacco farmer did, and far more than an indentured servant or a slave. In herds, a very few individual opinion leaders make decisions that influence the entire herd's grazing patterns. "Individuals in reality are continually gaining new information, and hence becoming informed," says Levin. "So for sure, any individual could become a leader (just look at elections in the United States). That does not mean that all men are created equal, but in terms of the ability to lead, they all have equal opportunity."

The idea that natural law and hard-wired biological instincts are somewhat synonymous forces lies at the very foundation of modern law. Remember *Calvin's Case*: "The law of nature is that which God at the time of creation of the nature of man infused into his heart, for his preservation and direction." In today's language, this is another way of saying "biological nature," or, more commonly, "human nature"—what drives us naturally, as reasonable creatures. What democracy did was to structure and channel those natural opinion dynamics—what we call natural law—for use in organizing society. Democracy is rooted in our biology.

Science, Art, and Creative Cultures

Science and art are intrinsically related and, in fact, were once one and the same. Both involve the detailed observation and representation of nature in its many aspects; both seek to capture and express some fundamental and perhaps ineffable truth. Both are concerned with the great questions of reality, of life, of an underlying order. Both require a sort of leisured study in a segregated place to maximize creativity, and both are driven forward by an intensely disciplined focus on the craft that can produce astounding bursts of creative insight. Physicists often talk about aesthetic qualities like beauty and symmetry, and indeed there is a long history of art apprehending the forms of nature later uncovered more explicitly by science. Great art and great science both produce a sense of wonderment, and the great artists and scientists are separated from the mediocre ones by the breadth of their minds and the originality of their ideas.

As with mathematicians, who are often also brilliant musicians, scientists frequently seek creative expression in other media such as painting, writing, or sculpture. The exploration of nature, the seeking of insight, the making of things, the importance of technique and finesse—all these drive both art and science.

As natural philosophy, science was considered part of the arts until around 1835, when the term "scientist," which had been in circulation for a few years, was adopted at the annual meeting of the British Association for the Advancement of Science in order to better describe what scientists did in both the philosophical and technical-analytical senses, using the term scientist "by analogy with artist."

Because of the close relationship between innate individual potential and creative expression, on the one hand, and the opportunity afforded by democracy, on the other, science has made the greatest advances in liberal (as in free and open) democratic societies that spread freedom and opportunity broadly, like fertilizer, through tolerance, diversity, intellectual and religious freedom, individual privacy, equal individual rights, free public education, freedom of speech, limited government authority, consideration of minority views, and public support for both arts and research that allow creative minds the opportunity to dream of new things. It is the cross-pollination of ideas that seems to have led to the greatest advances and the value is derived not from encouraging art or encouraging science, but from encouraging and supporting creativity.

These open, democratic societies that are supportive of—and attractive to—creative minds acknowledge the role of opinion leaders, but limit those leaders' actions closely to the feedback of the herd. In *The Science of Liberty*, Ferris shows how powerfully open societies have promoted the wealth and progress of nations, and how fundamentalist, theocratic, and totalitarian governments—in other words, authoritarian governments—have had comparatively few scientific advances. This resulted in those nations falling further behind the liberal democracies both economically and technologically.

This has occurred for three main reasons. The first is intellectual flight. Historically, the brightest and most creative minds have migrated to open societies, and, once there, have made discoveries and created works of art that advanced and enriched those societies. A classic example is the intellectual flight from fascist Europe in the years leading up to World War II. In the 1920s and early 1930s, Berlin was the world capital of science, culture, and art, and these aspects fed off one another. Persecution—particularly of Jews, homosexuals, and artists—spurred emigration that turned the United States into an intellectual mecca. The United States offered these intellectuals freedom, tolerance, egalitarianism, opportunity, and support for their work, and it had the military strength to protect those ideals. In return, the new immigrants gave the United States enormous breakthroughs in chemistry, biology, and physics, and helped shape Hollywood culture, which, together with advanced technology, became America's chief cultural export.

However, scientific leadership proceeds from not only openness but also the degree of opportunity available to creative citizens. By making education free and accessible to all, by stimulating cross-pollination and creativity with a diversity of views and languages and support for research and the arts, by financially supporting scientific research and artistic exploration, and by leveling the economic playing field to provide equal opportunity and freedom of inquiry, democratic societies have broadcast the intellectual fertilizer that helps talented people develop their creative potential wherever they may be— and that creative potential, in turn, benefits those societies.

Finally, these open societies with vibrant cross-fertilization between the sciences and the arts have historically produced innovations that have created new, previously unimagined economies, as well as profound technological breakthroughs that have led to the ability to project physical power over the natural world and against rivals. Personal computers were functional, but it took marrying science and technology with the art of design to make them into the ubiquitous and transformative tools that they became.

Combined, these three factors have had a stunningly powerful effect: even more than empowering individuals, they empower *ideas*. It is this mix of freedom, tolerance, creativity, talent, and diversity in science, in art, and in the social and intellectual interplay between the two that has spawned the great breakthrough cultures that produce new ideas and fresh insights.

A Nation of Thinkers—or Tinkerers?

Jefferson was certainly aware of some of this, and he heavily promoted science during the eight years of his presidency from 1801 to 1809, frequently writing and speaking of its value and importance to the nation and sponsoring major scientific expeditions such as that of Meriwether Lewis and William Clark. Befitting the great westward expansion, in the nineteenth century it was America's pioneer spirit and can-do attitude that produced the world's great inventors and implementers, the great trial-and-error engineers involved in communication, lighting, and power, including Eli Whitney, Samuel Morse,

Alexander Graham Bell, Thomas Edison, George Westinghouse, Nikola Tesla, and many others. But Europe was still the home of real *science* and the scientists—the curiosity-driven experimentalists and theorists—who made the fundamental basic-science breakthroughs, including Alessandro Volta, Michael Faraday, Andre Ampere, George Ohm, Charles Darwin, Marie Curie, James Maxwell, Gregor Mendel, Louis Pasteur, Max Planck, Alfred Nobel, and Lord Kelvin.

This focus on tinkering and engineering versus science and discovery (or, in some ways, applied science versus basic science) was partly because America lacked the well-established academies of Europe, and perhaps partly because of the American obsession with building a new country. But it also seemed to have something to do with the American social character itself. French political scholar Alexis de Tocqueville noted this focus on pragmatism and application when he toured America in 1831 and 1832, some fifty-five years after its birth. His report of what he learned, *Democracy in America*, contains a chapter titled "Why the Americans are More Addicted to Practical than to Theoretical Science." Tocqueville observed that free men who are equal want to judge everything for themselves, and so they have a certain "contempt for tradition and for forms." They are men of action rather than reflection, and hold meditation in low regard. "Nothing," he argued, "is more necessary to the culture of the higher sciences or of the more elevated departments of science than meditation; and nothing is less suited to meditation than the structure of democratic society. . . . A desire to utilize knowledge is one thing; the pure desire to know is another." Tocqueville argued that this relative disregard for basic, curiosity-driven science, on the one hand, and the focus on applied, objective-driven science, on the other, might eventually be the country's downfall. He related a striking cautionary tale that resonates powerfully today:

> When Europeans first arrived in China, three hundred years ago, they found that almost all the arts had reached a certain degree of perfection there, and they were surprised that a people which had attained this point should not have gone beyond it. At a later period they discovered traces of some higher branches of science that had been lost. The nation was absorbed in productive industry; the greater part of its scientific processes had been preserved, but science itself no longer existed there. This served to explain the strange immobility in which they found the minds of this people.

The Chinese, in following the track of their forefathers, had forgotten the reasons by which the latter had been guided. They still
used the formula without asking for its meaning; they retained
the instrument, but they no longer possessed the art of altering
or renewing it. The Chinese, then, had lost the power of change;
for them improvement was impossible. They were compelled at all
times and in all points to imitate their predecessors lest they should
stray into utter darkness by deviating for an instant from the path
already laid down for them. The source of human knowledge was
all but dry; and though the stream still ran on, it could neither swell
its waters nor alter its course.

In other words, China fell under a conservative, authoritarian intellectual fundamentalism that deeply honored tradition but lacked the substance,
freedom, and capacity to create anything new. The craft was there, but the
creativity, art, and science were gone.

Tocqueville concluded that the basic research that had the power to
change the future, to alter the course of the stream at will, was the product of
more liberal European thinking. His tale suggests the dangers posed by embracing the form of science at the expense of the process, of tradition and precedent
at the expense of openness and creativity, of applied research at the expense of
basic science, of fear at the expense of wonder, of utility at the expense of beauty,
and of insisting on financially quantifiable projections before an investment
is made—the idea of which runs contrary to the entire process of discovery
and creativity. Imagine, for example, an insistence on the promise of financial
return prior to Darwin's trips on the *Beagle* or Neil Armstrong's first steps on
the moon. They would never have happened. And yet it is hard to quantify the
enormous wealth that has spun off from those economically unsupportable
adventures into the unknown.

But Tocqueville was being a bit unfair in his assessment, or perhaps just
taking too short a view of history. Europe was the home of the greatest imperial
collapse in Western history, after all, and the collapse was, in many ways, all
about science and culture. Like sixteenth-century China, the Roman Empire
was the inheritor of centuries of scientific, artistic, and philosophical culture
from the ancient Greeks. The ruling classes of Rome were taught this Greek
knowledge and Greek culture.

But Roman intellectual and political culture was much more practical than theoretical—more applied than basic, more form than process. And with this approach Romans became increasingly anti-intellectual. There was an assumption that basic, curiosity-driven science just wasn't necessary. One can see this in the writings of Pliny the Elder. Gaius Plinius Secundus was the most prominent natural philosopher of his day. He published a famous book called *Natural History*, which was based on the much earlier observational, basic science of the Greek researcher Claudius Galen. But Pliny included several other theories in his book. Unlike Galen's work, they weren't based on observations of nature, and they turned out to be mostly wrong. Yet his book became a foundational textbook of the Roman Empire. As the Romans valued science and observation less and less, the artistic and scientific institutions that fed Roman culture began to weaken and decline.

Those who want to dismiss the arguments for basic research—thinking the private sector is the source of today's innovation and the public sector is a laggard—should consider the arguments by economist Mariana Mazzucato in her book *The Entrepreneurial State*. The Internet was a technological creation that came out of basic research. The Apple iPhone, while a creation of Apple as a design package, was based on technology that came out of government-funded basic research. Many universities today receive significant funding from licensing fees paid by private firms commoditizing their basic research. Private industry doesn't have the financial wherewithal to weather the risks that basic research imposes—namely, that a lot of it is wasted looking in the wrong places because of the trial-and-error nature of observational science. But when basic research hits, it hits big, creating entire new economies and transformative breakthroughs. We can't afford *not* to do basic research. The only thing we can be sure of is that, if we don't do it, we won't get the breakthroughs that solve global problems or make trillions of dollars. The private sector is timid by comparison, Mazzucato argues. It's the public sector that can be a catalyst for big, bold, problem-solving ideas, which is why the argument for science and democracy is so essential.

Despite his shortsightedness, it's possible that Tocqueville's general assessment of America may have been correct, and that the United States would have coasted off its vast natural resource exploitation until its economy eventually ran out of growth, but would have never really led the world, were it not for three major developments. The first grew out of Jefferson's insistence on public education, which over time did indeed provide opportunity to undiscovered talent. The second, also heavily encouraged by Jefferson, was the burgeoning

American university system, which was being built up to rival those of Europe. And the third was the American values of tolerance and freedom, which drew talented immigrants from elsewhere. By the first decades of the twentieth century, all three developments were beginning to pay major dividends for America—and particularly for Republicans.

When Conservatives were Pro-Science and Pro-Immigration

In today's Western culture, particularly throughout the former British colonies, conservatives have come to be aligned with vested economic and ideological interests, and have come to be seen as antiscience. Science itself, some conservatives have argued, has a liberal bias, a sentiment comedian Stephen Colbert echoed when he quipped that "reality has a well-known liberal bias" at the 2006 White House Correspondents' Dinner.

In fact, by its very nature, science is both progressive and conservative. It is conservative in that it is retentive of knowledge and cautious about making new assertions until they are fully defensible. But it is also progressive in that it is, and must always be, open to wherever observation leads, independent of belief and ideology, and focused on creating new knowledge.

It would thus be a mistake to characterize scientists as mostly Democrats or mostly Republicans, or mostly liberals or conservatives. They are mostly for freedom, exploration, creativity, caution, and knowledge—and not intrinsically of one or another party, unless a party or political orientation becomes authoritarian and begins to turn its back on evidence. In the early twenty-first century, the political orientation that most stands for freedom, openness, tolerance, caution, and science is the liberals. In the United States, this ideology is represented by the Democrats, which may explain why 55 percent of US scientists polled in 2009 said they were Democrats while only 6 percent said they were Republicans, compared to 35 and 23 percent of the general public, respectively. When one thinks about it, it becomes clear why this is currently the case. The conservative movement has largely become associated with—and financed by— old industry and traditional religion, both of which perceive an existential threat from new science. Rather than supporting exploration of wherever the evidence leads, they have invested big money in an authoritarian model of defending their values and business models, and that means denying science that contradicts those things. The rise of authoritarianism among the Republicans running in the 2016 US presidential elections is an example of this.

Early in the twentieth century this situation was almost reversed. It was the Southern Democrats, defending Jim Crow and traditional religion, who opposed science. Republican Abraham Lincoln had created the National Academy of Sciences in 1863. Republican Teddy Roosevelt, who had grown up wanting to be a scientist, became America's great defender of wildlife and the environment. Republican William McKinley, who would later be admired by Karl Rove, won two presidential elections, in 1896 and 1900, both times over the anti-evolution Democrat William Jennings Bryan, and supported the creation of the Bureau of Standards, which would eventually become today's National Institute of Standards and Technology. Bryan's strident anti-evolution campaigns, culminating in the 1925 Scopes Monkey Trial, helped to drive even more scientists toward the Republican Party.

As an exasperated Republican, the physicist and CalTech chair Robert A. Millikan, wrote in the leading journal *Science* in 1923, the year he won the Nobel prize,

> We have many people even here who hasten to condemn evolution without having the remotest conception of what it is that they are condemning, nor the slightest interest in an objective study *of the evidence in the case which is all that "the teaching of evolution" means*, men whose decisions have been formed, as are all decisions in the jungle, by instinct, by impulse, by inherited loves and hates, instead of by reason. Such people may be amiable and lovable, just as is any house dog, but they are a menace to democracy and to civilization, because ignorance and the designing men who fatten upon it control their votes and their influence.

Other prominent scientists noted the political divide. The great botanist Albert Spear Hitchcock, who would soon become principal botanist at the US Department of Agriculture, wrote the following spring in the same journal that "it is absurd for a scientist to shiver with fear if he sees a black cat cross his path or if he walks under a ladder. It is equally absurd to believe that all Germans or all democrats, or all Roman Catholics . . . are undesirables and a menace to society."

By the early twentieth century, the Democratic Party, which originally grew out of Thomas Jefferson's Anti-Federalists, had become dominated on the national level by Southern religious conservatives and was divided over culture-war issues like evolution, the prohibition of alcohol, restricting immigration, Jim Crow laws, the Ku Klux Klan, and the Catholic faith of Al Smith, the Democratic presidential nominee in 1928. Republicans, by contrast, were the party of Abraham Lincoln and Theodore Roosevelt, of progressive optimism and tolerance, of environmentalism and finance—the party of rationalism and national parks. And by the early 1930s, one of the most famous men in the world was a Republican scientist named Edwin Hubble.

Hubble, who was born in 1889, grew up in Marshfield, Missouri, and then, at the turn of the century, his father John relocated the family to Wheaton, Illinois, where he attended public school and was famous for his athletic prowess. At the time, science was considered a fanciful pursuit and a less-than-solid career path, much like the arts—something suited more for adventurers and wealthy "gentlemen scientists" than professionals. Hubble's father wanted him to be a lawyer, and when Hubble earned one of the first Rhodes Scholarships while a star student of Millikan's at the University of Chicago, he went to the Queen's College at the University of Oxford to study law, not physics.

Still, it was a time when great discoveries were being made in astronomy, which captivated Hubble's imagination. America was entering a golden age of science, propelled in no small part by the massive philanthropic investments of two Republican men: steel magnate Andrew Carnegie, who funded public libraries across the nation, helped found what is now Carnegie Mellon University, and funded basic scientific research through the Carnegie Institution of Washington (since renamed the Carnegie Institution for Science); and John D. Rockefeller Sr., who endowed the University of Chicago as well as Rockefeller University and Johns Hopkins University's School of Public Health. As Hubble began secretly studying astronomy on the side while at Oxford, the imagination of the American public was captured by the growing fame of a former Swiss patent officer with wild hair, an ever-present violin, a playful face, and some mind-blowing ideas—one Albert Einstein.

The Hoax of Relativity

Published in 1916 during World War I, Einstein's general theory of relativity had made the striking prediction that gravity could bend space and so disrupt the straight-line flow of light. On May 29, 1919, with the war over, the British astronomers Sir Arthur Eddington and Andrew Claude de la Cherois Crommelin set out to test the theory by traveling to the island of Príncipe near Africa, and carefully observing the way starlight behaved during a solar eclipse.

If Einstein was right, the sun's gravity would bend the light of stars that were in line with it, making them appear to be slightly offset. The eclipse, which lasted nearly seven minutes, was one of the longest of the twentieth century. It blocked enough sunlight that astronomers could see the stars and measure changes in their apparent locations. If they shifted, Einstein's theory would be proved.

The test's audacity drew the attention of scientists and journalists the world over. If Einstein was wrong, his reputation would be ruined. If he was right, he would be celebrated as a genius whose theory changed everything we thought about the universe. The results were dramatically presented at a November joint meeting of the Royal Society of London and the Royal Astronomical Society, and they confirmed Einstein's predictions spectacularly.

The popular press loved the drama, and Einstein became a household name—a little tramp of a professor who was also a bold genius, with his funny hair and beloved violin, not unlike Charlie Chaplin and his cane. In contrast to Americans' image of the snobby European intellectual, Einstein connected emotionally as an underdog, a trait that appealed to the antiauthoritarian aspect of the American spirit and was cited in press accounts of his "hero's welcome" when he first visited America in 1921, the year he won the Nobel Prize.

America's embrace of Einstein stood in stark contrast to the treatment he was getting at home in Germany. Even though Berlin was the world capital of culture, art, and science, right-wing relativity deniers were on the rise. Like modern climate-science deniers, relativity deniers mounted ad hominem attacks against Einstein, and loudly branded general relativity a "hoax," despite—or perhaps because of—its recent, dramatic scientific confirmation. They were led by an engineer named Paul Weyland, who formed a small but mysteriously well-funded group that held antirelativity rallies around Germany, denouncing the theory's "Jewish nature" and organizing a major event at the Berlin Philharmonic Hall on August 24, 1920. Einstein attended, only to suffer more personal attacks. The political animosity grew so bad that he decided to leave Berlin.

"This world is a strange madhouse," he wrote to a friend three weeks after the rally. "Currently every coachman and every waiter is debating whether relativity theory is correct. Belief in this matter depends on political party affiliation." His words would be echoed decades later by mystified climate scientists.

Even prominent German physicists were getting into relativity denialism, largely along political lines having to do with nationalism and rising anti-Semitism, which, paradoxically, was occurring as Germany was awash in a new liberalism. The winner of the 1905 Nobel Prize in Physics, Philipp Lenard, who had previously exchanged flattering letters with Einstein, had since become bitter about Jews and jealous of the popular publicity Einstein's theory was receiving. He now called relativity "absurd" and lent his name to Weyland's group's brochures. As a Nobel laureate, he worked behind the scenes to try to deny Einstein the prize.

The Greatest Triumph of Satanic Intelligence

At the same time, antiscience had been growing in the United States in reaction to the perceived evils of the theory of evolution. Like relativity, evolution was seen by social conservatives as undermining moral absolutes—in this case, biblical authority. The movement was led in part by an attractive, charismatic sex-symbol and revivalist named Aimee Semple McPherson, the founder of what may have been the first American evangelical megachurch. The 5,300-seat Angelus Temple in Echo Park, Los Angeles, was equipped with radio towers to broadcast her sermons, and McPherson filled it to capacity three times a day, seven days a week. She stressed the "direct-experience" approach to religion, not unlike the empirical spirituality of the Puritans that had been so central to the creation of Western science. Like the Puritans, she considered the mainstream Protestant churches too orthodox—but unlike the Puritans, her complaint was that the mainstream churches were not authoritarian *enough*.

This revivalist spirit was propelled by the wave of immigration that followed World War I, which many Americans found disconcerting; the recovery from the traumatic flu pandemic of 1918, which had killed millions; and the return of millions more from the war, many of them still with untreated "shell shock," the condition we now describe as post-traumatic stress disorder. Fueled by new optimism and cheap labor, the stock market boomed. Moral restrictions were loosening and the country needed to blow off some steam. Materialism soared during what F. Scott Fitzgerald called the Jazz Age, a subject he explored in his 1925 novel *The Great Gatsby*.

But the line between liberalism and running amok depends upon one's psychological keel, as *Gatsby* showed. For many, this powerful mix of materialism, diversity, and newfound tolerance was simply too much, and they began to lose their moral bearings—or to feel that other Americans were losing theirs. McPherson was among the latter group. She set about working to bring order to society, and her moral fierceness offered a bulwark in the storm to many.

By the mid-1920s, McPherson had become a household name. She was made an honorary member of police and fire departments across the country, and, at 10,000 members, she ran the largest Christian congregation in the world. She purchased one of the first three radio stations in Los Angeles, and eventually claimed more than 1,300 affiliated churches that preached her "Foursquare Gospel" of literalist Bible interpretation. From this great platform, she took up a campaign against the two most profound evils threatening America at that time: the drinking of alcohol and the teaching of evolution in public schools.

McPherson was not alone in this holy campaign. William Jennings Bryan, the former secretary of state, had been the Democratic candidate for president for a third time in 1908. He had spoken throughout the United States in favor of Prohibition and against the teaching of evolution, which he believed had led to World War I. If we were in fact "descended from a lower order of animals," he professed, then there was no God and, as a consequence, nothing underpinning society. Like Thomas Hobbes, he felt that, without an absolute authority, society would fall into decay.

Darwin himself had not seen it this way. He had written to John Fordyce about the issue in 1879, saying, "It seems to me absurd to doubt that a man may be an ardent Theist & an evolutionist," though Darwin himself had by then given up his own Christianity. In 1880, he wrote to the young lawyer Francis McDermott that "I am sorry to have to inform you that I do not believe in the Bible as a divine revelation & therefore not in Jesus Christ as the son of God," a view that only became known when the letter was sold at auction in 2015.

Following Bryan's fiery stump speeches warning of the moral decay that teaching evolution would wreak on society, several states passed laws banning the practice. The most notable of these laws was Tennessee's Butler Act, signed into law on March 21, 1925. By April, the American Civil Liberties Union had recruited a substitute teacher named John Scopes to break the law in Tennessee, after which the organization would pay for his defense to challenge it. On the other side, Bryan was asked to represent the World Christian Fundamentals Association, defending the law at the resulting trial, and, with it, his personal reputation and political future.

McPherson was a strong supporter of Bryan during the trial. He had been her guest at the Angelus Temple and had watched her preach that social Darwinism had corrupted students' morality. The teaching of evolution was "the greatest triumph of satanic intelligence in 5,931 years of devilish warfare against the Hosts of Heaven. It is poisoning the minds of the children of the nation," she had said. During the trial, McPherson sent Bryan a telegram, which read, "Ten thousand members of Angelus Temple with her millions of radio church membership send grateful appreciation of your lion hearted championship of the Bible against evolution and throw our hats in the ring with you." The confrontation at the trial between Bryan and Scopes's "sophisticated country lawyer" Clarence Darrow, also a Democrat, was the climax of one of the nation's earliest major scientific-political-religious controversies. Though Darrow lost the case in the Bible Belt state of Tennessee, the accordant publicity turned American public opinion in support of teaching evolution in public schools.

The Vatican stayed out of this debate, partly because its healthy network of parochial schools meant it had little skin in the game—state laws concerning public-school curricula were of little concern. Even today, in Georgia, the joke is, "If you want your kids to learn about evolution, send them to Catholic school, because they won't learn it in public school."

The event also marked a curious milestone: Evangelical Protestants and Roman Catholics had now nearly reversed their respective popular positions with regard to science. While Protestants had once embraced it as Catholics had found themselves at odds, now it was Protestants who were rejecting science and Catholics who were beginning to more fully embrace it—a reversal that astronomer Edwin Hubble would soon help to accelerate.

The Largest Scientific Instrument Known to Man

It was into this hothouse climate that the Protestant-raised Hubble, adorned with the cape, cane, and British accent he had acquired while a Rhodes Scholar at Oxford, returned after the war, having traveled Europe and formed friendships with several of its leading astronomers. He arrived at the Carnegie-funded Mount Wilson Observatory outside Pasadena, California, insisting on being called "Major Hubble." He quickly made enemies among the other scientists with his pompous airs and his self-aggrandizing tall tales. Looking through the great Hooker Telescope—at 101 inches in diameter and weighing more than one hundred tons, it was by far the largest and most powerful scientific instrument in the world—Hubble was able to view the universe with the light-gathering capacity of more than two hundred thousand human eyes.

Despite his propensity for stretching the truth, Hubble was a very strict Baconian observer when it came to science, limiting his statements only to what he observed and what could be strictly concluded from those observations, as John Locke had prescribed. Despite these conservative precautions, or perhaps because of them, what Hubble saw changed humanity's view of the universe forever—and would further roil the controversy over science's role in defining the origins of creation. Hubble photographed a small blinking star in the Andromeda nebula that he identified as a Cepheid variable. Like Galileo's view of Venus, Hubble's observation of a Cepheid in Andromeda would become iconic in its power.

The Human Computer That Opened the Heavens

Hubble's work relied heavily on that of another astronomer, Harvard College Observatory's Henrietta Leavitt, who had in 1912 shown something remarkable about Cepheid variable stars, which change from dim to bright to dim again over a period ranging from a few hours to about a month. Scientists were trying to figure out a way to measure the distance to stars. It was impossible to tell if a star appeared dim because it was far away, or because it didn't emit as much light, so this was a difficult task.

As a woman, Leavitt was not allowed to be part of the scientific staff; she was a "computer"—one of several women hired merely to identify and catalog stars and calculate light curves for the male scientists. Leavitt began to suspect that there might be a relationship between the brightness of a variable star and the length of its period. She reasoned that all stars in the Small Magellanic Cloud were roughly the same distance from Earth and so their apparent brightness could be compared to one another. She then created a graph showing the maximum luminosity of each Cepheid variable compared to the length of its period, and found that there was indeed a relationship. The longer the period, the brighter the star actually was at maximum luminosity.

Danish astronomer Ejnar Hertzsprung seized on Leavitt's insight. Using inductive reasoning, Hertzsprung determined that if two Cepheid variable stars had similar periods but one was dimmer than the other, it was probably farther away. He then searched for and found a Cepheid variable close enough to Earth to measure the distance to it using parallax. Knowing this distance, he measured its apparent brightness and used Leavitt's graph to reverse engineer its actual brightness. The resulting formula, called the Distance-Luminosity relation, allowed scientists to measure the distances to all Cepheid variables, and thus also to the stars and other nearby formations. The blinking stars became "standard candles" throughout the heavens. Leavitt was paid a premium rate of thirty cents per hour over the usual two bits because of the high quality of her work.

The standard candle measurement of space was an immense discovery. In 1915, American astronomer Harlow Shapley, a Democrat, used it and Mount Wilson's sixty-inch telescope to map the Milky Way in three dimensions. Shapley's measurements expanded the known size of the Milky Way severalfold and showed that the sun was not at the center of the galaxy, as had been thought until that moment, but was in fact located in a distant outer arm. This overturned the concept of the centrality of humans yet again, and Shapley was celebrated as the greatest astronomer since Copernicus—a title he himself helped promote—for having achieved the "overthrow" of the heliocentric universe.

The Great Debate: A Cautionary Tale

In what would become an important cautionary tale for science—and, by extension, democracy—Shapley became blinded by his belief that the Milky Way was the entire universe—or maybe by his hubris, in wanting to believe that he had mapped the whole shebang. He argued that the spiral nebulae seen in the heavens were simply wisps of gas and clouds of dust within the Milky Way, rather than entire "island universes"—that is, galaxies—of their own, as fellow astronomer Heber Curtis posited. Shapley debated this point with Curtis at a meeting of the US National Academy of Sciences in April of 1920, in an event famously called the Great Debate.

The debate ended in a draw because there wasn't yet enough observational data to draw firm conclusions. But this was partly because Shapley, without realizing it, had stopped using Locke's and Bacon's inductive reasoning to build knowledge from observation. Instead, he was trying to prove his point with a *rhetorical* argument—an a priori, top-down, Cartesian approach of first principles that had him arguing more like an attorney than a scientist. When his assistant, Milton Humason, showed him a photographic plate that seemed to indicate the presence of a Cepheid variable in Andromeda, Shapley shook his head and said it wasn't possible. Humason had unimpressive formal credentials—he had been elevated to assistant from a mule driver and had only an eighth-grade education—but it was Shapley's a priori ideas that occluded his vision. He took out his handkerchief and wiped the glass plate clean of Humason's grease pencil marks before handing it back. No one realized it at the time, but Shapley's career as a major scientist ended in that moment.

Soon after, Shapley moved on to run the Harvard College Observatory while his tale-telling rival, Edwin Hubble, took over the telescope. Adopting the uncredentialed but brilliant Humason as his assistant and adhering to his strict Baconian observational methodology, Hubble soon identified Cepheids in Andromeda and used them to show that the spiral "nebula," as he called the galaxy, was not part of the Milky Way at all. In fact, it was nearly a million light-years distant—more than three times farther away than the diameter of Shapley's entire known universe. The Great Debate was settled, and Hubble became an overnight sensation.

A Republican Expansion

In 1929, Hubble and Humason followed this accomplishment up by showing that there is a direct correlation between how far away a galaxy is and its redshift—the degree to which its light waves are shifted to the red end of the spectrum. Light waves are emitted at known frequencies. Redshift is caused by a star's light waves stretching out, apparently due to the star's rushing away from Earth, making them lower in frequency and thus shifted toward the red end of the visible-light spectrum.

Scientists had already established that the redshift suggested that light waves were subject to the Doppler effect. We notice the Doppler effect in everyday life when the sound of a train's whistle or a police siren lowers in pitch as it races away from us. Astronomers believed redshift could be used as a measure of the speed at which a star appears to be moving away from us. Hubble correlated that redshift with distance, and then showed via painstaking observation—performed mostly by Humason, whom Shapley had recommended for promotion to the scientific staff—that the farther away a star is, the greater the redshift. The odd couple of the liar and the mule driver found this to be uniformly true in every direction of the sky. This suggested that the universe itself was probably expanding at an even rate.

To picture this, imagine blowing up a perfectly round balloon until it is no longer flaccid but not yet taut. Now take a marker and mark spots in a grid pattern over the entire surface of the balloon, each spot exactly one inch from the next. Now finish blowing up the balloon and watch what happens. As the balloon expands, every dot moves farther away from every other dot. The space between each pair of dots expands. In addition, the dots that are, say, five dots apart from each other move apart five times faster than the dots that are only one dot apart do because the surface is expanding uniformly. Five times the distance, five times the expansion. This is a close analogy to what Hubble and Humason saw happening in three dimensions in the universe.

This fundamental velocity-distance relationship came to be known as Hubble's law, and it is recognized as one of the basic laws of nature. It was a liar who, ironically, using the tools and methods of science, discovered some of the universe's most fundamental truths. But his work also implied something even more momentous.

A Catholic Priest's Big Bang

Georges Lemaître was a pudgy, pinkish Belgian Jesuit abbé—a Catholic priest—and also a skilled astronomer. Like the Puritans, Lemaître was interested in reading the Book of Nature, something the Catholic Church has come to support in the ensuing centuries, funding major astronomical observatories. He was also a reasonably good mathematician, and he had noticed that Einstein's general theory of relativity would have implied that the universe was expanding but for a troublesome little mathematical term called the cosmological constant, which Einstein had inserted into his equations. Lemaître saw no convincing scientific or mathematical reason why the cosmological constant should be there. In fact, Einstein himself had originally calculated that the universe was expanding, but he was a theoretician, not an astronomer. When he turned to astronomers for verification of his theory, he found that almost all of them held the notion that the universe existed in a steady state, and that there was no motion on a grand scale. In deference to their observational experience, Einstein adjusted his general theory calculations with a mathematical "fudge factor"—the cosmological constant—that made the universe seem to be steady.

Lemaître worked independently off the same mathematical principles that Einstein had originally laid out. Nine years later, in 1927, he wrote a dissenting paper in which he argued that the universe must be expanding, and that, if it was, the redshifted light from stars was the result. This redshift had been observed by a number of astronomers, but until then there had been no consensus on the cause.

Lemaître saw Hubble's self-evident observations and clear logic and immediately realized that Hubble's work confirmed his math—and refuted Einstein's general theory of relativity. Furthermore, he deduced, if the universe was expanding equally in all directions, it must have initiated in a massive explosion from a single point. This meant that the universe is not infinitely old; it has a certain age, and that the moment of creation—which British astronomer Fred Hoyle later mockingly called the "big bang"—was analogous to God's first command: *Let there be light.*

Hubble's meticulously reported observations and ironclad, self-evident conclusions convinced Einstein that he may have been wrong to insert the cosmological constant. He made a pilgrimage from Germany to Mount Wilson Observatory outside of Pasadena, where he joined Hubble, Humason, Lemaître, and others to speak with Hubble and examine the stars through the one-hundred-inch telescope. Then he held a news conference. Standing with Hubble, Humason, and the other scientists, he made a stunning public announcement. Unlike Shapley, Einstein changed his mind based on the evidence, and removed the cosmological constant from his general theory of relativity, later calling it "the greatest blunder of my life." The universe was indeed expanding.

Science Rock Star

This dramatic mea culpa by Einstein, who was perhaps the most famous man in the world, drew even more attention to Hubble and the striking depictions of the immense universe coming from the astronomers atop the 5,715-foot Mount Wilson. Breathless newspaper headlines screamed about the gargantuan distances and the millions of new worlds Hubble was discovering. He began to lecture on science to standing-room-only crowds of five thousand people, and he, too, became one of the most famous men in the world for redefining our ideas about our origins.

Decades later, the Hubble Space Telescope would be named in his honor, but, in the 1930s, Hubble's work captured the public interest like that of few scientists before him. He became one of the first great popularizers of science with his traveling and speaking, even delivering, as part of a scientific lecture series, a ten-minute national radio address heard by millions during the intermission of a New York Philharmonic broadcast. He and his wife Grace were the special guests of director Frank Capra at the March 4, 1937, Academy Awards ceremony, where Capra, the academy's president that year, won best director for *Mr. Deeds Goes to Town*. Hubble became the toast of Hollywood, and a long line of actors and directors made the journey up Mount Wilson to peer through the lens of his telescope.

Unlike the admiration that his papers produced, this activity engendered the wrath of his fellow scientists, who scorned him as an arrogant egotist and shameless self-promoter. His protégé, Allan Sandage, discoverer of quasars and further mapper of the universe, said that Hubble "didn't talk to other astronomers very much, but he was certainly not arrogant when he was in the company of other people." Regardless of his temperament, he had the talent to back up his celebrity. And, in part because of the press coverage, the transparency of his work, and his popular speaking, the public felt in on his discoveries, embracing them rather than becoming suspicious.

Among the many celebrities who came to visit Hubble on the mountain was Aimee Semple McPherson. Milt Humason, who was a famous womanizer, told Sandage that, in 1926, during a month-long disappearance in which McPherson claimed to have been kidnapped, tortured, and held for ransom in Mexico, the attractive radio evangelist had actually been up on Mount Wilson, enjoying Humason's special attentions in the Kapteyn Cottage. If true, this would seem an example of the phenomenon of preachers and politicians who attempt to impose rules on society in areas in which they themselves have weaknesses, perhaps seeking to control their own overpowering and unacceptable urges.

Hubble maintained a tolerant but skeptical relationship toward all religions. But according to Sandage (a Democrat), when it came to politics Hubble was a staunch Republican who colluded with other Republican scientists to schedule known Democrats for telescope time on Election Day to prevent them from voting.

In 1951, Pope Pius XII gave a momentous speech in which he addressed Hubble's work and the big bang theory, stating that the big bang proved the existence of God by showing there was a moment of creation, which meant there must be a creator. A friend of Hubble's read the text of the pope's speech in the *Los Angeles Times* and wrote to him,

> I am used to seeing you earn new and even higher distinctions; but
> till I read this morning's paper I had not dreamed that the Pope
> would have to fall back on you for proof of the existence of God.
> This ought to qualify you, in due course, for sainthood.

Hubble, heralded by scientists as the greatest astronomer since Galileo, and loved by the public and the press for his indefatigable popularization of astronomy, had managed to bring the relationship between science and the Roman Catholic Church full circle.

Chapter 5

GIMME SHELTER

Turning and turning in the widening gyre
The falcon cannot hear the falconer;
Things fall apart; the centre cannot hold;
Mere anarchy is loosed upon the world,
The blood-dimmed tide is loosed, and everywhere
The ceremony of innocence is drowned;
The best lack all conviction, while the worst
Are full of passionate intensity.

—William Butler Yeats, 1919

An Intellectual Weapon

Science took an important leap in public consciousness during World War II, when it transformed from an exploration of nature into a means to win the war for democracy and against the tyranny that had overtaken Germany, Italy, and Japan. Radar and the atomic bomb were both Allied inventions that had major impacts on the war's outcome, as did sonar, synthetic rubber, the proximity fuse, the mass production of antibiotics, and other key wartime innovations, with many of the efforts led by emigrants from an increasingly antiscience Third Reich.

The war didn't start out that way, though. In fact, during the 1930s, Adolf Hitler was an early adopter of the latest science and technology, which he used to great political advantage. He forbade smoking around him because German scientists had shown a link between smoking and lung cancer. He based his politics of white supremacy on ideas he appropriated from early research into genetics. He barnstormed twenty-one cities by airplane—the first politician to use an airplane to campaign on that level—in his 1932 race for president against Paul von Hindenburg, an effort the campaign called *"Hitler über Deutschland."*

The Nazi Party mounted gramophones—at the time a relative novelty—on vehicles, using the public's attraction to them to broadcast a uniform political message. Hitler lost the presidential election, but won enough support to be named chancellor in 1933. That year, the Third Reich introduced another weapon with which to spread mass Nazi ideology: the *Volksempfänger*, or "people's receiver," which was offered to the public at low cost and with great success. It had no international shortwave bands, only domestic, which the Nazis filled with propaganda and patriotic music. The world's first regular television broadcast was instituted in Germany beginning in March 1935, with similar goals, and the Third Reich pioneered the use of the classroom filmstrip to inculcate uniform Nazi ideas about politics and racial pseudoscience in students. In short, Hitler placed science and technology in service of politics, leveraging its new power in ways no one had before.

As Hitler's minister for armaments, Albert Speer, recounted at his trial in Nuremberg after the war,

> Hitler's dictatorship differed in one fundamental point from all its predecessors in history. It was the first dictatorship in the present period of modern technical development, a dictatorship which made complete use of all technical means for the domination of its own country. Through technical devices like the radio and the loudspeaker, eighty million people were deprived of independent thought.

Science and technology were employed as tools to spread authoritarian ideology and whip up extreme partisanship and nationalism. The German suspicion of government-conducted science, and the desire of citizens to have greater control over it, is likely a reaction to this misuse of science for ideological ends, and motivates European—and particularly German—attitudes toward science to this day.

Germany also made great strides in mechanized warfare and developed key technological advancements to the submarine and the ballistic missile. But the intolerance of the Nazi regime, and the elevation of authoritarian ideology and propaganda over knowledge and science, began to backfire. Berlin may have become a scientific and cultural capital in the late 1920s and early 1930s, but the Nazis considered the city's artistic and scientific cross-pollination degenerate. As they elevated rhetoric and ideology over science and tolerance,

Germany's intellectuals began to either conform to Nazi authoritarianism or flee. Within a decade, German scientific and technological progress ground to a halt as the Third Reich lost many of its most creative minds to the United Kingdom and the United States.

Presiding over the American science war effort was Edwin Hubble's boss, Vannevar Bush, an engineer and the president of the Carnegie Institution of Washington. There had been a lack of cooperation between the Europe-friendly science enterprise and the US military during World War I, and administrative barriers to the military's adoption of new technologies, that Bush was anxious to avoid repeating—particularly with the vast influx of talent into the United States as a result of the expansion of right-wing totalitarianism across Europe. Albert Einstein was the most famous of these immigrants, but there were many others—some of them gay, many of them Jewish, most of them creative intellectuals from both the sciences and the arts. The entire Frankfurt School decamped and reconstituted itself as part of the University in Exile, a home for German and Italian intellectuals dismissed from their teaching jobs in Europe that was created at the New School in New York City. Later, the university also took in many leading French intellectuals at its École libre des hautes études, or Free School for Advanced Studies. Other universities similarly benefited, as did the US economy as a whole. Patents in the fields the émigrés studied increased by 31 percent over prior years, and the result was an innovation contagion. The émigrés' arrival increased US innovation by attracting a new group of US researchers to their fields, rather than by increasing the productivity of incumbent inventors, according to Stanford economist Petra Moser. US inventors who collaborated with émigré professors began to patent at substantially higher levels in the 1940s and continued to be exceptionally productive in the 1950s, her study found.

The same was true of Southern California, where many of the giants of European cinema fled from Prague and other cities, breathing innovation and creativity into America's fledgling storytelling industry, transforming Hollywood into the world's leading cultural powerhouse. Writers, dramatists, architects, dancers, musicians, and philosophers—the gays, Jews, artists, gypsies, and intellectuals rejected by the Nazi jackbooters—similarly enriched US and UK culture with a flood of new ideas and innovations that created much of the West's postwar culture.

Vannevar Bush saw this growing influx of talent and believed that science and technology would lead to military superiority for whichever country best exploited them. After the Germans invaded Poland in September 1939, Bush became convinced of the need to establish a federal agency that would coordinate US research efforts. He scheduled a hasty meeting in June 1940 with President Franklin D. Roosevelt, who approved the agency in less than ten minutes.

The National Defense Research Committee (NDRC), the forerunner to today's National Science Foundation, was established on June 27. The open society, the wartime esprit de corps, the federal dollars, and the marshaling of talented citizens and émigrés organized the American science enterprise into an intellectual weapon unlike any seen before. Under the auspices of this and a related agency, Bush initiated and oversaw the development of the atomic bomb (until it was taken over by the military), as well as the development of radar, sonar, and numerous other inventions critical to the war effort, in addition to several significant medical advances, including the mass production of penicillin.

The End of Innocence

One of the four top scientists Bush would appoint to lead the NDRC was Harvard president James B. Conant, who was initially in charge of chemistry and explosives. When the NDRC took on the goal of making an atomic bomb before the Germans could, Conant recruited a former Harvard chemistry major, the charismatic and popular University of California, Berkeley, theoretical physics professor J. Robert Oppenheimer, who was recommended by his friend and fellow Berkeley physicist Ernest Lawrence. It was to be physics' finest hour, and Oppenheimer, the poetic son of German Jewish immigrants, who read the Bhagavad Gita in Sanskrit and studied philosophy under Alfred North Whitehead, threw himself into the problem with abandon, assembling a crack team of the best minds in physics, including some of his own top students and several European immigrants. In September 1942, the project was turned over to the military under the command of engineer and brigadier general Leslie Groves. Groves recognized Oppenheimer's brilliance and ambition and appointed him scientific director of what was now code-named the Manhattan Engineer District, or, more simply, the Manhattan Project. The work was "without doubt the most concentrated intellectual effort in history," wrote William Laurence, science reporter for the *New York Times*. Science was to be America's greatest defense against tyranny.

But then, in the blink of an eye, everything changed. The project succeeded, and on August 6, 1945, the United States dropped Little Boy, the first of two of its new bombs of light, on the Japanese city of Hiroshima. On August 9, Fat Man fell on Nagasaki. The bombs proved the power of knowledge once and for all, and Oppenheimer, as the director of the project, was the first public spokesman for the awesome power of science in a new era.

After the euphoria of winning the war had ebbed, the idea that the United States had used science to kill an estimated 110,000 Japanese civilians without any warning—with another 230,000 dying from radiation injuries over the next five years (a side effect the United States at first officially denied)—weighed heavily on Oppenheimer's conscience, and on the American public's collective conscience as well.

In addition to his moral unease, Oppenheimer, like many other leading scientists, had a mounting strategic concern that the Soviet Union, with its vast uranium deposits, would engage the United States in an arms race.

Up to this point, the Allies had regarded themselves as fighting the good fight—honorable, fair, and true, with one hand tied behind their backs, like Superman. The obliteration of two cities of civilians avoided what would surely have been a bloody invasion against a radicalized nation that was using suicide bombers, but it also exposed the dark side of the power that science could unleash, and the horrific consequences that can arise when ethics lag behind knowledge. Mainstream Americans, who had been largely proscience during the 1920s and 1930s and through World War II, now became deeply ambivalent. Was it right, what America had done? Was it honorable? And could it come back to hurt them?

Science, and with it democracy, was growing up, and with increased power came the dawning of a new age of responsibility. Seven weeks after the Hiroshima and Nagasaki blasts, Laurence, whom the War Department had contracted to be the atomic bomb's official historian, characterized this visceral feeling:

> The Atomic Age began at exactly 5:30 Mountain War Time on the morning of July 16, 1945, on a stretch of semi-desert land about fifty airline miles from Alamogordo, NM, just a few minutes before the dawn of a new day on this earth. . . . And just at that instant there rose from the bowels of the earth a light not of this world, the light of many suns in one.

There was a sense that scientists had unlocked a power whose use crossed an ethical boundary—that this act had soiled science and might even destroy humanity. Oppenheimer, the poet-physicist, who thought of a verse from the

Bhagavad Gita upon seeing the first atomic detonation at Trinity test site in New Mexico—"I am become Death, the destroyer of worlds"—spoke of his growing misgivings at the American Philosophical Society in November:

> We have made a thing, a most terrible weapon, that has altered abruptly and profoundly the nature of the world. We have made a thing that by all standards of the world we grew up in is an evil thing. And by so doing, by our participation in making it possible to make these things, we have raised again the question of whether science is good for man, of whether it is good to learn about the world, to try to understand it, to try to control it, to help give to the world of men increased insight, increased power.

Albert Einstein, who had played a key role in alerting President Roosevelt to the possibility of making such a bomb, shared Oppenheimer's misgivings. He sent a telegram to hundreds of prominent Americans in May 1946, asking for $200,000 to fund a national campaign "to let the people know that a new type of thinking is essential if mankind is to survive and move toward higher levels. . . . This appeal is sent to you only after long consideration of the immense crisis we face." The telegram contained what has become one of the most famous quotes in science:

> The unleashed power of the atom has changed everything save our modes of thinking and we thus drift toward unparalleled catastrophe.

Ethical Infants

Today, the idea that everything has changed "save our modes of thinking" might refer not only to the bomb but also to climate change, biodiversity loss and habitat fragmentation, ocean trawling, geoengineering, synthetic biology, genetic modification, mountaintop removal mining, chemical pollution, pollinator collapse, the sixth mass extinction, the deployment of killer robots on the battlefield, and a host of other science-themed challenges we now face. Science and technology have delivered awesome power to governments and industry, but they have not granted us total awareness of the consequences of this power or sustainable mechanisms for its use. We are ever like teenagers being handed the car keys for the first time.

After the war, this feeling—that our scientific ability had outstripped our moral and ethical development as a society, perhaps as a species—was not limited to physicists. The Austrian Jewish biochemist Erwin Chargaff emigrated to the United States to escape the Nazis in 1935. His work would lead to James Watson and Francis Crick's discovery of the double-helix structure of DNA. Chargaff's autobiography described his changed feelings about science:

> The double horror of two Japanese city names [Hiroshima and Nagasaki] grew for me into another kind of double horror: an estranging awareness of what the United States was capable of, the country that five years before had given me its citizenship; a nauseating terror at the direction the natural sciences were going. Never far from an apocalyptic vision of the world, I saw the end of the essence of mankind—an end brought nearer, or even made possible, by the profession to which I belonged. In my view, all natural sciences were as one; and if one science could no longer plead innocence, none could.

Military leaders shared a similar concern. Omar Bradley, the first chairman of the US Joint Chiefs of Staff and one of the top generals in North Africa and Europe during World War II, gave blunt voice to this cultural angst in a 1948 Armistice Day speech:

> Our knowledge of science has clearly outstripped our capacity to control it. We have many men of science, but too few men of God. We have grasped the mystery of the atom and rejected the Sermon on the Mount. Man is stumbling blindly through a spiritual darkness while toying with the precarious secrets of life and death. The world has achieved brilliance without wisdom, power without conscience. Ours is a world of nuclear giants and ethical infants.

The Endless Frontier: From Wonder to Fear

In November of 1944, Roosevelt had asked Vannevar Bush to consider how the wartime science organization might be extended to benefit the country in peacetime—to improve national security, aid research, fight disease, and develop the scientific talent of the nation's youth. After the war was won, Bush submitted his report to President Harry S. Truman. *Science, the Endless*

Frontier, made the case that the creation of knowledge is boundless in its potential. The report is widely credited with laying the groundwork for the second golden age of Western science, during which governments, rather than wealthy philanthropists, became the principal funders of scientific research in peacetime as they had been in war.

In his report, Bush argued that science was of central importance to freedom, an argument that was powerfully underscored when, in August 1949, the Soviet Union detonated an atomic bomb of its own, as Oppenheimer had feared it would. The sense of impending doom over the power scientists had unleashed by splitting the atom turned into the fear of a clear and present danger: nuclear war.

In less than a year, a bill creating the National Science Foundation (NSF) was signed into law, and science began to undergo a subtle but profound change in its relationship to Western culture. For two centuries, it had been motivated by a sense of *wonder* on the part of noble idealists and adventurers, wealthy visionaries, civic-minded philanthropists, and scrappy entrepreneurs. But it was now largely driven by government investments that were, in no small part, motivated by the public's sense of *fear*.

Atomic Terrorism

This fear would impact the world for the next fifty years. Western troops developed a gallows humor when referring to the nuclear weapons they monitored and strapped to the bellies of airplanes. Canadian troops stationed at Zweibrücken, a NATO air base in West Germany, called the bomb "a bucket full of sunshine."

Stateside, American defensiveness bordered on hysteria. Americans knew what a nuclear weapon could do—they'd done it. And now the possibility that it could boomerang back on them was very real. They'd sacrificed and died and put off love, children, and careers in order to beat back the authoritarian threat, and now suddenly it was back in a different way, and what was at risk was society's most precious asset: their children. The baby boom.

The Federal Civil Defense Administration determined that the country that would win a nuclear war was the one best prepared to survive the initial attack. Achieving this required a homeland mobilization on an unprecedented scale, and children needed to know what to do when nuclear war came. The government commissioned a nine-minute film called *Duck and Cover* that showed Bert the Turtle pulling into his shell to survive a nuclear explosion that burns everything else. The film exhorted millions of schoolchildren to "duck and cover" like Bert by covering the backs of their heads and necks and ducking under their desks if they saw a bright flash. The film didn't mention that the gamma-ray burst, which carries most of the lethal radiation, arrives with the flash, nor the fact that school desks were hardly sufficient protection against the flying shrapnel of broken glass and building materials. As the film said,

> Now, we must be ready for a new danger, the atomic bomb. First, you have to know what happens when an atomic bomb explodes. You will know when it comes. We hope it never comes, but we must get ready. It looks something like this: there is a bright flash! Brighter than the sun! Brighter than anything you have ever seen! If you are not ready and did not know what to do, it could hurt you in different ways. It could knock you down hard, or throw you against a tree or a wall. It is such a big explosion it can smash in buildings and knock signboards over and break windows all over town. But, if you duck and cover, like Bert, you will be much safer. You know how bad sunburn can feel. The atomic bomb flash can burn you worse than a terrible sunburn.

The children were reminded regularly that, because a nuclear attack could happen at any time, they, like soldiers in a combat zone, needed to maintain a high level of alertness, forever vigilant to the possibility of attack without warning, ready to duck and cover. As Hiroshima City University nuclear historian Bo Jacobs put it,

> This is the narrative about nuclear war, about the Cold War, and about childhood that millions of American children, the Baby Boomers, received from their government and from their teachers in their schoolrooms: a tale of a dangerous present and a dismal future. Ducking and covering is, after all, a catastrophic pose, one in which the emphasis is on avoiding head injury at the expense of bodily injury: it is the desperate posture of an attempt at bare survival. To duck and cover is to fall to the ground and hope that you live to stand back up. As we watch each setting of childhood succumb to the bright flash of death and destruction in the film, no grown-ups are in sight; it is up to the children to survive the world

that their parents have made for them—a world seemingly without
a future, where survival is measured day to day, minute to minute.

They did drills in school, and participated in citywide mock Soviet atomic-bomb attacks. Many were given metal dog tags so their burned bodies could be readily identified by their parents after a nuclear explosion burned them beyond recognition. "While adults perceived a threat to the American way of life—to their health and wellbeing and those of their families—their children learned to fear the loss of a future they could grow into and inhabit. These kids of the Atomic Age wondered if they might be the last children on Earth," a worry that Jacobs says "had unforeseen and profound effects on the Baby Boomer generation."

At the same time that the federal government was promoting *Duck and Cover*, the Office of Civil and Defense Mobilization (OCDM) was broadly distributing public-service pamphlets whose intent was to instill in everyone a sustained alertness to danger, the better to prepare the country to survive the first wave of a nuclear attack. These pamphlets were bundled with vinyl recordings of survival instructions, such as this one from Tops Records:

Our best life insurance may be summed up in four words: be alert; stay alert. This will take some doing on your part. It will take ingenuity; it will take fervor; it will take the desire to survive. . . . We might label our nuclear weapons "instant death." There is no doubt about it: if you live within a few miles of where one of these bombs strike, you'll die. Instantly. . . . It may be a slow and lingering death, but it will be equally as final as the death from the bomb blast itself. You'll die, unless you have shelter. Shelter from the intense heat, and the radiation that is the by-product of a nuclear explosion. . . . Let's assume bombs fall before you have time to prepare a shelter, or while you wait, in the belief atomic war will never come. We can always hope that man will never use such a weapon, but we should also adopt the Boy Scout slogan: be prepared. . . . It may be safe for you to leave your house after a few hours, or it may be as long as two weeks or more. Two weeks with very little food or water . . . tension . . . unaccustomed closeness. Two weeks with sanitary facilities most likely not operating. No lights. No phone. Just terror.

Fallout shelters were built around the country. New commercial buildings had them. New homes had them, and residents stocked them with water, canned goods, candles, blankets, and tranquilizers. Public buildings had them installed. The possibility of sudden nuclear annihilation at the hands of the Communists became part of everyday life. Many families rehearsed and planned for living for extended periods in dark and rancid basement shelters, and for being separated from one another indefinitely if an attack came when the children were in school.

That the trauma of unending fear and the need for hypervigilance could have psychological or neurological impacts on the developing baby-boom generation didn't occur to psychologists until the social erosion of the 1960s and 1970s was becoming apparent. Yale psychiatrist Robert Jay Lifton, founding member of the International Physicians for the Prevention of Nuclear War, which was awarded the Nobel Peace Prize in 1985, conducted a series of interviews in the late 1970s with people who had been children during the Cold War. Writer Michael Carey was his assistant, and reported on it in the January 1981 issue of the *Bulletin of the Atomic Scientists*.

> This generation had America's only formal and extended bomb threat education in its schools, and that education—along with the lessons about the bomb from government, the media and the family—were well-learned. This generation has a collection of memories, images and words that will not disappear, even for those who profess not to be troubled.

Pulitzer Prize–winning Harvard psychiatrist John Mack was part of a 1977 task force for the American Psychiatric Association that sought to understand the effects of the sustained nuclear threat on children's psyches.

> We may be seeing that growing up in a world dominated by the threat of imminent nuclear destruction is having an impact on the structure of the personality itself. It is difficult, however, to separate the impact of the threat of nuclear war from other factors in contemporary culture, such as the relentless confrontation of adolescents by the mass media with a deluge of social and political problems which their parents' generation seems helpless to change.

Analyzing the results of studies in Canada, the United States, Sweden, Finland, and the USSR, together with her clinical experiences with families and young patients, Canadian psychiatrist Joanna Santa Barbara found that nuclear-age youth were "profoundly disillusioned," and that this affected their capacity to plan for the future. She urged adults to help in young people's efforts to overcome their sense of betrayal.

Betrayal, cynicism, absurdity, and a profound mistrust of authority were overriding themes not only in the books of such popular writers as Ken Kesey and Kurt Vonnegut, but also in results from studies and commentary on the issue at the time. Jacobs writes on the psychological history and effects of nuclear war:

> While children were supposedly being trained to physically survive an atomic attack, *Duck and Cover* also delivered a subtle message about the relationship of children and their world to the world of their parents. "Older people will help us like they always do. But there might not be any grown-ups around when the bomb explodes," the narrator somberly reminds them. "Then, you're on your own." Duck and Cover was designed to teach children that they could survive a surprise nuclear war even in the absence of adult caretakers, conveying a powerfully mixed assurance. The film leaves no doubt that the threat of attack is always imminent and that the key to the survival of these children is their constant mental state of readiness for nuclear war: "No matter where we live, in the city or the country, we must be ready all the time for the atomic bomb. . . . Yes, we must all get ready now so we know how to save ourselves if the atomic bomb ever explodes near us." But the film also reveals that the world children take for granted, the safe world of their childhood, could dissolve at any moment. And when that debacle happens, the adults will be gone; the youngsters will be on their own.

We now know that, in some people, the constant amygdala stimulation of such hypervigilance in childhood can produce long-term effects in the brain that lead to post-traumatic stress disorder. The classic fight-or-flight reaction to perceived threat is a reflexive nervous-system response that has obvious survival advantages in evolutionary terms, say psychiatrists Jonathan Sherin and

Charles Nemeroff, who have published on the neurological changes underlying PTSD. However, the constant stimulation of the systems that organize the constellation of reflexive survival behaviors following exposure to perceived threat can cause chronic dysregulation of these systems. This, in turn, can effectively rewire brain pathways, causing certain individuals to become "psychologically traumatized" and to suffer from post-traumatic stress disorder, which can lead to other mental-health issues, impulsivity, and self-medication with drugs, alcohol, sex, or other forms of addiction.

Later studies showed an increased incidence of mental-health problems among the baby-boomer generation that sharply diverged from prior and subsequent generations, and in many ways fit the PTSD profile. Drug and alcohol use ran at rates far higher than any other generation, as did rates of divorce. Baby boomers also experienced a sharp increase in the adolescent suicide rate versus previous or subsequent generations, a problem that, like drug use and divorce, continued to plague the generational cohort into their later years. Suicide rates among baby boomers continue to run as much as 30 percent higher than other age cohorts, and divorce rates in 2010 among those aged fifty and older ran at twice the rate of the prior generation (measured in 1990) while the overall US divorce rate had fallen 20 percent during the same time.

As the largest age demographic, the baby boomers soon began to rule the cultural conversation across the Western world, and particularly in the United States, with what many commentators called an overweening narcissism. During the sixties, when baby boomers were in their teens, they rebelled and self-medicated with sex, drugs, and rock 'n' roll. In their twenties, as they were getting jobs and finding their identities, they became the me generation. In the 1980s, as they settled down, had families, took up professions, switched to cocaine, and began gentrifying urban neighborhoods their parents had abandoned, they became yuppies: young urban professionals. In the 1990s, as they began investing for retirement, they created the tech bubble. And in the 2000s, as they realized they were mortal, there was a resurgence of big religion in suburban megachurches, and a sharp increase in libertarian, "no new taxes" politics whose goal was to "shrink government until it's small enough to drown in the bathtub." Through it all, the generation has largely maintained its mistrust of—and antipathy toward—both government and science. Over the ensuing decades, the baby boomers felt vindicated in this attitude again and again as government officials from the president on down were discredited, and as environmental science began to expose the damage being wrought by the civic, industrial, and agricultural application of pesticides and other chemicals developed during World War II. The world was a mess, thanks to science.

From Duck and Cover to Run Like Hell

We do not have enough data to say that the threat of imminent nuclear annihilation, and the resulting hypervigilence and learned helplessness, are wholly responsible for any of the social or epidemiological characteristics of boomer culture. But historical records suggest it was a significant contributing factor. Whatever the long-range effects, the threat of nuclear war in the 1950s represented terrorism on a new scale. Imagine that ISIS were in charge of a country the size of the Soviet Union and had nuclear weapons trained on the United States; one can get a sense of the era's fear. Americans knew what these weapons could do, and knew they could be used again. The only option was to plan for an attack on American soil, which was regarded as inevitable. This knowledge changed American culture and its relationship to science in some surprising ways.

For example, it has long been the prevailing opinion that American suburbs developed as a result of the increased use of the car, GI Bill–funded home construction, and white flight from desegregated schools after the 1954 Supreme Court decision in *Brown v. Board of Education of Topeka*. But in reality the trend started several years before *Brown*.

The idea had first been pushed by a utopian short film called *To New Horizons*. Shown at the 1939 World's Fair in General Motors' "Futurama" exhibit, which imagined a city of 1960, the film first introduced the idea of a network of expressways. "On all express city thoroughfares, the rights of way have been so routed as to displace outmoded business sections and undesirable slum areas whenever possible," the film said. But the notion didn't really get traction until 1945, when the *Bulletin of the Atomic Scientists* began advocating for "dispersal," or "defense through decentralization," as the only realistic defense against nuclear weapons. Federal civil defense officials realized this was an important strategic move. Most city planners agreed, and the United States adopted a completely new way of life by directing all new construction "away from congested central areas to their outer fringes and suburbs in low-density continuous development," and "the prevention of the metropolitan core's further spread by directing new construction into small, widely spaced satellite towns."

General Motors, a major US defense contractor, heavily supported the idea, as did other auto, tire, glass, concrete, oil, and construction companies that stood to gain. GM's president, Charles Wilson, became Dwight Eisenhower's secretary of defense in 1953, and, in his Senate confirmation hearing, made the famous blunder of saying that "what was good for our country was good for General Motors and vice versa." The fifty largest US corporations accounted for a quarter of the country's gross national product that year, with GM sales alone exceeding 3 percent. Extending the war footing by redirecting resources toward suburban development was good for business and good for the economy, as well as for national defense.

Nuclear safety measures, supported by business interests, drove the abandonment of US cities. After being told that "there is no doubt about it: if you live within a few miles of where one of these bombs strike, you'll die" and "We can always hope that man will never use such a weapon but we should also adopt the Boy Scout slogan: be prepared," moving far away from the "target" city seemed wise. Those who could afford to left. Those who remained were generally less affluent, and minorities made up a disproportionate share of the poor.

A far worse development for American urban minorities came in 1954, when the federal Atomic Energy Commission realized that, with the advent of the vastly more powerful hydrogen bomb, "the present national dispersion policy is inadequate in view of existing thermonuclear weapons effects." The dispersion strategy was akin to "matching a sleeping tortoise against a racing automobile." By then, however, it was too late; the suburbs were growing rapidly, but offices were still largely downtown. A new strategy was needed— one that had been laid out by General Motors in *To New Horizons*. President Eisenhower promoted a program of rapid evacuation to rural regions via expressways. As a civil defense official who served from 1953 to 1957 explained, the focus changed "from 'Duck and Cover' to 'Run Like Hell.'"

Cities across the United States ran nuclear-attack drills, each involving tens of thousands of residents, practicing clearing hundreds of city blocks in the shortest possible time. It became clear that this would require massive new transportation arteries in and out of cities. The resulting National Interstate and Defense Highways Act of 1956 was the largest public-works project in history. It created a system that provided easier access from the suburbs into cities, as well as a way to more rapidly evacuate urban areas in case of nuclear war. The new freeways had to be built in a hurry and were routed through the cheapest real estate, which usually meant plowing through vibrant minority communities, displacing "outmoded business sections and undesirable slum areas whenever possible" and uprooting millions of people. Although poverty had been concentrated in these neighborhoods, so was a rich culture and a finely woven fabric of relationships, and the neighborhoods' destruction ripped apart the social networks that had supported minority communities for years, leading to a generation of urban refugees.

These defense accommodations—with the encouragement and involvement of what Eisenhower would later regretfully refer to as the "military-industrial complex"—brought about immense changes, altering everything from transportation to land development to race relations to energy use to the extraordinary public sums that are now spent on building and maintaining roads. This created social, economic, psychological, and political challenges that are still with us today—all because of science and the bomb.

The Protection Racket

The fear that was changing the nation kicked up another notch with the Soviet launch of *Sputnik 1*, the first Earth-orbiting satellite, on October 4, 1957. Its diminutive size—about that of a beach ball—made it perhaps the most influential twenty-three-inch-diameter object in history. Traveling at roughly eighteen thousand miles an hour, the shiny little orb circled the planet about once every hour and a half, emitting radio signals that were picked up and followed by amateur radio buffs the world over—but nowhere more closely than in the United States.

Sputnik shocked America in ways that even the 1949 Soviet nuclear test had not. For the first time, the Commies were not just catching up—they were ahead. The fear was that North America stood at risk of being overrun by an authoritarian society. The little orb focused this amorphous fear and placed the entire continent in danger, at least psychologically. In the United States, since the 1949 Soviet nuclear test, debates had been swirling about the need to invest more in education, particularly science, technology, engineering, and mathematics (often referred to as STEM), because of their critical importance to national defense. But until *Sputnik*, these discussions had foundered on the shoals of congressional indifference. Now those debates came into sharp focus. As historian JoAnne Brown put it:

> The struggle for federal aid may have been won in the sky, but it was
> fought in the basements, classrooms and auditoriums, as educators

adapted schools to the national security threat of atomic warfare
and claimed a proportionate federal reward for their trouble.

Within a year, the National Defense Education Act of 1958 was passed,
with the goals of improving education in defense-related subjects at all grade
levels and bolstering Americans' ability to pursue higher education. The NSF's
budget, which had been quite low, jumped dramatically in 1957 and continued
to grow. Science would become a major issue on the presidential campaign
trail in 1960. If Americans didn't recommit to science and technology, it was
argued, they might lose the Cold War. Their entire way of life, perhaps their
very survival, was at stake, and it all hinged on what they could do to protect themselves by reinvesting in science and technology to beat "those damn
Russkies." American public opinion about science, which, for the twelve years
since the bombings of Hiroshima and Nagasaki, had been one of great moral
ambivalence, began a new relationship with it almost overnight. Scientists
might be sons of bitches, but they were American sons of bitches.

When Science Walked Out on Politics

By this time, it was clear that science was the answer to the twin threats of the
arms race and *Sputnik*—and that America was, in fact, in a science race, as
Vannevar Bush had essentially argued in *Science, the Endless Frontier*. Science
had become one of the primary weapons in a new kind of war. The nation that
invested the most in science and engineering research and development would
lead the world—and perhaps, find safety.

In the span of two short decades, science had attained sacred-cow status
enjoyed by few other federal priorities. Gone were the days of scientists need-
ing to reach out to wealthy benefactors to justify and explain their work in
order to get funding. The adoption of science as a national strategic priority
changed the relationship between science and the public. Over the course
of a single generation, government funding allowed scientists to turn inward,
away from the public and toward their lab benches, at the very time that the
public had developed a love-hate relationship with science.

This love-hate relationship came with the conflicting emotions of need
and resentment. Though their work is by nature antiauthoritarian and somewhat artistic, scientists became figures of authority in white lab coats—bland,
dry, value-neutral, and above the fray. This new image of science, implanted
in baby boomers by hundreds of classroom filmstrips, couldn't have been less

inspiring—or further from the truth. Scientists are very often passionate and curious, interested in many things. They often are world travelers and lovers of the outdoors and the arts. These are the very qualities that typically motivate their interest in science—the exploration of creation—to begin with. But very little of that characteristic passion and curiosity would be communicated to the general public for the next fifty years. Science came to be regarded as a culture of monks: intellectual, quietly cloistered, sexually and creatively dry.

With tax money pouring in from a vastly expanding economy and the public respect afforded the authority of the white lab coat, two generations of scientists instead had only to impress their own university departments and government agencies to keep research funds coming their way. But they no longer had to impress the public, which was growing increasingly mistrustful of science.

At the same time, science was becoming less accessible even to other scientists. As knowledge mounted and research became increasingly specialized, no one could keep up with all the latest findings. There was simply too much information. With scientists unable to follow each other outside their own fields, reaching out to the public seemed a hopeless exercise. What mattered was not *process*, but *results*. University tenure tracks rewarded the scientists who had successful research programs and multiple professional publications that attracted large sustaining grants, which, in turn, attracted and funded the top graduate students. But tenure gave no similar consideration to science communication or public outreach. "Those who can, do," the attitude of scientists became, "and those who can't, teach." It was a horrible mistake.

Locked in a subculture of competitive, smart, and passionate people focused on their own research, scientists forgot that they were responsible to—indeed, a part of—the community of taxpayers that funded much of their work and so deserved a say in what they did. Scientists became notoriously cheap donors of both time and money, and withdrew from civic life in other ways. Giving back and participating in the greater civic dialog just wasn't part of their culture or value system. As in any cloistered society, attitudes of superiority developed within the science community—attitudes that ran counter to the fundamentally antiauthoritarian nature of scientific inquiry.

Many scientists, for example, came to view politics as something dirty and beneath them. Arguing that they did not want to risk their objectivity, they eschewed voicing opinions on political issues. So while science was entering its most dizzyingly productive and politically relevant period yet, very little of this creativity was being relayed to the public. Only the results were publicized. From the public's perspective, the science community had largely withdrawn into its ivory tower and gone silent. This proved to be a disaster.

Public Sentiment Is Everything

In democracy, there is a mistaken idea that *politics* is the lowly part of the business—what we have to put up with in order to enact *policy*—but, in fact, the opposite is true. This mistake is made especially often by scientists, who view politics as tainted. Abraham Lincoln eloquently illustrated this point when he debated his opponent Stephen Douglas in the 1858 Illinois campaign for the US Senate. Lincoln lost that election, but he forced Douglas to explain his position on slavery in a way that alienated the Southern Democrats. That set Lincoln up to defeat him in the race for president two years later.

"Public sentiment," Lincoln said, "is everything. With public sentiment, nothing can fail; without it, nothing can succeed. Consequently he who molds public sentiment goes deeper than he who enacts statutes or pronounces decisions. He makes statutes and decisions possible or impossible to be executed."

Thus politics, which moves the invisible hand of democracy, is more important than policy. It reflects and shapes the will of the people. It is the foundation on which policy is based. Lincoln's thinking in this regard echoed that of Thomas Jefferson. It was also a view that industry would soon adopt with a vengeance.

Scientists were certainly smart enough to realize this, but the structure put in place under Vannevar Bush's grand vision worked against it. Who now had to worry about shaping public sentiment? As president of the Carnegie Institution of Washington, Bush was familiar with the time and resources that fund-raising required, and his goal was to lift the onus of obtaining research funding off scientists and universities to propel the nation forward in a more coordinated way. Other nations—both East and West—quickly followed suit. But the need to sell the worth of one's work to the public and donors, to converse about new discoveries and their meaning, and to inspire and excite laypeople may be the only thing that keeps the public invested and supportive in the long term—support that, in a democracy, is critical to sustained effort. Bush may have done his job too well. The shift to public funding changed the incentive structure in science.

This might not have been a problem if scientists had valued public outreach, but, by and large, they didn't. As economists are quick to point out, people often adjust their behavior to maximize the benefit to themselves in any given transaction, and the economics of the new structure rewarded research but not public outreach or engagement. As a result, most scientists ignored it. Science coasted off the taxpayers' fear of the USSR, even as public mistrust was building.

The Two Cultures

The growing divide between science and mainstream culture was famously articulated by British physicist, novelist, and science advisor C. P. Snow, a man who straddled many worlds, like the scientist/artist/statesmen of old. In a famous 1959 lecture titled "The Two Cultures and the Scientific Revolution," Snow warned that the widening communication gulf between the sciences and the humanities threatened the ability of modern peoples to solve their problems:

> A good many times I have been present at gatherings of people who, by the standards of the traditional culture, are thought highly educated and who have with considerable gusto been expressing their incredulity at the illiteracy of scientists. Once or twice I have been provoked and have asked the company how many of them could describe the Second Law of Thermodynamics. The response was cold: it was also negative. Yet I was asking something which is about the scientific equivalent of: "Have you read a work of Shakespeare's?"
>
> I now believe that if I had asked an even simpler question—such as, "What do you mean by mass, or acceleration," which is the scientific equivalent of saying, "Can you read?"—not more than one in ten of the highly educated would have felt that I was speaking the same language. So the great edifice of modern physics goes up, and the majority of the cleverest people in the Western world have about as much insight into it as their neolithic ancestors would have had.

Scientists didn't see this as a warning or an invitation to reach out; rather, they viewed it as a criticism of the willful ignorance and snobbishness of those practicing the humanities. To a certain extent, their view was justified: intellectuals weren't giving their work its due. The fast-growing importance of the sciences was garnering scientists considerable funding and public regard in exchange for the new powers and freedoms they were giving society. Yet that same society's highbrows, especially in academia, still refused to acknowledge their work's significance.

But the threat of nuclear war made survival the priority and relegated other important things to the realm of luxuries. Suddenly, citizens didn't have the luxury of indulging wonder, or the humanities, to the extent they once had. And science had adeptly proven its utility to society, as Snow argued. Although he criticized scientists who could scarcely make their way through Dickens with any understanding of its subtleties, he saved his harshest criticism for British universities—which had underfunded the sciences to the benefit of the humanities, despite the former's contributions—and for the snobbishness of literary intellectuals. "If the scientists have the future in their bones," he said, "then the traditional culture responds by wishing the future did not exist." It was a statement that could just as easily describe the US Congress, or the Canadian parliament, some fifty years later.

The lecture was printed in book form and widely debated in Britain, as well as in Canada and the United States. It has been declared one of the one hundred most influential Western books of the last half of the twentieth century.

For a solution, Snow envisioned the emergence of a third culture of people schooled in both the sciences and the humanities. But that is not what took place. A great change had begun in Western universities, and humanities professors felt themselves slipping from the top spots and being supplanted by scientists, who generally seemed as if they couldn't have cared less about the humanities. Why bother with all the reading and writing and talking when science was actually *doing* things? But this was equally shortsighted, and in this shift the West let go of something precious: a grasp on the classics that had informed Western culture. Since scientists couldn't be bothered with civics, democracy continued to draw its elected leaders primarily from the humanities, creating a culture war that is with us to this day, threatening the ability of Western democracies to solve their problems—just as Snow feared.

Democracy is, as we now know, rooted in science, knowledge, and the biology of natural law. But most of our elected leaders have not had significant training in science, or, more importantly, in how the foundational ideas of modern law and democracy relate to, and grew out of, science. In the middle years of the twentieth century, this was beginning to pose a problem.

The twin threads of fear and resentment created a growing sense that science might be outpacing the ability of a democracy to govern itself. The situation was alarming enough that it compelled President Eisenhower to warn the

American people about it. On January 17, 1961, in his farewell address to the nation, he famously warned of the dangers of the emerging "military-industrial complex." Ike blamed the rise of this behemoth on the federal government's growing funding of science, and he complained that the solitary inventor was being overshadowed by teams of scientists in cloistered labs, hidden from the watchful eye of the public and awash in taxpayer money. "[I]n holding scientific research and discovery in respect, as we should," Ike warned, "we must also be alert to the equal and opposite danger that public policy could itself become the captive of a scientific-technological elite."

How far the United States had come from the days of the first State of the Union address, when George Washington told Congress that "there is nothing which can better deserve your patronage than the promotion of Science and Literature." Democracy itself had been created by a scientific-technological elite that had included Thomas Jefferson, Benjamin Franklin, George Washington, Benjamin Rush, and other Founding Fathers. Elitism had been something to aspire to. Now, thanks to its association with a cozy cabal of military officers and allied contractors, science had become something to be feared.

Chapter 6

SCIENCE, DRUGS, AND ROCK 'N' ROLL

Doubt has replaced hopefulness—and men act out a defeatism that is labeled realistic. The decline of utopia and hope is in fact one of the defining features of social life today. The reasons are various: the dreams of the older left were perverted by Stalinism and never recreated; the congressional stalemate makes men narrow their view of the possible; the specialization of human activity leaves little room for sweeping thought; the horrors of the twentieth century, symbolized in the gas-ovens and concentration camps and atom bombs, have blasted hopefulness. To be idealistic is to be considered apocalyptic, deluded. To have no serious aspirations, on the contrary, is to be "toughminded."

—The Port Huron Statement of the Students
for a Democratic Society, 1962

Curiosity-Driven versus Goal-Driven Science

There is a dynamic tension between the two types of science—basic, curiosity-driven science, and applied, goal-driven science—that began to emerge out of both industry and the war effort in World War II. They aren't so much two types as two ends of the spectrum. Basic science was the realm of the gentleman and philanthropist-funded explorers, and, after World War II, the recipient of about half of the spending on government-funded research centers and university science labs. Applied science was the realm of engineering-oriented American entrepreneurs like Edison, Tesla, and Bell, and later of the major corporate research programs like pharmaceutical research, as well as goal-driven government projects like the Manhattan Project and the majority of other military-funded research.

Much of the public's skepticism toward science derives from the narrow focus of military- and industry-funded applied research, and, in the early years, the industrial applications of war-effort-funded research such as the development of pesticides—research that, too often, was applied without regard to its wider consequences, and that has historically been weakly regulated.

And then there were the public-private research programs of the military-industrial complex. Despite the fact that the bulk of both applied and basic government research has historically been about improving health, military-funded applied research began to drive the political conversation about almost all of government science.

The Graduates

Eisenhower's warning about science as part of the military-industrial complex fed into the momentous changes afoot in American, British, and broader Western culture. Traumatized by a dozen years of high alert to the threat of nuclear holocaust, the public's patience was wearing thin. The enormously powerful hydrogen bomb had made "duck and cover" a ludicrous farce. Advances in government- and industry-funded applied science were viewed with increasing skepticism. The baby boomers found their generational power by questioning the authority of the government and, by extension, government science. Instead of the solution, government was seen as the problem, as Ronald Reagan would point out two decades later.

That their parents were incapable of providing safety, that their world might end at any moment, that the government—their government—had brought this threat into their lives, that their parents' generation had managed to screw things up so badly while at the same time celebrating the "victory culture" of the postwar years, that the stated ideals of democracy didn't seem to extend to blacks or, during the war, to Japanese, who had been held in American concentration camps, that the military-industial complex seemed to be increasingly captivating public policy for its own profiteering ends—all this combined to increase the cynicism, rebellion, mistrust, and antigovernment sentiment that fueled the baby boomers' late adolescence and early twenties. This gave rise to a new counterculture. Adults and the government had lost their moral authority, hedonism was justifiable since death might come at any moment, and the young ruled the cultural conversation. These conditions were described in the seminal document of the new left movement that came to define the counterculture, the 1962 Port Huron Statement of the Students for a Democratic Society:

As we grew . . . our comfort was penetrated by events too troubling to dismiss. First, the permeating and victimizing fact of human degradation, symbolized by the Southern struggle against racial bigotry, compelled most of us from silence to activism. Second, the enclosing fact of the Cold War, symbolized by the presence of the Bomb, brought awareness that we ourselves, and our friends, and millions of abstract "others" we knew more directly because of our common peril, might die at any time. We might deliberately ignore, or avoid, or fail to feel all other human problems, but not these two, for these were too immediate and crushing in their impact, too challenging in the demand that we as individuals take the responsibility for encounter and resolution.

While these and other problems either directly oppressed us or rankled our consciences and became our own subjective concerns, we began to see complicated and disturbing paradoxes in our surrounding America. The declaration "all men are created equal . . ." rang hollow before the facts of Negro life in the South and the big cities of the North. The proclaimed peaceful intentions of the United States contradicted its economic and military investments in the Cold War status quo.

Scientists and engineers as a whole were suddenly seen as associated with the military-industrial complex, as were organized religion and other sources of authority, all of which had failed this young generation. The culture at large began rejecting science and tradition and the government itself—and with it, government contractors like Honeywell—and began moving more toward nature, hedonism, anarchy, and spiritualism.

And why not? Faced with the collapse of the mainstream culture's moral authority but lacking the power to change it, many baby boomers either raged in anarchistic riots or tuned in, turned on, and dropped out. World-renowned British mathematician Bertrand Russell captured the dour pessimism in a 1963 *Playboy* interview:

The human race may well become extinct before the end of the present century. Speaking as a mathematician, I should say that the odds are about three to one against survival. The risk of war by accident—an unintended war triggered by an explosive situation such as that in Cuba—remains and indeed grows greater all the time. For every day we continue to live, remain able to act, we must be profoundly grateful.

With towering intellectual figures like Russell—who co-wrote the iconic *Principia Mathematica*, won the Nobel Prize in Literature, and championed logic and freedom at the beginning of the twentieth century—making these sorts of dire pronouncements in a hedonistic outlet like *Playboy*, living for the moment seemed like an entirely rational idea. If everything is going to hell in a handbasket, why not?

The Dark Side of the Moon

A new president was attempting to turn the tide of fear and to restore a sense of wonder to science, not out of vision so much as desperation. By May 1961, John F. Kennedy had been in office for a little more than three months and had already stumbled into deep trouble, seemingly justifying the public's growing skepticism of the government. The recession of 1958—the worst since the end of World War II—had been quickly followed by another that began during the presidential campaign of 1960 and lasted into 1961. The Bay of Pigs had already occurred, resulting in an embarrassing failure for the new administration. Five days before the thwarted invasion, the Soviets had sent the first human into orbit, pulling the rug out from under Kennedy's campaign rhetoric about besting them in the space race. Aiming to take advantage of the growing social upheaval in the West, Soviet premier Nikita Khrushchev was testing Kennedy in every way he could, a pattern that would continue all year long and include the building of both the Berlin Wall and the nuclear-missile sites that would result in the Cuban Missile Crisis. His credibility on the line at home and abroad, Kennedy looked weak, inexperienced, and outmaneuvered. He needed a way to assert his leadership—and America's. He turned to the moon—not for science's sake, but to beat Khrushchev. He later admitted as much in a November 1962 meeting with NASA administrator James Webb. "I am not that interested in space," he told Webb. The main reason he wanted Apollo was its importance in the Cold War rivalry with the Soviet Union.

The Russians had led with *Sputnik 1* in 1957 and, a month later, had sent a dog into space on *Sputnik 2*. Two years after that, they had crash-landed *Luna 2* on the moon. Now, on April 12, 1961, they had beaten the United States yet again by sending the first human, Yuri Gagarin, into space. On May 25 of that year, Kennedy addressed a joint session of Congress and laid out several "urgent national needs," among them getting America back into the space race:

> Recognizing the head start obtained by the Soviets with their large rocket engines, which gives them many months of lead-time, and recognizing the likelihood that they will exploit this lead for some time to come in still more impressive successes, we nevertheless are required to make new efforts on our own. For, while we cannot guarantee that we shall one day be first, we can guarantee that any failure to make this effort will make us last.

He laid out a bold agenda, a desperate and visionary agenda, to regain the military and ideological lead, and, at the same time, to turn around the economy by landing a man on the moon and returning him safely to Earth. The effort would require a peacetime science mobilization on par with the Manhattan Project, requiring the building of entire cities to support it. At its peak, the Apollo program would employ some four hundred thousand people.

There were just two problems: the United States didn't have a clue how to do it, and, with the country in a recession and federal tax revenues down, it didn't have the money either. To make matters worse, Kennedy's inspirational ideas for domestic policy were getting shot down. He had painted a grand vision of the New Frontier and the War on Poverty, but he couldn't get Congress to pay for a major expansion of social programs—at least not yet. If a program were tied to the Cold War, however, he thought he could get Congress to support it, as it had the National Interstate and Defense Highways Act five years earlier and the National Defense Education Act of 1958. Apollo could be a bold new vision and a jobs program in one, an economic stimulus with benefits. The cultural anxiety was so high by then that the very idea of Russians crawling all over the moon caused a visceral reaction in many Americans. Kennedy thought he had a winner.

But once the budget numbers came back, they showed that the program would cost almost $20 billion over eight years, eating up all the discretionary funds that Kennedy needed for his War on Poverty. If he wanted Apollo, he would have to sacrifice every other goal of his presidency. He began looking for a way out.

He realized that, if he took away the Cold War justification, he'd lose the support of fiscal conservatives, and he could use that loss as an excuse to move the deadline back indefinitely. So, feigning friendship, he reached out to Khrushchev, his worst enemy, at the Vienna summit. Over lunch, he suggested that they bury the space-race hatchet and go to the moon together as a cooperative venture. Surprised, Khrushchev said no. Kennedy looked at him silently and, finally, Khrushchev said, "All right, why not?" But as they talked further he thought about what it was going to cost and changed his mind again, now saying that disarmament was a prerequisite for cooperation in space. Both sides were hoping to avoid the costly space race.

Kennedy took this nugget of an offer and sold the idea in a speech at the University of California, Berkeley, in March 1962, and another at the United Nations in September 1963. Fiscally conservative Democrats who had backed the program now saw Kennedy's support wavering and started jumping ship. The Senate responded to the changing political climate by proposing to cut NASA's funding and scrap the Apollo program. Kennedy's political escape plan was working.

In general, Americans were more concerned about domestic issues like poverty, race relations, and the economy, and suspicion of science was growing. In 1962, marine biologist Rachel Carson's book *Silent Spring* came out and made a permanent impact on the national psyche, shocking Americans already suspicious of science into an awareness of chemical pollution, reaffirming Eisenhower's warnings about the scientific-technological elite, and launching the field of environmental science and the modern environmental movement. That year, only about 35 percent of Americans thought Apollo was worth the cost.

Democracy's Most Severe Test

At the same time, opposition to Apollo was also building in the scientific community. By 1963, the Senate Committee on Aeronautical and Space Sciences was holding hearings, at which scientists were testifying that human spaceflight was inordinately expensive and risky, since its paramount objective was not research but bringing astronauts back alive. Unmanned spaceflight with robots, they said, would cost much less and return much more information, since we didn't have to worry about feeding or protecting the lives of robots, or about bringing them back, cutting fuel costs by more than half.

The nation was adrift, and the public hardly thought of the space program as a good thing. Why couldn't we spend our money on putting our best scientific minds to work solving issues here on Earth? The cost of funding the newly created NASA and the Apollo program was incredible. By 1966, it would reach 4.5 percent of the federal budget, an astronomical figure compared to today, when it is about one-ninth of that.

But then, in November 1963, everything changed. Kennedy was assassinated in Dallas, and the next year the elections went overwhelmingly to the Democrats. The new president, Lyndon Johnson, committed to the Apollo program and the Great Society in honor of Kennedy's memory. Very few could stand opposed. Enacting Kennedy's dream had become a national cause célèbre, and, in a dramatic turnaround, both initiatives were funded.

Still, the funding of Apollo stood in contrast with the mainstream culture's attitude. In 1965, Ralph Lapp, the former head of the nuclear physics branch of the Office of Naval Research, captured this growing fear when he published *The New Priesthood*, in which he reiterated Eisenhower's argument that the "scientific elite"—people who understood how science and technology work—were starting to supplant the country's elected leadership. Lapp's argument reflected an emerging and critically important idea: that "democracy faces its most severe test in preserving its traditions in an age of scientific revolution."

Carpe Diem

The live-for-today ethos birthed by the bomb and the Cold War was changing the way the West approached living, consumption, and finance. The change was particularly evident among the baby-boom generation. "I hope I die before I get old," sang the Who on *My Generation*, their 1965 debut album. "Don't trust anybody over thirty" cautioned Berkeley free-speech activist Jack Weinberg. These phrases became watchwords. With the older generation having lost its moral credibility, the boomers felt older people had no right to tell young people anything. "Today there is great concern among my generation that an era of permissiveness has resulted in unrest among our young people," said California governor Ronald Reagan on May 20, 1967. "But just to keep things in balance there is a widespread feeling among our young people that no one over thirty understands them."

Feeling powerless, these baby boomers needed an outlet for their anger and distrust of science, government, and the older generation. They adopted the protest songs of folk music. Singer-songwriter Bob Dylan became an overnight sensation, the "poet to a generation." Satire became a dominant cultural art form, lampooning all kinds of authority for its hypocrisy and failure. *The Graduate, Catch-22, One Flew over the Cuckoo's Nest, MAD* magazine,

Dr. Strangelove, Saturday Night Live, and many other satirical cultural touch-stones emphasized the absurdity of the times. The hilarious, humanitarian, anti-scientific, pessimistic novels of Kurt Vonnegut became runaway hits. In 1970, Vonnegut gave a commencement address at Vermont's Bennington College in which he famously said, "Everything is going to become unimaginably worse and never get better again." He cautioned the baby boomers that "we would be a lot safer if the government would take its money out of science and put it into astrology and the reading of palms. I used to think that science would save us. But only in superstition is there hope." American kids loved his cynical pessimism because he gave voice to their fear and anger. In 2006, a year before he died, Vonnegut showed the same satirical mix of anger and regret when he penned a "Confetti print," which he sold in numbered silk screens:

> Dear future generations:
> Please accept our apologies. We were roaring drunk on petroleum.
> Love, 2006 A.D.

By that time, however, the baby boomers were in charge, and few were still listening.

Ladies and Gentlemen, Step Right Up!

Into this inferno of cynicism went NASA. Because of the nationalist, political, and competitive nature of its mission, NASA, more than any other US science agency, had come to understand Lincoln's dictum on the importance of public sentiment. In a way, the whole Apollo program was a big show of American technological prowess and innovation—political theater intended to influence public sentiment—that just happened to spin off immense benefits. But the showmanship, wonder, and politics were lost on most other scientists, who saw no value—or worse, a negative value—in showing personality or telling engaging stories about their work. Hubris, after all, is what had ruined Harlow Shapley's career; it was the gateway to the path of a priori principles.

From a purely scientific viewpoint, human spaceflight was wasteful. But from a public-engagement (and thus funding) viewpoint, it was sheer genius. It gave the public protagonists starring in an epic narrative about science. It was, in some sense, an example of the third culture C. P. Snow had hoped for—a marriage of literary resonance and "doing the big things," as Kennedy had urged in his United Nations address.

The remainder of the 1960s and 1970s, however, brought increasing dis-illusionment with mainstream culture and the military-industrial complex with which science had become closely aligned. NASA battled this growing antiscience backlash with its own powerful narrative, which culminated in the dramatic 1968 Apollo 8 mission. Humans left Earth's orbit for the first time and orbited the moon in both a scientific mission and a public-relations move with literary panache. In a Christmas Eve television broadcast that became, at the time, the most-watched program ever, the astronauts read the first ten verses of the book of Genesis—another attempt to tie the narrative of science and exploration back to the fundamental human search for meaning. The mission returned to Earth on December 27, the day on which Darwin's *Beagle* had departed England 137 years before.

This unification of science with religion in a shared sense of wonder at the great questions of the universe and in humanity's sense of curiosity and exploration, captivated the nation. Science was accessible through storytelling, and it was epic—an attitude that no doubt later inspired science-fiction film-makers like George Lucas and Steven Spielberg, whose films often sought to recreate a similar sense of awe. Pop culture put astronauts everywhere. The *Peanuts* comic-strip character Snoopy, a beagle, wore a spacesuit on *Snoopy the Astronaut* pins and was the most popular cartoon character of the day—so much so that the pin became the symbol of the Apollo program. Apollo 8 set the stage for the dramatic climax of Apollo 11, less than seven months later, when American astronauts actually walked on the Moon, fulfilling Kennedy's goal before the decade was out.

A Question of Priorities

But in the months leading up to this grandest of human adventures, uniting science and literature in a way that had rarely been achieved, voices of concern were heard anew. Despite the incredible successes of the Apollo program and the spiritual sense of wonder it generated, the quasi-military organization and culture that NASA had adopted to get the job done recalled, for many, Eisenhower's warning of eight years before.

This distrust in the American science enterprise was particularly sharp in the nation's debates over race and poverty, which had been fueled by the civil-rights movement, the destruction of many minority communities by highway construction, and the increasing disparity in wealth between the predomi-nantly black urban cores and the new white suburbs.

The Reverend Ralph Abernathy, who had cofounded the Southern Christian Leadership Conference with the Reverend Dr. Martin Luther King Jr., carried on King's focus on economic justice for the poor of all races after his assassination on April 4, 1968. In 1969, Abernathy led a national march against the Apollo 11 moon launch, criticizing the incredible spending as an inhuman priority at a time when so much suffering existed in the nation. "One-fifth of the population lacks adequate food, clothing, shelter, and medical care," he argued.

> A society that can resolve to conquer space; to put man in a place where in ages past it was considered only God could reach; to appropriate vast billions; to systematically set about to discover the necessary scientific knowledge; that society deserves both acclaim and our contempt . . . acclaim for achievement and contempt for bizarre social values. For though it has the capacity to meet extraordinary challenges, it has failed to use its ability to rid itself of the scourges of racism, poverty and war, all of which were brutally scarring the nation even as it mobilized for the assault on the solar system.

NASA administrator Thomas O. Paine walked out to where the demonstrators were gathered and met with Abernathy on the eve of the Apollo 11 launch. Paine said that if he could alleviate human suffering by not pressing the launch button he would, but that sending three men to the moon was "child's play" compared to "the tremendously difficult human problems" Abernathy and his people were discussing. Incredibly, the exchange ended amiably, with Abernathy agreeing to pray for the lives of the astronauts.

There is a very good chance that, had Kennedy lived, the United States would not have put a man on the moon by 1969, since the costs and the politics were tilting so heavily against it. The conflict between social and science spending lives on to the current day.

The Apollo program's triumph of both technology and narrative inspired pop culture in profound ways. Its adventurous mix of science and religion seemed, in spite of the times, to reflect the best of humanity, and it inspired songs like David Bowie's "Space Oddity," Deep Purple's "Space Truckin," Elton John's "Rocket Man," Joni Mitchell's "Willy," Pink Floyd's "Brain Damage" and "Eclipse" from *The Dark Side of the Moon*, and many more. It also inspired many films, including the ambivalent *2001: A Space Odyssey* and, later, the more optimistic *Star Wars*, *ET: The Extraterrestrial*, and *Apollo 13*. It brought America together, briefly, in a moment of shared pride, wonder, and imagination. And it spun off scientific and technological advances that laid the groundwork for the digital revolution.

The Seven Stages of Technological Adaptation

But space was the exception. In other ways, science was taking it on the chin. As the United States was roiled by antiauthoritarian demonstrations against the Vietnam War, against poverty, against racism, against "the man," science came to be seen as a part of "the man"—the corrupt, jingoistic, largely white and male cultural elite about which Eisenhower had warned the nation, dominated and corrupted by business interests too closely aligned with government in a sort of neofascism. Things seemed out of control and America's youth felt trapped, like Jack Nicholson, in a "cuckoo's nest" of insanity in which it was the authorities and caretakers who were brutal and insane.

By the 1970s, compounding the problems of racial inequality, nuclear weapons, poverty, sexism, and the war, the destructive environmental consequences of object-oriented science and technology were becoming abundantly clear. Much of this science was applied research undertaken by industry, as opposed to the basic, curiosity-driven science of academia and government. During a more naive and optimistic time, what had been viewed as great basic-science breakthroughs in the nuclear, chemical, and agricultural sciences had been rushed into broad application. But these were beginning to boomerang back with broad and sometimes horrible environmental consequences. What had been a manageable impact in the smaller scale of the past was a devastation on the vastly larger scale that science and technology now made possible. As was the case with the nuclear bomb, the consequences of making a mistake had become exponentially amplified—and, at the same time, the probability of such a mistake being made had become compounded.

In 1971, *The Lorax*, a children's book by Dr. Seuss, was published. A parable about the environmental costs of industrialized technology, the book echoed the atomic-bomb theme of scientific and technological power outpacing moral and ethical development. Year after year, more and more environmental disasters seemed to be bear this attitude out. Unregulated and widespread uses of industrial chemicals and processes were found to have unanticipated and sometimes catastrophic environmental consequences, establishing a pattern that became the all-too-familiar hallmark of industrial science and muted the euphoria over the moon shot. In the coming years, conservative businesspeople became proscience and liberal environmentalists became antiscience. The process followed seven stages, over and over:

1. DISCOVERY

A new process or tool (for example, a chemical or, today, a nanotechnology or genetic technology) is discovered that vastly expands utility, power, convenience, or efficiency.

2. APPLICATION

Industrial applications are quickly developed and commercialized, often increasing productivity and lowering costs. But the science of biocomplexity and ecology—of how the process or tool will affect and be affected by its broader context, from the human body to the environment—lags behind.

3. DEVELOPMENT

Industries grow up around the new application. Major capital investments are made and its use intensifies.

4. BOOMERANG

A tipping point is reached at which the application has noticeable negative effects on health or the environment. Fueled by growing public outcry, scientists study the degree of the systemic effect to determine what to do.

5. BATTLE

Regulations are proposed to minimize the negative effects, but vested economic interests sense a potentially lethal blow to their production systems and fight the proposed changes by denying the environmental effects, maligning and impeaching witnesses, questioning the science, attacking or impugning the scientists, and/or arguing that other factors are causing the mounting disaster. A battle ensues between the adherents of old science and those of new science. This stage is caustic to science's credibility, because science becomes a rhetorical tool and facts are cherry-picked to win arguments, pitting real science against clever public relations campaigns, with the public unsure about what and whom to believe.

6. CRISIS

Evidence continues to accumulate from the emerging science until the causation becomes irrefutable, often through dramatic deaths or disasters (or in the case of climate disruption, extreme weather events) that draw increased public scrutiny and outrage, finally tipping the politics in the direction of reform.

7. ADAPTATION

Regulations are passed or laws are changed to stop or modify use and to mitigate the effects. The industrial approach grudgingly shifts to take into account the relationships between the application and its environmental and/or physiological context. Or this does not occur, in which case the process returns to stage 3.

This pattern has been repeating since the 1960s with escalating stakes. The same drama of object-oriented science and the development of technological solutions has led to the denial of consequences over and over again, from cigarettes to DDT to asbestos to acid rain to Love Canal to the hole in the ozone layer to Three Mile Island to the Dalkon Shield to toxic-shock syndrome to lead paint and leaded gas to atrazine to the Vioxx scandal to emerging battles over microbeads and other nanotech—and eventually to the granddaddy of them all (so far), global climate disruption.

Wonders upon Wonders

By the late 1970s, science had become known for environmental and health debacles, most of them coming out of industry applied science, all playing out against the nuclear backdrop, and even NASA's victories couldn't stem the growing antiscience sentiment. A 1979 NBC/AP poll showed that just 41 percent of Americans thought the benefits of the space program outweighed its costs. Philip Handler, the president of the National Academy of Sciences, cautioned that scientists needed to find a way to reverse the trend of increasing public skepticism toward science.

Witnessing all this, American astronomer Carl Sagan took up Hubble's mantle and joined a small group of scientists—including his friend the evolutionary biologist Stephen Jay Gould—who were trying to turn things around. The discoveries Hubble and his contemporaries made about the universe had inspired Sagan the moment he had first learned of them as a young boy at the 1939 New York World's Fair. "I was a child in a time of hope," he said. "I wanted somehow to immerse myself in all that grandeur. I was gripped by the splendor of the Universe, transfixed by the prospect of understanding how things really work, of helping to uncover deep mysteries, of exploring new worlds—maybe even literally."

But now, with the exception of NASA, science had gone silent and become disengaged from the broader culture. Its justification for funding had shifted from that special wonder Sagan had felt as a child to the smaller-minded fear and pragmatism of business and defense. In the growing divide between science and society, Sagan saw the rise of a "demon-haunted world" as baby boomers aged away from science and toward the ESP, New Age mysticism, faith healing, miracles, and UFO abductions that increasingly occupied the popular press.

In Sagan's view, this threatened the nation's fabric:

> I have a foreboding of an America in my children's or grand-children's time—when the United States is a service and information economy; when nearly all the key manufacturing industries have slipped away to other countries; when awesome technological powers are in the hands of a very few, and no one representing the public interest can even grasp the issues; when the people have lost the ability to set their own agendas or knowledgeably question those in authority; when, clutching our crystals and nervously consulting our horoscopes, our critical faculties in decline, unable to distinguish between what feels good and what's true, we slide, almost without noticing, back into superstition and darkness.

In the public's growing mysticism and antiscience, Sagan saw a hunger for the lost sense of wonder and adventure that science had once provided. The world had become much smaller since then, more materialistic and mechanistic. Skepticism toward the environmental and health failures of industrial science, coupled with fear of the atomic bomb and the horrors of the Vietnam War, like Agent Orange and napalm, meant that the public had had enough. For a generation, public funding of science had been driven by fear, and Ronald Reagan was running for president saying that "government is the problem." Sagan set out to change the growing public apathy and distrust and created the 1980 television series *Cosmos*—"the greatest media work in popular science of all time," as Gould would call it.

With *Cosmos*, Sagan sought to counter the fear and distrust of applied science, and to inspire the kind of basic science wonder that Hubble's lectures and the moon landing once had. The series was enormously successful. For the first time since Hubble, a huge audience was engaged in exploring the grand questions about life, nature, the structure of the universe, and what it might all mean. *Cosmos* examined how our search for meaning through science was the greatest of all human quests.

The show was seen by an estimated five hundred million people (about a ninth of the world's population) in sixty countries—nearly as many as the moon landing. This was an enormous viewership, and by far the largest for any science show ever. And yet Sagan was later denied admission to the National Academy of Sciences by his peers, who voted against his nomination on the grounds that his research work as a scientist was not strong enough to justify admission. But this was likely a stalking-horse for the real reason: the animosity scientists felt toward Sagan's status as a TV star and as a spokesman for their work. The scientist who nominated Sagan, origins-of-life researcher Stanley Miller, described the hostility he perceived. "I can just see them saying it: 'Here's this little punk with all this publicity and Johnny Carson. I'm a ten times better scientist than that punk!'"

Following this rejection, and Sagan's failure to secure tenure at Harvard, scientists developed a new term: the Sagan effect. One's popularity with the general public was considered inversely proportional to the quantity and quality of one's scientific work, a perception that, in Sagan's case at least, was false. He published, on average, once monthly in peer-reviewed publications over the course of his thirty-nine-year career—a total of five hundred scientific papers. More recent research suggests that scientists who engage the public tend to be better academic performers as well.

In the years following Sagan's drubbing—by the very National Academy whose president had called for increased science outreach just a short while before—the public's perception of science continued to erode. By 1999, less than half of all Americans—just 47 percent—said that scientific advances were one of the country's most important achievements. By 2009, that number had fallen to 27 percent.

Sagan's rejection became a poignant and symbolic example of how America's most prominent scientists had lost their appreciation of the value of their relationship to society—a relationship that was critical to their future and the future of the country, but that was slipping through their fingers even as they voted against Carl Sagan, like the sands of an ancient streambed, forgotten and now run dry.

Purity

Throughout the 1970s and 1980s, as the political schism over science began to widen, ambivalence toward science was expanding throughout much of the

Western world. Religious fundamentalists were getting organized; industry was mounting antiscience PR campaigns; and people were becoming more aware of the unanticipated effects of chemicals on the environment and the human body.

At the same time, the use of synthetic chemicals in everyday products was exploding, "oppressing" those who did not have the training to understand the language. Take, for example, butylated hydroxytoluene, a common food additive. It takes knowledge of organic chemistry and biology to understand how a chemical that is commonly used as an additive in jet fuel, rubber, electrical transformer oil, and embalming fluid can possibly be good for one to eat. In food, it serves as an antioxidant, keeping baked goods made with oil and butter from going rancid. But do we really understand all of its effects on the human body?

Now multiply that question by 3,968—the number of additives currently approved in the United States for use in foods, which are listed in the FDA's "Everything Added to Food in the United States" list, more commonly known as the EAFUS list. From a health perspective, science seemed alien, out of step with mainstream culture, and it seemed as though the applied science of agribusiness might be unwittingly poisoning people in the same way it had unwittingly weakened the shells of bald eagles' eggs with the "miracle" pesticide DDT. People who did not speak the language of organic chemistry, and even many who did, began to worry about unanticipated biocomplex effects—those arising from the complex biological, chemical, physical, and behavioral interrelationships between living organisms and their environment. This spawned a massive resurgence of the organic-food movement, and a suspicion of foods derived from organisms that had been bred using genetic modification. But genetic modification is the same thing humans have been doing for millennia, just with more precision. It actually changes less of a plant's organic genome than hybridization.

Many GM opponents worry that it increases the use of herbicides in the environment and chemical residue on foods; some plants have been engineered to be resistant to glyphosate, which the chemical company Monsanto markets under the brand name Roundup, an herbicide that kills broadleaf plants. Studies have shown that, similar to broad antibiotic use causing the evolution of "superbugs," the broad use of the herbicide glyphosate has increased the prevalence of glyphosate-resistant "superweeds." A recent study sampled waterways in 38 states and found glyphosate and its residue in the majority of ditches, rivers, streams, and wastewater treatment plants tested. It was also found in about 70 percent of rainfall samples, though not in concentrations high enough to be a health threat.

Such hybrid uses that are tied to the broader application of a chemical herbicide pose complex environmental, legal, health, and public-policy issues that extend far beyond the science-based question of whether genetically modified foods are safe to eat (the answer to which is a very clear yes). However, this sort of hybrid use with chemical pesticides is only one of the ways we are using plant genetic engineering. Other GM crops are engineered for improved flavor, longer shelf life, abundance (requiring less land and fewer resources, and thus offering higher sustainability and efficiency), nutrient content (e.g., golden rice, which contains beta-carotene, a precursor to vitamin A that helps prevent blindness and gives the rice a carrot-like orange color), hardiness, drought resistance (requiring less use of groundwater), virus resistance (e.g., the Rainbow papaya, which is resistant to the ringspot virus that nearly wiped out the papaya industry), salinity resistance, extreme-temperature resistance, or other factors, which have nothing to do with the use of chemicals. Crops are also being engineered to use in the production of recombinant medicines and other health products, such as monoclonal antibodies, vaccines, biofuels, and plastics that don't require drilling for oil.

As for the science, the scientific consensus on the safety of eating GM foods is even stronger than that for the existence of human-caused global warming. A 2015 Pew Research Center/AAAS study found that 88 percent of all AAAS-member scientists said that genetically modified foods are safe to eat, compared with just 37 percent of the general public. The fifty-one-point gap makes this the largest opinion difference between scientists and the public. This enormous difference highlights the public's insecurity over the use of herbicides, pesticides, and artificial ingredients being added to food, even though that is not what genetic modification is. Genetic modification is a more precise form of plant breeding, whereby food made from the child plants have the same nutrient content and ingredients as food made from their non-GM parent plants. However, few members of the public understand that.

There are other issues with GM that pose legitimate policy questions, such as corporate ownership of genomes, monoculture farming and its effects on pollinators, economic justice, farmers' inability to harvest their own seeds, and the possible spread of GM pollen. These are issues we should be discussing, but the evidence does not support health concerns.

Tin Hats

Others grew worried about high levels of exposure to electromagnetic fields as people were increasingly bombarded by electromagnetic waves from electronic devices. Like understanding chemical additives, understanding the risks of electromagnetic fields requires an understanding of science—in this case, physics, chemistry, and biology—that many people do not have.

Beginning in 1989, the *New Yorker* magazine ran a series of stories about electromagnetic pollution by staff writer Paul Brodeur that was similar to the excerpts it once ran from *Silent Spring*. They eventually formed the basis for Brodeur's popular 1993 book *The Great Power-Line Cover-Up*. By the time the book was published, belief was widespread that electromagnetic fields from power lines and household appliances like microwave ovens were linked to childhood leukemia and other cancers. But there was little or no evidence of this in broad epidemiological studies.

In January 1993, a man named David Reynard appeared on the television show *Larry King Live*. Reynard was suing the nine-year-old cell-phone industry, insisting that his late wife's brain cancer had been caused by her cell phone. "She held it against her head, and she talked on it all the time," he said. The interview fed into the public's growing distrust of science and had an electrifying effect. Reynard's suit seemed to corroborate what Brodeur had been writing about, and this new fear continued to spread even after broad epidemiological studies showed no links at all between cell-phone use and an increased incidence of brain cancer. In 2015, the number of mobile-phone subscribers exceeded five billion, or nearly 70 percent of the world's population, yet the brain-cancer rate has not increased.

Physics explains why no links were found. The microwaves used in cell-phone transmissions do not have enough energy to break the chemical bonds of DNA, which is how cell mutations occur and cause cancer. How do we know this? Light and other parts of the electromagnetic spectrum—including microwaves, radio waves, infrared waves, and ultraviolet light waves—are all forms of radiation. A single unit of radiation is called a photon. A photon can be thought of either as a particle or as a wave. A century ago, Albert Einstein showed that the energy (E) of a photon can be calculated as Planck's constant (h) times the frequency (v) of its wave form, or $E=hv$. This formula was set out in one of his early, most famous papers, which was published in 1905 and is one of the reasons he won the 1921 Nobel Prize in Physics. Photons with low frequencies are at the red end of the spectrum. They include radio photons, whose waves can be as long as a football field and thus fly past us with low frequency and low energy. Microwaves are slightly stronger, followed by the infrared radiation your skin is giving off as you sit reading this. You are probably emitting somewhere around ninety watts. Then, after infrared, we get into visible light waves—what our eyes use to see the world. As we climb through visual light from its red end to its blue, the frequency of radiation continues to climb. Waves at the high end of the visual spectrum fly past us at much faster frequencies (and thus have more energy) and appear more blue. The spectrum then moves into ultraviolet light, followed by the even more energetic X rays, followed by gamma rays, whose waves can be shorter than the diameter of the nucleus of an atom, and are thus very high frequency and very highly energetic.

Microwaves are slightly more energetic than radio waves, but far less energetic than even the infrared radiation that our skin gives off (which is how we can be seen by someone wearing infrared night-vision goggles). Both visible light and microwaves can be used to cook food and heat up your coffee by concentrating them in very large amounts, such as in a solar oven or a microwave oven. But this process doesn't work because of radiation in the way many people think. The concentrated waves excite the molecules in a manner that increases their vibration, and the friction this produces increases their temperature, thereby cooking them. But they still don't have anywhere near enough energy to break chemical bonds. If they did, the food would turn into goop.

Microwaves are much weaker than visible light, though, so it takes a lot more of them to make an oven work than it does to make a solar oven work, which is why microwave ovens are such electricity hogs. It's not until we travel into the red end of the visible-light photons, then on to the yellow emitted by incandescent bulbs, then to the blue that illuminates many fluorescent bulbs, and then on past the visible spectrum into ultraviolet light, that we get photons with a high enough frequency, and thus enough energy, to knock an electron out of a carbon atom, thereby ionizing it, altering how it chemically bonds, and potentially altering our DNA. But photons at this frequency have about a million times more energy than microwave photons. It's like the difference between getting hit by a pea from a pea shooter and getting hit by a BMW Z4, which is about a million times the mass of a single pea. But these ultraviolet photons still don't have enough energy to penetrate us very deeply, so they can only give us skin cancer, unless we block them with sunscreen.

Electromagnetic radiation above this level grows increasingly dangerous. X rays are sometimes stopped by our skin, but if we are bombarded by enough of them they have sufficient energy to penetrate through us. A few of them are absorbed by skin and muscle, many more by our bones (which are denser). And many shoot that cue ball of a photon clear through us, which is why we can use them to make images of the insides of our bodies. X rays can and do cause cancer, but our bodies can almost always stop it if the exposure is low enough. Gamma rays are so energetic that they can penetrate us, kill cells, and cause cancer very easily. In high exposures, they cause radiation poisoning, which kills much more quickly by damaging bone marrow and gastrointestinal tracts.

To the general public, the word "radiation" came to mean danger, even though our bodies give it off. Physicists like Bob Park tried to point this out and show that not all radiation is bad for us. Some kinds of radiation, like light and heat, are even necessary for life. Tsunamis kill, but streams nourish—the difference is in the amount and energy of the water, just as with radiation.

But, by then, science's credibility had been damaged by a constellation of silence, failure to anticipate biocomplex consequences, environmental mismanagement, corporate public-relations spin, and political rhetoric. When that confidence has been shattered, all that is left to rely on is superstition, magic, religion, opinion, partisan pundits, television celebrities, and "prudent avoidance"—all "ways of knowing," but none of them *objective knowledge*. A new form of antiscience skepticism was taking root.

A Congressional Lobotomy

Despite all the growing cultural ambivalence toward science, it was protected until the 1990s by the same motivation that had held sway since the Russians exploded their first atomic bomb in 1949—fear. But in 1991 the Soviet Union collapsed. With the collapse, the West won the Cold War and lost its major scientific competitor. The Western science enterprise, without realizing it, suddenly lost the rationale Vannevar Bush had used to get government funding.

Two generations of scientists had largely depended upon government largesse. But now science, which had was seen as a means to the end of maintaining security, no longer had a specific mission. What's more, it was increasingly clear that science—particularly, but not exclusively, applied industry science—had produced a lot of messes.

Physicist Leo Kadanoff described the dour public mood in an October 1992 *Physics Today* piece titled "Hard Times":

> Today when the public thinks of the products of science it is likely to think about environmental problems, an unproductive armament industry, careless or dishonest "scientific" reports, Livermore

cheers for "nukes forever" and a huge amount of self-serving noise on every subject from global warming to "the face of God."

Over the course of four decades of disconnection, specialization, environmental and health disasters, and tunnel vision, science had transformed in the eyes of the public from noble savior to troublesome spoiler—or, worse, a tool of oppression. Science—its values, priorities, methods, and the effects of its advances—was perceived to be out of touch with shared values and priorities, and beholden instead to the interests of a small but powerful minority. Funding began to fall off, and science, which had been relatively silent for nearly two generations, began to lose its standing.

By 1994, the value of science had fallen so steeply in the eyes of policymakers that newly elected House Speaker Newt Gingrich, a self-professed science supporter, felt compelled to propose eliminating funding for Congress's own science and technology advisory body, the Office of Technology Assessment (OTA), in order to sell his budget-cutting package. The proposal was accepted. In fairness to Gingrich, he was able to protect and even increase science funding in other areas, ultimately doubling the research budget of the National Institutes of Health with the help of Congressman John Porter (R-IL). But the elimination of the OTA was a strategic blow that would, many argued, affect science policy for the next generation. In saving a relative pittance, Congress gave itself "a lobotomy," as former congressman and physicist (later CEO of the AAAS) Rush Holt put it. When crafting major policies, congressional staffers instead turned to lobbyists and the Internet for science information of dubious quality. The truth, forever nebulous in the eyes of baby boomers, became even more fungible.

Michael Halpern of the Union of Concerned Scientists says that, as a result of losing the OTA, Congress wasted billions of dollars on policies like the fence along the Mexican border that OTA scientists could have told them would not work. "It was penny-wise and pound-foolish," Halpern says. "Without the OTA, it all became rhetoric."

Poison

By the late 1990s and through to today, nonscientists had become increasingly unable to discern between knowledge and opinion. With a subjective news media and a niche for everything online, the American public dialogue was getting hopelessly confused. Something had gone terribly wrong. The world was a mess. And science was poisoning us. Ruining our environment, causing cancer— and now delivering a new threat.

Children—whom baby boomers protect with a level of ferocity never before seen, perhaps prompted by their Cold War fear of the bomb—were in grave danger. Now science was taking *children* from us too, by giving them autism. But there was a lone scientist on our side, telling the truth, fighting the man: the British surgeon Andrew Wakefield.

In 1998, Wakefield published in the prestigious British medical journal the *Lancet* a scientific paper that linked childhood measles, mumps, and rubella (MMR) vaccines to autism. The paper gave Wakefield instant celebrity because it crystallized the amorphous public fear of poison by science. Autism, after all, was inexplicably on the rise, and science had earned a reputation for this sort of thing: the spread of poisonous chemicals in the environment; the insertion of chemicals into foods; the destruction of the ozone layer; the obfuscation of links between smoking and cancer; and, in the early twentieth century, the injection of syphilis, hepatitis, cancer, and other diseases into patients (often children), federal prisoners, and African Americans—without their knowledge—so researchers could study the diseases.

Wakefield became a media star. He began touring and speaking on the subject, and coauthored a 2010 book. The only problem was that it wasn't so. Wakefield, it later turned out, had doctored his evidence to fit his a priori conclusion. The paper has since been discredited as "fraudulent" and was withdrawn by the journal. The General Medical Council, which oversees British doctors, said Wakefield's "conduct in this regard was dishonest and irresponsible."

But the damage was done. Scared parents—the vast majority of them liberal and well-educated—bought into the "science is poisoning us" meme that had been building all their lives since *Silent Spring*. They began refusing the MMR and other vaccines for their children. This was especially true in Britain, where Wakefield is from, and where MMR vaccination rates fell from 90 percent to just 54 percent within a year. In some regions, the rate fell below 50 percent. By 2013, thousands of teens whose parents hadn't vaccinated them as younger children were spreading measles in outbreaks that raged across the United Kingdom.

The next year, the situation was made worse when the United States Food and Drug Administration found that children might be receiving as much as 187.5 micrograms of ethylmercury within the first six months of life, mostly from the preservative thimerosal that was used in vaccines. It had been used since the 1930s to prevent contamination of vaccines by bacteria and fungi, after a horrible 1928 tragedy in which a multiuse vial of the diphtheria vaccine became contaminated with bacteria and killed twelve of the twenty-one children who were injected with doses. Thimerosal had helped ensure public safety for nearly seven decades since, without any measurable negative effects. But now, suddenly, health-conscious, organic-food-shopping, industrial-science-mistrusting public opinion turned against the preservative.

The American Academy of Pediatrics responded to the FDA findings and public worry, and recommended removing thimerosal from all vaccines given to young infants. In the midst of the Wakefield autism scare, public worries over poison had suddenly expanded—without any evidence—to include thimerosal in vaccines as another possible cause of autism. For the first time in decades, large numbers of frightened parents began talking about not vaccinating their children. Like the word "radiation," the word "mercury" had come to mean "poison" to the semi-science-literate, and no distinction was made between the toxic form of mercury (methylmercury) and the ethylmercury that was safely used in vaccines. Antimercury advocacy groups sprang up, even though the symptoms of mercury poisoning are nothing like autism and there was no evidence to support any link.

The situation was exacerbated when talk-show hosts Oprah Winfrey and Larry King gave antivaccine advocate and charismatic, attractive former *Playboy* model Jenny McCarthy a platform on their shows. Like many of us, McCarthy is well-meaning, passionate, and concerned about her child, but she has no background in science. With an a priori conclusion and a skepticism of science, she unintentionally did harm by promoting the nonexistent link between vaccines and autism. She was not alone. Robert F. Kennedy Jr., another nonscientist celebrity on the political left, authored a widely distributed, well-intentioned 2005 article in *Rolling Stone* and on Salon.com (since removed from the site), arguing that he was "convinced that the link between thimerosal and the epidemic of childhood neurological disorders is real."

Even after thimerosal was removed from vaccines, the antivaccine movement continued to gain momentum, fueled by the same skepticism of science that had propelled the electromagnetic-field scare, the cell-phone scare, the GM scare, and the poison-meme antiscience backlash as a whole.

This backlash is not related to lower education levels or economic status. On the contrary, according to a 2004 study from the Centers for Disease Control and Prevention (CDC), in the United States, unvaccinated children tend to be white and to have upper-middle-class, college-educated parents who express concerns about the safety of the vaccines. A 2011 study found similar results. These people are indeed organic-food shoppers, and alternative-health-care and nutritional-supplement consumers. There are large pockets of them in affluent liberal communities, like Marin County, California, where more than 50 percent of students were unvaccinated, and where, in 2014 and 2015, epidemics of whooping cough and measles broke out. They know enough about science to be skeptical, but not enough to allay their fears, and they have too little trust in science to take the risk. In Eugene, Oregon, rated the "Best US City for Hippies," there were schools with vaccination exemption rates over 60 percent.

According to the same CDC study, some seventeen thousand American children were going without childhood vaccines annually. The largest numbers were in liberal counties in California, Illinois, New York, Washington, Pennsylvania, Texas, Oklahoma, Colorado, Utah, and Michigan. States that allow "philosophical exemptions" to laws that mandate certain vaccinations for children entering school had more communities with unvaccinated children, creating pockets of risk.

Most parents who exercise these exemptions, a 2011 study found, express concerns regarding the safety of vaccines, or are suspicious that vaccines may cause autism. "Additionally," wrote the authors, "some parents may claim a personal exemption in response to their disdain for government organizations and pharmaceutical companies who are perceived as benefiting financially from mandatory vaccination requirements. The reasons for philosophical exemptions may vary, but interestingly there seems to be at least one commonality: emotionally-based negative conjecture despite overwhelming positive scientific evidence."

The danger of this is that unvaccinated pockets can form living petri dishes where nearly eradicated viruses can gain new footholds and endanger the wider public's health, much as cholera metastasized in the broader Haitian population in 2010. This has already happened. In 2000, the CDC declared that measles had been eliminated in the United States. But by 2008, after ten years of antivaccine misinformation, the United States had more measles cases than it had had since 1996, two years before the Wakefield paper was published. By 2014, the number had skyrocketed to more than six hundred cases in twenty-three separate outbreaks.

As a result, California passed a new law in 2015, requiring vaccinations for children to attend school, and Oregon, the state with the highest vaccine exemption rate, considered a similar measure. "California Gov says yes to poisoning more children with mercury and aluminum in manditory [*sic*] vaccines. This corporate fascist must be stopped," tweeted actor (and Jenny McCarthy's former boyfriend) Jim Carrey on June 30, 2015, after Governor Jerry Brown signed the bill into law.

But the antivaccine scare also creates pockets of disease in certain religious and immigrant communities. The school with the highest 2013–2014 exemption rate in Oregon, for example, was the Roman Catholic St. Thomas Becket Academy in Veneta, with a rate of 71.84 percent. And in December of 2010, an unlicensed and discredited Wakefield visited with members of the Minnesota Somali community. Referred to by Somalis as "little Mogadishu," Minnesota has the largest Somali community in the United States. Wakefield held three such visits, reportedly telling the immigrants to "get vaccinated, but do your homework and know the risks." The Somalis were concerned because autism appears to run at twice the national average in their community, but they did not recall high autism rates in Somalia or in Kenyan refugee camps. As a result of the fear, MMR vaccination rates among the Minnesota Somali population fell from more than 90 percent in 2004 to just 54 percent in 2010, the last time community members invited Wakefield to speak.

Then, in February 2011, an unvaccinated Somali infant returned from a trip to Kenya infected with measles. By April, fourteen cases had been reported, all but one traceable back to the unvaccinated infant. Half were Somali children, six of whom were unvaccinated and one who was not old enough for shots. The state had reported zero or one case of measles per year for most of the past decade.

Dr. Abdirahman Mohamed, a Somali physician in Minneapolis, told the Associated Press that Wakefield has caused global hysteria that has cost lives. He said he has warned the Somali community to stay away from him. "He's using a vulnerable population here, mothers looking for answers," Mohamed said. "He's providing a fake hope."

As with other wars on science, the war on vaccines is one of shifting goalposts. After the antivaxxers' claims that the MMR vaccine caused autism were debunked, they argued that the thimerosal in vaccines caused autism. After this was debunked, new claims surfaced that the cause was the vaccines being given too close together and too early, somehow shocking a child's immune system and resulting in autism. This, too, was debunked. Science shows that even though the number of shots has risen over time, the actual load on the immune system has decreased because today's vaccines are better engineered.

Before 1991, the whooping-cough vaccine had three thousand different antigens. Today's whooping-cough vaccine has no more than five particles—just as effective, but much easier on the immune system. After the "too many, too soon" myth was debunked, antivaxxers began claiming that the MMR vaccine was "triggering" autism in children who were somehow genetically predisposed to it. That, too, was debunked.

Conservative Antivaxxers

Recently, the antivaccine movement, traditionally the domain of educated liberals, started spreading into conservative politics through two different vectors. The first was human papillomavirus (HPV). There is currently a global HPV epidemic, with hundreds of millions of people infected with the sexually transmitted disease. It is the most common sexually transmitted disease in the United States, with almost every sexually active person getting it at some point in their lives. It is also the main cause of cervical cancer, and the fourth most common cause of cancer in women, with 266,000 deaths and 528,000 new cases in 2012. A large majority (around 85 percent) of the global burden occurs in less developed regions. In men and women alike, it can also cause anal and genital cancer, head and neck cancers, and genital warts. But it can be prevented by vaccination of girls and boys before they become sexually active.

In 2012, HPV vaccination became a major subject of debate during the US Republican presidential primaries. Candidates Michele Bachmann and Ron Paul criticized their rival Rick Perry for becoming the first governor in the nation to sign a Texas bill requiring the vaccination of prepubescent girls against HPV. Religious conservatives oppose HPV vaccination, believing it may increase promiscuity among girls by removing one of the punitive consequences (cervical cancer and genital warts) of having sex outside of marriage. The conservative backlash against Perry was so strong that he reversed course, calling the bill "a mistake" and allowing the legislature to overturn it.

In 2014, Texas Republicans placed the antivaccine plank in their party platform, stating, "All adult citizens should have the legal right to conscientiously choose which vaccines are administered to themselves, or their minor children, without penalty for refusing a vaccine. We oppose any effort by any authority to mandate such vaccines."

The second antivaccine vector emerged in the 2016 US presidential race, again on the Republican side. But this time it had moderated into language coded to appeal to both religious conservatives and antigovernment libertarians, with candidates advocating for parental choice over the mandates of an intrusive big government. On February 1, 2015, in the midst of a major measles outbreak linked to Disneyland, President Obama told the *Today* show's Savannah Guthrie, "I understand that there are families that in some cases are concerned about the effect of vaccinations. The science is, you know, pretty indisputable. We've looked at this again and again. There is every reason to get vaccinated, but there aren't reasons to not."

The next day, New Jersey governor Chris Christie was asked by reporters in London about the president's comments. He said, "The concern would be measuring whatever the perceived danger is by a vaccine, and we've had plenty of that over a period of time, versus what the risk to public health is and you have to have that balance."

"I'm not anti-vaccine at all, but particularly, most of them ought to be voluntary," Senator Rand Paul (R-KY), a medical doctor, said on Laura Ingraham's radio show the same day. Paul has long been a member of the Association of American Physicians and Surgeons, a Tucson, Arizona, organization that advocates conservative and free-market solutions in health care and other political issues, expresses doubt about the connection between HIV and AIDS, and opposes mandatory vaccinations.

Later in the day, former Hewlett-Packard CEO Carly Fiorina (to whom Craig Barrett had turned in 2008 to help connect me and the ScienceDebate effort with the McCain campaign) told Buzzfeed, "I think there's a big difference between—just in terms of the mountains of evidence we have—a vaccination for measles and a vaccination when a girl is 10 or 11 or 12 for cervical cancer just in case she's sexually active at 11. So, I think it's hard to make a blanket statement about it. I certainly can understand a mother's concerns about vaccinating a 10-year-old. . . . I think vaccinating for measles makes a lot of sense. But that's me. I do think parents have to make those choices. I mean, I got measles as a kid. We used to all get measles. . . . I got chicken pox, I got measles, I got mumps." A 2016 study showed that measles depresses immune response to other deadly diseases for up to three years, helping to explain why measles vaccination reduces childhood deaths across the board.

Ben Carson, a retired neurosurgeon, took a stand against the emerging Republican position, sounding a voice for reason on vaccinations even as he continued to deny evolution, climate change, and the big bang. "Although I strongly believe in individual rights and the rights of parents to raise their children as they see fit, I also recognize that public health and public safety are extremely important in our society," Carson told the *Hill* in a statement. "Certain communicable diseases have been largely eradicated by immunization policies in this country and we should not allow those diseases to return by foregoing safe immunization programs, for philosophical, religious or other reasons when we have the means to eradicate them." Carson's proscience stance in this instance was perhaps not surprising, since he is chairman of the board of a company called Vaccogen, which is working on a vaccine against colon cancer.

Donald Trump presented an unusual mix of conservative and liberal concerns. He didn't like vaccines given "in 1 massive dose," he tweeted in 2014, relaying one of the more recently debunked antivaxxer tropes. He said the syringe used on the child of an employee was of the size used on a horse, a story he repeated on the campaign trail and in debates over the coming year (although he made no mention of what he was doing at an employee's child's medical appointment, nor of the fact that childhood vaccines are not given all at once in a single shot). "Tiny children are not horses," he tweeted, and said he was convinced the shot gave the child first a fever and then autism—a condition he thought might have been avoided if vaccines were spread further apart and not given with such big syringes. "I'm not against vaccinations for your children, I'm against them in 1 massive dose," he tweeted. "Spread them out over a period of time & autism will drop!" Vaccinations are already spread out over several years. As well, the US Centers for Disease Control and Prevention, the US National Institutes of Health, the American Academy of Pediatrics, the World Health Organization, the Institute of Medicine, and other leading worldwide health and medical organizations all point out that, according to scientific evidence, vaccinations do not cause autism, whether they are spread out or bunched together (as in the MMR vaccine), and, again, that the actual immune-system load in today's vaccines is much lighter.

Canadian Conservative prime minister Stephen Harper, who was widely criticized for clamping down on scientists' ability to speak to the media and for defunding some of the country's science investments, reacted strongly against the right-wing political pandering in the United States at an event with Bill Gates announcing a $22.5 million Canadian investment to fund vaccination programs in poor countries. "We in the educated, advanced, medically advanced, sophisticated part of the world, we have a responsibility when it comes to this, not just a responsibility to vaccinate our children, which I think every parent has a responsibility to do, and not just a responsibility to encourage that widespread vaccination so we're not putting other kids at risk," Harper said. "But we have a responsibility to set an example, for God's sake. We know these medical interventions work and as an advanced, educated society it is completely irresponsible of people in this society to communicate anything other than that anywhere else in the world."

He chided parents worried about the supposed health risks of vaccinations, saying, "Get the facts from the medical and scientific community and if you're not a doctor or a scientist yourself, listen to the people who are. It's that simple." Newly-elected Liberal prime minister Justin Trudeau took a similar position.

As in liberal communities in the United States, there is a long tradition of antivaccination propaganda in the Canadian alternative-medicine community, and naturopaths and chiropractors have been implicated in the spread of misinformation. Canadian supplement and vitamin sellers market Influenzinum, for example, as a homeopathic alternative to the flu vaccine. It is essentially water, and has no scientific merit.

Similarly, in Australia, conservative prime minister Tony Abbott reacted to the international dialogue by announcing that the government would stop childcare and family tax payments—worth up to $15,000 per child—to "conscientious objector" parents of unvaccinated children. "Parents who vaccinate their children should have confidence that they can take their children to childcare without the fear that their children will be at risk of contracting a serious or potentially life-threatening illness because of the conscientious objections of others," Abbott said.

In the United Kingdom, however, a group called Arnica has begun working to spread antivaccination propaganda and a belief in "natural immunity" across Europe. "When we compare a vaccine's possible adverse reactions, and possible longer-term health problems, against complications from childhood disease in healthy children, we feel that in most cases the non-vaccinated child is healthier," the group's website argues, without presenting any scientific evidence to support this conclusion.

The British Invasion

Antivaccine scares are not new. Their history is almost as long as that of vaccination, and, like the modern Wakefield scare, it has its roots in the United Kingdom.

In 1870, antivaccination demonstrations in England drew thousands of people in reaction to an 1853 law requiring vaccinations against smallpox. British businessman and leading antivaxxer William Tebb organized the First

International Anti-Vaccination Congress in Paris. The group appointed a committee to further spread its antivaccine message, made up of members from France, Germany, Austria, Holland, Belgium, Switzerland, Russia, Sweden, England, Canada, and the United States.

Tebb had moved from England to Blackstone, Massachusetts in 1852, as an emissary of the Vegetarian Society of Great Britain, whose members often opposed smallpox vaccinations because they used calf lymph (cowpox conferred immunity to smallpox without being communicable in humans). Similar to the language used by Arnica a century and a half later, the Vegetarian Society members felt that both vaccines and eating animals adulterated the "natural purity" of human blood.

Tebb moved his family back to London in the 1860s—ironically, to escape a malaria outbreak. In 1879, he returned to the United States to speak about vaccines, and the Anti-Vaccination Society of America was founded following his visit. Several subsequent US groups waged court battles to attempt to repeal vaccination laws in a number of states, including California.

The first US Supreme Court case concerning antivaxxers and the power of the states in enacting public-health laws followed a 1902 smallpox outbreak. The board of health in Cambridge, Massachusetts, had mandated that residents be vaccinated against smallpox. Otherwise, they would be fined $5 (over $100 in today's money). A resident named Henning Jacobson refused, claiming that he and his son had had bad reactions to earlier vaccinations and arguing that the law violated his civil liberties. The city filed criminal charges against him. Jacobson lost, but he appealed the decision to the Supreme Court. In 1905, the court ruled that states could enact compulsory laws to protect public health.

By the 1950s, future Republican US Congressman John Porter was a young boy growing up in a Christian Science family in Evanston, Illinois. The Church of Christ, Scientist had been chartered in New England by Mary Baker Eddy in 1879, the same year William Tebb visited New England. Eddy was a believer in homeopathy and experimented with placebos. Her 1875 book *Science and Health* argued that sickness is an illusion that can be corrected by prayer alone. But Porter's father walked with leg braces because of polio that hadn't been cured by prayers—his father's or his own. Then, in 1955, the Salk vaccine was licensed. Porter watched in amazement and anger as several of his Christian Science classmates refused to get it. He thought of what the vaccine would have meant to his father's life, and he cut his ties with the church. "That is not what I would call moral thinking," he says. The polio vaccine's success was profound.

In 1952, the year Jonas Salk first began to test it in humans, the United States had 57,879 cases. By 1957, two years after the vaccine was approved for broad use, the number had fallen to 5,485, and by 1964 it was 122.

Ten years later, in the mid-1970s, the United Kingdom exported another international antivaccine scare, this one over the safety of the diphtheria, pertussis, and tetanus (DPT) immunization. Three researchers from the Great Ormond Street Hospital for Sick Children in London published a paper alleging that thirty-six children suffered neurological damage following DPT immunization. As in the case of the US cell-phone scare, television documentaries and newspaper reports drew public attention to the controversy. An advocacy group, the Association of Parents of Vaccine Damaged Children, sprang up, warning the public about the supposed risks and consequences of the DPT vaccine. By 1977, pertussis vaccine rates had dropped from 77 percent to 33 percent, and as low as 9 percent in some districts. American media attention to the controversy helped fuel the spread of the movement in the United States, until government agencies like the CDC were able to spread enough vaccine education to quell the panic. But by then the fear had spread across Europe, Japan, the Soviet Union, Australia, and Canada.

As a congressman in the 1990s, Porter saw the public's vulnerability to misinformation and worked with Gingrich to double the budget for the National Institutes of Health. After retiring, he lobbied Congress for medical-research investments as chair of Research!America, a health-research advocacy group. He says the silence by scientists over the last two generations has contributed to the problems in American policymaking—not just around vaccines and health care, but on a host of issues related to science.

"What members of Congress need to hear is passion," Porter says. "When scientists and engineers talk to a member they need to say, 'Look, this is really important.' The scientific community is very, very timid about asserting themselves. When President Bush was undermining scientific integrity and rewriting scientific studies to make religion preeminent, the science community did nothing publicly except for the Union of Concerned Scientists, and that silence was a shameful travesty. It's not just about 'Give me funding.' They've got to step up and say, 'This is wrong!' They need to impact the public in a way that really highlights how critical science policy is to our lives."

Chapter 7

THE RISE OF THE ANTISCIENCE
NEWS MEDIA

Every time we proceed to explain some conjectural law or theory by
a new conjectural theory of a higher degree of universality, we are
discovering more about the world, trying to penetrate deeper into its
secrets. And every time we succeed in falsifying a theory of this kind,
we make an important new discovery. For these falsifications are most
important. They teach us the unexpected; and they reassure us that,
although our theories are made by ourselves, although they are our
own inventions, they are none the less genuine assertions about the
world; for they can *clash* with something we never made.

—Karl Popper, 1972

Deaf, Dumb, and Blind

As science sat on the sidelines and antiscience rose on both the left and the right,
the intellectual erosion of the 1970s and 1980s worsened. In August 1987, during
the administration of President Ronald Reagan, the Federal Communications
Commission (FCC) abolished the Fairness Doctrine in a historic 4–0 vote.

The doctrine had its roots in the Radio Act of 1927 and had been a formal
policy of the FCC since 1949, when television went mainstream. It required
those who held licenses to broadcast over the public airwaves to present programs on controversial issues of public importance—and to present them in a
way that was, in the FCC's view, honest, equitable, and balanced.

After the policy was abolished, Congress recognized the danger and tried
to codify the doctrine into law, but President Reagan vetoed the legislation. As
a result, broadcasters were unburdened from the requirement to present balanced news coverage, and the age of yellow journalism was reborn.

Chief among the early beneficiaries were angry, opinionated baby-boomer talk jocks like Rush Limbaugh, who began broadcasting political rants that charged up listeners' amygdalae with outrage, attacking political correctness, government regulations, and other enemies of cultural conservatives, and driving audience numbers sky-high.

It was as if the *National Enquirer* and *Star* magazine had bought up the nation's broadcast media. For science, the problem was that the nation began to lose its common sense of reality—the assumption on which science is predicated. The balanced programming broadcast on public airwaves had long been held up as the voice of objectivity. While network news shows often got facts wrong, they were nevertheless scrupulous—one could almost say scientific—in their efforts to be accurate. What was said on the air was taken to be generally true and impartial. This is why Orson Welles's 1938 radio play about an alien invasion, *The War of the Worlds*, could cause a nationwide panic.

To better understand the effect of the Fairness Doctrine's repeal, consider a bit of educational psychology: the cognitive styles of field-dependent and field-independent personalities. Field-dependent personalities are more socially oriented and require externally defined goals and reinforcements. Field-independent personalities are more analytical and tend to have self-defined goals and reinforcements. Educators have developed strategies for teaching people with each learning style. Field-dependent citizens in the population, who valued authority and looked to it for guidance, needed the protection of the Fairness Doctrine to give them a fair grasp of reality. In the doctrine's absence, many field-dependent people have drifted under the influence of slanted conservative commercial media, creating one of the most divided political climates in history.

It made sense that this would happen. Field-dependent personalities need more externally supplied goals and structure than field-independent types, and talk radio provided them. By freeing broadcasters from the requirement of fairness and thus the need to ground their messages in established knowledge and fact, the repeal plummeted the country into a partisan public dialogue. One-sided rhetorical arguments backed by outrage and wattage could now hold sway over facts and reason. It was a further undoing of Locke's ideas about knowledge, upon which the founding principles of democratic governance are based. By removing the goal of objectivity, it set modern politics up for endless arguments between warring pundits.

Proponents of the repeal, who have since opposed congressional attempts at correction, argued that the market forces of the invisible hand will unleash freedom and competition in the marketplace of ideas, which will stimulate broad-ranging and higher-quality discussion. But, in fact, just the opposite has occurred, with discussion becoming less diverse and more polarized. Without the attempt at an objective standard of knowledge, listeners turned into opinion "dittoheads," as Limbaugh's followers happily describe themselves. Diversity of thought has been quashed in favor of an uncritical, authoritarian, and vehemently partisan allegiance to a political tribe—a new conservative identity politics that rejects the left-wing identity politics of the postmodern era. The talk jocks act as chorus masters, conducting the audience members' political opinions in an us-versus-them narrative that maximizes audience share and directs the anger and political contributions of millions of listeners like so many sports fans.

This is the opposite of the protection of diverse viewpoints that was intended by the Founding Fathers, and it has inevitably run up against science on several occasions. The most current and prominent example is the complete rejection of climate science as a manufactured political project of socialists bent on a global takeover and the government-funded scientists who enable them.

What Marketplace?

But the argument for doing away with the doctrine was based on a faulty assumption: that ideas exist within a social dialogue akin to a marketplace in which journalistic truth has market value—that people want to hear it enough to pay extra, and that media with a variety of viewpoints will compete to deliver the highest-quality journalism for the best price. This is nonsense.

In fact, ideas do not exist in anything akin to a marketplace, and journalism's function in a democracy is often to tell people what they *don't* want to hear but is important to know anyway. Its role is to report the news, meaning the facts of recent events, in order to provide the public with a common ground for debate and discussion.

The incorrect assumption that there exists some sort of marketplace of ideas where truth is the gold standard is destroying mainstream news and, with it, the balanced, moderate political weltanschauung of the countries that practice this approach. This is especially true in the United States, where media outlets are now divided into opposing groups of partisans whose primary goal is to manufacture conflict and produce the sort of instant drama that helps win the battle for ratings. This approach applies to the news the principles of building drama instead of those of journalistic reporting on facts, and so it is authoritarian and ultimately leads to the tyranny of might-makes-right, evidence of which we have seen growing throughout the media in the recent decades.

Hollywood producers will immediately tell us why this idea could never have worked. There is no marketplace of ideas; there is, instead, a marketplace of emotions. Given the choice, the majority of people want drama, sex, violence, and comedy—the four horsemen of entertainment. These four elements have motivated plays, paintings, and stories for all of history. But they are not *news*. They are "but faith, or opinion, but not knowledge." News, on the other hand, is knowledge.

The only way we can have a market-driven model for news is with a nonfinancial reward structure. You could do this with a peer-review system like that used in scientific publishing, but review by outside peers is difficult in the extremely short time frame of competitive newsgathering. You could also have broadcasters perform an abbreviated form of the peer-review process internally, which is what the Fairness Doctrine did. By repealing the policy, the FCC forced news programs into financial competition with entertainment— the marketplace of emotions. Broadcasters were forced to turn to the tools that deliver audience share, and those tools are the four horsemen. It was the success of highly partisan, emotional commentators like Limbaugh that took market share away from broadcasters who presented the news more evenhandedly. Without enforced standards, news was cut loose from knowledge and the emotions of outrage and comedy were increasingly relied upon to sell the news.

Imagine what would happen in the scientific community if peer review were abolished and scientists could, in the name of freedom of speech and market economics, publish anything they wished with only a commodity regard for the truth. The major journals would compete for market share and give a portion of the proceeds to the scientists. A free-for-all would eventually ensue, with scientists competing for the most dramatic papers. Some would reflect good science, but many would not. Some scientists would argue for the value of truth while others would see the rewards of emotion. Scandals might even be manufactured in order to get more market share—say, a study manipulated to link the MMR vaccine to autism. Satirists would arise, pointing out how ridiculous it all is. Serious scientists would soon have to dress their findings in the enticing clothing of sex or drama, much as some of them try to do now, in an effort to be relevant to the public. Many would object to the new situation, arguing that by substituting market economics for standards of truth and quality we were losing something precious: a common ground upon which to build knowledge, which is essential to the functioning of the scientific community. But others would argue that this was a healthy development because it opens science up to a marketplace of ideas and draws in a wider readership from the general public. The journals *Science* and *Nature* would likely increase their circulations, but eventually science itself would be lost.

This would not be freedom; it would be its opposite. Science would descend into darkness, and tyranny would again reign. For the sake of freedom and the public good, we must not use market economics as the sole arbiter of value in matters of public knowledge such as science and journalism. We need a reward system tied to objective truth. The Fairness Doctrine was a tool of self-governance that encouraged peer review and helped ensure an attempt at objectivity.

The New News Crisis

Meanwhile, over the last two decades, the move to free online news has been driven by the founding principle of the Internet: the hyperlink, a trail of context-dependent endnotes, based on a concept first developed by Vannevar Bush in a 1945 *Atlantic* magazine article in which he presciently describes the modern personal computer. The model of free and linkable news stories completely upended the revenue structure of newspapers, making it very difficult to remain solvent and propelling them to cut costs. But putting news stories behind a paywall destroyed their linkability and so eliminated them from the dialogue occurring on the Internet. To save money, newspapers began by cutting expensive endeavors like investigative reporting and specialty divisions like science. Thus, an important social faculty—the capacity for broad, critical self-assessment and data-based reflection—was suddenly eliminated.

As newspapers were grappling with obsolescence, yellow journalism had already spread from AM talk radio into TV with the advent of cable news. This trifecta—talk radio, the Internet, and cable news—combined to devalue the factual reporting that once kept society balanced, supplanting it with the opinion wars of the new media. Having trained at postmodernist universities, many emerging leaders in journalism didn't recognize this as a problem. It wasn't their role to discern the reality of things, they believed. Truth was subjective, a matter of one's perspective. Thus a reporter's role was to provide both sides of the story fairly and with balance, but also without judgment, which, in a world cut loose from knowledge, could be deemed an insertion of a politically motivated bias, and no one would be educated enough to refute the charge.

Marcia McNutt, then director of the United States Geological Survey (USGS), describes the mind-numbing, tail-chasing public-policy effects this "marketplace of ideas" produces:

There doesn't seem to be much accountability. Remember when the "sand berm" issue for Louisiana was such a big deal? [It was proposed as a way to prevent the flood of oil from the 2010 Deepwater Horizon spill from reaching the shore.] Governor [Bobby] Jindal and the parish presidents got pretty much 24/7 coverage on CNN, Fox, and everywhere else, saying what a bunch of losers anyone and everyone in the government was who wasn't rushing to put in their project, even though the USGS scientists and the Louisiana scientists said it was a stupid idea. But did the scientists get any airtime? No, and in fact with the way the media played it up, they were fearing for their lives. So the government reluctantly granted the permits and tried to make the best of it, because the public outcry in favor of the berms was so loud after all of the media attention. Then after the fact, when every prediction of the scientists proved to be right, where was CNN? Fox? Oh yeah—they did show up—to say what on earth was the government thinking to build the sand berms! "Another example of hasty government action!"

The result is a crisis throughout journalism. Serious journalism is being forced into small outlets on the Web, many of them nonprofit, or is being wrapped in the guise of one of the four horsemen of entertainment. Thus we see serious news being covered on Comedy Central's *The Daily Show* and Canada's *Naked News* and in theatrical documentaries such as *A Sea Change* and *Inside Job*, while cable news peddles scandal and outrage. Public broadcasting, one of the few news sources left that attempt to stick to reporting the facts, is attacked as politically motivated. As Hollywood screenwriting expert Robert McKee says, "Story is not about facts. It's not about intellect. It's about values and emotion." Forcing journalists to deliver the high degree of values and emotion necessary to win in a crowded entertainment marketplace forces them to become yellow journalists, substituting emotions for facts and depriving citizens of the fourth estate's crucial contribution to a democracy.

Losing Touch with Reality

The effects of the "there is no such thing as objectivity" school of journalism could be seen early in 2015, as NBC News president Deborah Turness sent an e-mail to staff announcing that, effective immediately, Brian Williams had been suspended without pay as managing editor and anchor of *NBC Nightly*

News, a position he had held since 2004. Williams ruled in TV news ratings, but, while covering a ceremony honoring retiring soldier Tim Terpak, who provided ground security for Williams when he was in Iraq twelve years previously, Williams's account veered away from what had actually happened.

"The story actually started with a terrible moment a dozen years back during the invasion of Iraq when the helicopter we were traveling in was forced down after being hit by an RPG," Williams said on the broadcast. "Our traveling NBC News team was rescued, surrounded, and kept alive by an armor mechanized platoon from the US Army third Infantry."

The only problem was that it didn't happen. Another chopper had come under fire about an hour before Williams even arrived in the area. Because of an impending sandstorm, the helicopter that Williams and his camera crew were traveling in later landed next to the one that had taken fire. Soldiers took to social media to debunk Williams's story. Their posts were picked up by the media and the gaffe went viral.

Williams responded to their social media posting with an apology.

> You are absolutely right and I was wrong. In fact, I spent much of the weekend thinking I'd gone crazy. I feel terrible about making this mistake, especially since I found my OWN WRITING about the incident from back in '08, and I was indeed on the Chinook behind the bird that took the RPG in the tail housing just above the ramp. Because I have no desire to fictionalize my experience (we all saw it happened the first time) and no need to dramatize events as they actually happened, I think the constant viewing of the video showing us inspecting the impact area—and the fog of memory over 12 years—made me conflate the two, and I apologize.

Media pundits were quick to blame Williams's character for the gaffe. Others suggested that he was the victim of false-memory syndrome. But it wasn't "the fog of memory." The soldiers saw Williams and his camera crew photographing the damaged chopper that day, and said they recalled a news story at the time in which Williams made the same claim. The NBCUniversal online archive indeed shows that the network broadcast a news story on March 26, 2003, with the headline "Target Iraq: Helicopter NBC's Brian Williams Was Riding In Comes Under Fire." The segment's long description reads, "NBC's Brian William [sic] recounts being shot at by a rocket propelled grenade while riding along in

Chinook helicopter on a mission over Iraqi airspace." Williams had lied to the public at the time, and now he was doing it again, by blaming it on "the fog of memory."

How could this be? Did he think no one would remember the story, or later check the archives? Why would he possibly risk his career? Was it true that Williams had "no desire to fictionalize . . . and no need to dramatize events"? What could have driven this colossal blunder?

A few reporters remembered other stories Williams had told, such as of his dramatic experiences covering Hurricane Katrina. "My week—two weeks—there was not helped by the fact that I accidentally ingested some of the flood-water," he told Tom Brokaw in a 2014 interview. "I became very sick with dysentery. Our hotel was overrun with gangs, I was rescued in the stairwell of a five-star hotel in New Orleans by a young police officer. We are friends to this day."

Other stories were even more heartrending. "When you look out of your hotel room window in the French Quarter and watch a man float by facedown, when you see bodies that you last saw in Banda Aceh in Indonesia and swore to yourself that you would never see in your country," Williams said. "I beat that storm. I was there before it arrived. I rode it out with people who later died in the Superdome."

But this didn't happen either. The French Quarter stands on high ground and never flooded. "I can tell you that at no time did any of my people report any sightings of any bodies," Myra DeGersdorff, the manager of the French Quarter Ritz-Carlton during Katrina, said in 2015. "He may have simply misremembered. But I can tell you no one broke out in the hotel with dysentery." Dr. Brobson Lutz, a former city health director who manned an EMS trailer in the French Quarter, said, "We were never wet. It was never wet." As to Williams's claims of fighting dysentery, Lutz said, "I saw a lot of people with cuts and bruises and such, but I don't recall a single, solitary case of gastroenteritis during Katrina or in the whole month afterward." He added that his dogs drank the floodwater "and they didn't have any problems."

Ten days later, after defending Williams, conservative Fox News host Bill O'Reilly was himself accused, in a *Mother Jones* article, of embellishing accounts of his time in war zones. "I've covered wars, okay?" conservative commentator Tucker Carlson quoted him as saying in a 2003 book. "I've been there. The Falklands, Northern Ireland, the Middle East. I've almost been killed three times, okay." Then, in a 2004 column, O'Reilly wrote, "Having survived a combat situation in Argentina during the Falklands war, I know that life-and-death decisions are made in a flash." He spoke often of his experiences, offering graphic details. But these stories, too, were shown to be false. O'Reilly's former colleagues at CBS, where he worked at the time, now disputed his claims. "It was not a war zone or even close. It was an 'expense account zone,'" said Eric Engberg. Several other journalists made similar comments, debunking O'Reilly's depictions of dramatic events that never occurred.

Fair and Balanced

What could be happening to US journalism? Was this a coincidence? Do these men—both of them top television journalists with enormous audiences, albeit very different ones—simply have poor characters? Or could there be another explanation for their embellishments? There are several possible ones, of course, starting with the simple fact that people have been exaggerating the drama of their experiences since the dawn of time. But doing it publicly, with one's credibility as a television news anchor at stake, was breathtaking. Something more unique was happening—there was a clear record that could easily be checked, and the exaggerations happened with such high stakes, making the theory that they were simply liars untenable. Some journalists theorized Williams and O'Reilly were victims of false-memory syndrome. Others wondered about their mental health. But few looked in a more obvious direction: the subjective approach that has taken over modern journalism. Occam's razor suggests that the simplest explanation that accounts for all of the evidence is often the correct one. It is possible I am not being fair to hold these men out as examples of a problem in the field, but it's also possible that these men's judgment was clouded by their career-long exposure to the view running institutionally throughout journalism, taught in journalism schools since the 1970s, that there is no such thing as objectivity. It is possible that, in a high-stakes professional environment like television journalism, they were simply caught doing what others do.

Journalism students are taught that every story is subjective, that it is impossible for a reporter to filter out their own biases, and that responsible reporters will acknowledge this. In fact, they are told, to present a story as objective is fundamentally dishonest. This notion is widespread. Publications' reporter guidelines contain it. "There is no such thing as objectivity," the former NBC journalist Linda Ellerbee wrote. "Any reporter who tells you he's objective is lying to you." Students are taught that the best they can hope to achieve is to be fair and balanced. The Society of Professional Journalists dropped "objectivity" from its Code of Ethics in 1996. Even Nick Gillespie, the editor of the libertarian, objectivist website Reason.com, repeats on the speaking circuit that there is no such thing as objectivity.

But the notion is mistaken. Of course we are each subjective in our perspective, but there is such a thing as objectivity: a statement about reality that stands independent of our subjective qualities and is verifiable by others. And such objectivity is attainable in reporting. The belief in, and search for, objective truth might have motivated journalists such as David Gregory to have the confidence that it was "our role" to push President Bush to produce evidence of the existence of weapons of mass destruction in Saddam Hussein's Iraq before the United States invaded. Is that a partisan position? No. Would it have been political? Yes, and that is what responsible journalists do: they hold the powerful accountable. Today, this distinction is more important than ever, as the stories journalists present to the public have a great deal to do with objective knowledge from science, and reporters must find a new way to sort out their thinking on it.

Physicist Lawrence Krauss, a cofounder of the ScienceDebate effort, wrote presciently about this in a 1996 *New York Times* opinion piece, in which he described how presidential candidate Pat Buchanan had recently professed that he was a creationist. Although journalists questioned other Buchanan campaign planks like trade protectionism and limits on immigration, Krauss said, there were no major articles or editorials declaring the candidate's views on evolution to be counterfactual.

> Why is this the case? Could it be that the fallacies inherent in a strict creationist viewpoint are so self-evident that they were deemed not to deserve comment? I think not. Indeed, when a serious candidate for the highest office of the most powerful nation on earth holds such views you would think that this commentary would automatically become "newsworthy."
>
> Rather, what seems to have taken hold is a growing hesitancy among both journalists and scholars to state openly that some viewpoints are not subject to debate: they are simply wrong. They might point out flaws, but journalists also feel great pressure to report on both sides of a "debate."

This erosion of our capacity to tell knowledge from opinion, and the press's culpability in it, began to have profound effects that aided and abetted the rise of the antiscience right.

The Authority of Subjectivity

And here's where the problem comes to a head. If all reporting is subjective, all reporters must in a sense become gonzo reporters—the intellectual offspring of Hunter S. Thompson. They must tell or tweet the story from the subjective point of view; in one way or another acknowledging that they are somehow *in the story. Embedded.* Like Brian Williams, their reporting therefore must come to be about *their experiences.* This is called "firsthand" reporting. If the highest authority is in the immediacy of the subjective story, a reporter who doesn't personally experience a story is not regarded as having the same authority as one who did. It follows, then, that if one wants the most authority, one must put one's self into stories intensely and personally, especially in television. This is, some may argue, what led to Williams and O'Reilly's exaggerations: the substitution of the subjective eyewitness participant for the journalist. In the climb to the top of American journalism, imagine the pressure to *be in the story* in order to speak with the most authority and create the most compelling TV. Imagine the little slips of reason that could result.

This approach to reportage is backwards, as we now know from science: the eyewitness account is among the *least* reliable. Science has shown it to be "shockingly inaccurate." Let's grant that Williams and O'Reilly did not consciously lie, but had the sort of vagary of thought common to many eyewitnesses. With time, things fade, and the stories we remember are the ones charged with emotion, and that confirm our identity, not necessarily the ones derived from fact. Scientists deal with this all the time. Bacon wrote about it at the dawn of the scientific revolution: "For what a man had rather were true he more readily believes." Memory, bias, emotion, identity and desire are powerful forces over which we do not always have full control, and science shows that emotions play a considerable role in the formation and organization of memory. Knowing this, one can appreciate that turning to the eyewitness, subjective view as a primary vehicle for reporting on matters of national import is not wise. This is the opposite of science's idea of knowledge. But it sells in the marketplace of emotions.

In a world in which gonzo journalism has become mainstream, who's to say that Williams didn't *feel* attacked in that helicopter? That it wasn't his *experience*? It certainly was his memory of it. Except, objectively, the attack happened to another chopper in a different place and time. Who's to say that he didn't feel "something get dislodged that changes the usual arm's length relationship between me and the stories I cover," as his interviews about his eyewitness reporting of Hurricane Katrina indicated, from seeing that dead body floating outside his hotel in the French Quarter ("These are *Americans*," he said, "my brothers and sisters. And one of them was floating by"), or that he wasn't terrified when the hotel was overrun by gangs? That it wasn't his *experience*? Certainly that's the powerful sense that informed his memory of the events. Except that, objectively, there was no flooding in the French Quarter, no dead body, no gangs overrunning his hotel. When these objective facts were exposed, reporters threw up their hands, shocked to learn that Williams, and later O'Reilly, would have said such things. But is it credible to think that no other reporters engage in this, given our understanding of eyewitness accounts and the way memories are formed, and given the elevation of subjectivity and first-hand reportage present throughout journalism and its schools of higher learning?

Were these distortions the result of a character flaw? Did these men suffer from false-memory syndrome? Or is it more likely that these errors are a creation of journalism itself? That almost any journalist placed in a high-stakes situation—the pressure, the money, the ego, the reputation, the competitors nipping at one's heels—would be likely to misspeak, solely out of the industry's imperative to capture the eyewitness's firsthand experience?

That, of course, is precisely the kind of bias that science—and, once upon a time, journalism—would endeavor mightily to strip away.

Rethinking the Journalistic Method

During the short time since the 1970s that journalists have been arguing about objectivity, scientists and engineers have completely transformed our world using objective knowledge. Climate change is objectively measurable, no matter which thermometer records one looks at. Electricity is objectively measurable, no matter who plugs the multimeter into the wall socket. Earth objectively goes around the sun, no matter who makes the telescopic observations. Radiometric dating always comes up with the age of Earth at about 4.54 billion years old, no matter who runs the mass spectrometer. Vaccines have been objectively shown to prevent the spread of deadly diseases, no matter who administers them. Evolution has been objectively verified thousands upon thousands of times by people of many cultures. Scientists have labored, sacrificed their careers, and died to cull this and other objective knowledge from our many biases and failings, using a system devised to do just that. They have done so in order to build a literature of objective knowledge that has, in turn, improved the lives of the rest of us.

Let's be clear: scientists have egos and biases. They recognize that, and the scientific method is designed to help filter those qualities out. Objective does not mean complete or perfect or apolitical; it means stripped of personal, religious, political, emotional, cultural, sexual, referential, and other biases, which is what the process of science works to achieve via repeated testing, confirmation, and peer review. If knowledge is falsifiable but holds up regardless of who does the testing, it is said to be reliable; i.e., objective. A scientific conclusion is always provisional, because knowledge is never complete, and it is always political, because new knowledge always threatens the status quo, but it is also increasingly reliable as it is tested and survives. Further, industries and careers built on today's new discoveries will become the vested interests of tomorrow that are, in turn, upended by new discoveries. Science is a process. But it is, with each passing confirmation and refinement, increasingly objective.

It is also true that journalists have egos and biases that can affect their reporting, and that it can also be a problem when journalists purport to be "objective" while failing to acknowledge their biases, a case NYU journalism theorist Jay Rosen points out quite eloquently. For example, a 2015 study by the Canadian Centre for Policy Alternatives found that news media coverage can cause cynicism by emphasizing policy failures on science topics like climate change, or inspire action by providing a compelling narrative about an everyday hero working for solutions. A study by Rutgers University communications researcher Lauren Feldman found that the *New York Times*, the *Wall Street Journal*, the *Washington Post*, and *USA Today* often did just this, by framing climate-change actions as unsuccessful or costly instead of manageable or effective.

The problem some journalists run into is that they confuse objectivity with completeness—as in "the full story." They then apply their experience interviewing subjective witnesses—and noting their own subjectivity as reporters—to all reporting, and conclude that because objectivity is not always possible it is never possible, or, worse, that passing news off as objective is a lie, as Ellerbie said. That is, ironically, when they become vulnerable to manipulation.

So how can journalists account for both subjectivity and objectivity, and not one or the other alone? Can we devise a comparable "journalistic method" that seeks to strip away biases and leave verifiable knowledge? Some may say it's hubristic to even talk about it, but the problems are real and demand an adjustment.

Obviously, statements have to be supportable—that is, tied back to the evidence of primary sources that confirm one another. Those sources, in turn, cannot be falsely balanced in the story between those who are relaying knowledge and those who are simply voicing an opinion. This is where it gets sticky, because that's a judgment call that must be made by the reporter, and that gets into objectivity, bias, criteria, and editorial review. But a good reporter can use his or her judgment to determine which view is most supported by knowledge, i.e., able to be tied back to independent evidence, and can then place the weight on that view, properly balancing the story and removing himself or herself in the process. However, this is only possible as long as that goal is the paramount value in the newsroom, reinforced from the publisher through the editors on down—which, as we have seen, is not always, and perhaps not often, the case.

One way to build in a further effort at objectivity by a subjective reporter is to seek, as a sort of peer review, a meta-consensus from fellow reporters who have different perspectives. Then one would present the story as a more or less objective assertion of what happened, reviewed and reliable to the best of our ability to determine, but also clearly identifying at the bottom of the story, as a scientist does, possible contraindications. If one really believes that one should leave it to the reader, such an approach would at least make an attempt at objective reporting while preserving transparency about how the subjective reporter might be wrong (and where he or she isn't). But such a novel approach could only flourish in a newsroom where objectivity is a stated value and reporters are encouraged to work as a team. This is expensive and slows the reporting down. It is also hard to imagine in the Twitterverse of immediate and gonzo political reporting, where it seems quaint to the point of naivety. One can easily imagine a scenario in which an otherwise conscientious reporter will take the path of least resistance and simply acknowledge his or her own subjectivity (on Twitter that's implied on an ongoing basis). The problem is that such a laissez-faire approach in a news story in no way provides the full picture, and since it is only an admittedly narrow window, the resulting lack of rigor can suspend a certain level of critical faculty, allowing bias to slip in to a much higher degree, since the consequences for not eliminating it are so low.

Such a low standard becomes problematic in an age when major political issues have considerable scientific dimensions. Very often, there is, in fact, objective knowledge that is readily available, and the misapplication of a reporter's well-meaning view that there is no such thing as objectivity can become a recipe for disaster. Cumulatively, this view becomes a danger to democracy, because it makes reporters vulnerable to easy manipulation by public-relations campaigns. We are left with mainstream journalists (i.e., not those on talk radio or other purposely slanted news outlets) who often simply present "both sides" of controversial issues, which gets us nowhere, doesn't help the public make informed decisions, and plays into the hands of powerful vested interests. Yet this has become the expectation and politicians become distrustful of journalists who don't use such a he-said she-said approach.

When a television news program presents a split screen with a scientist on one half representing the knowledge accumulated from tens of thousands of experiments performed by thousands of scientists, and then presents a charismatic advocate with an opposing opinion on the other half, as if the knowledge and opinion carry equal weight, this creates false balance. It skews democracy toward extremes by giving equal weight to both opinion and knowledge. But if reporters don't believe that objective knowledge exists, or that claims of objectivity can be honestly made, one can understand why they have been slow to discern the problem.

Diving Deeper

Another way to avoid the problem and improve the quality of reporting at the same time is to go deeper into politically contentious science topics. Often, the political arguments, particularly those mounted by antiscience public-relations campaigns, only function on the surface of a discussion, and when a journalist can disaggregate that "tarball," things can automatically become less polarized and more interesting—and more focused on the actual knowledge from science.

Stephanie Curtis is the senior producer of *Climate Cast*, a weekly program covering climate change on Minnesota Public Radio News that has both news and call-in/online-posting components. The show is nation-leading as a major broadcast weekly news program that goes in depth on climate change, featuring the world's leading climate scientists as regular guests. It is the brainchild of MPR News host Kerri Miller and the network's meteorologist Paul Huttner. "I'm not aware of any other radio stations that have a weekly climate show like this," says climate scientist Michael Mann. "Kerri and Paul certainly stand out."

What makes the program interesting to this discussion is that the hosts and producer decided early on not to entertain the question "Does climate change exist?" which has trapped so many other journalists into false balance reporting about science. Producer Stephanie Curtis explains their thinking and approach:

> When you're taking it seriously enough that you're looking at it every week, you can't just set up things and say "This is what the debate is." You have to step back and ask, "Is that really what the debate is?" You have to look smaller, and be more specific. Not what is all of climate change, but "What is El Niño and how does it affect things in our region?" Or "How are changes in the water cycle affecting rivers?" It's in the specificity that you get away from the broad-brush politics, and it also makes it much more interesting for your audience. There's just so much to talk about about the real science—the effects on the economy and people's lives—that we don't need to take the "Is it really happening?" question. That's been solved. It's easy to use the conflict frame, that's too easy. The more you learn about climate change, when you have an environmental reporter like Elizabeth Dunbar and a meteorologist like Paul Huttner that have the knowledge, the less you are going to need to fall back on the easy hot-button issue. And the more you know about it, the more fascinating things there are to cover, like "What's going to be happening with the water supply in California and how does that affect water in Minnesota?" because it does. And it's really driven by Kerri's confidence in saying, this is it. Just because it's a question in some people's minds doesn't mean we need to entertain that. But that takes a certain level of journalistic confidence.

Curtis says that the focus on specificity helps the program automatically avoid falling into the usual negative talking points that callers might get from some science-denial source. "We'll be talking about a narrow point of view about how climate is affecting rivers and they'll ask about the troposphere. Obviously a canned, classic denier talking point, but it's off-topic so I just say no, that's not helpful to the conversation. So being very specific allows for that clarity."

That doesn't mean that the show doesn't allow debate. "Absolutely," Curtis says, "We want debate. Just not canned denial talking points." But there are very sincere people who for whatever reason haven't been able to get their head around climate change and have real questions, like "What difference does two degrees make? I can't feel a two-degree difference in my office." But that's a specific, sincere question and Curtis says they expect their guests to answer those. "Because they really are wondering, and if we don't answer people's real, sincere questions we become part of the politics of it. There are plenty of believers in climate change who are simply ideological believers and they can't talk to the specifics of it, and taking an approach like that as journalists wouldn't be helpful."

The topic is so large and impactful—and so under-covered—that Curtis thinks journalists, and more specifically meteorologists—who as a group have often been in the "climate-denier" camp—are missing an enormous opportunity. "I really sing the praises of Paul Huttner here. I wish this became a more normal discussion for all meteorologists, so your local meteorologist could cover this every day. There is so much to explore to enrich their coverage, and their jobs. It could make their jobs almost news anchor level."

PART III

The Three-Front War
on Science

Chapter 8

THE IDENTITY POLITICS
WAR ON SCIENCE

The human understanding is no dry light, but receives an infusion from the will and affections; whence proceed sciences which may be called "sciences as one would." For what a man had rather were true he more readily believes. Therefore he rejects difficult things from impatience of research; sober things, because they narrow hope; the deeper things of nature, from superstition; the light of experience, from arrogance and pride; . . . things not commonly believed, out of deference to the opinion of the vulgar. Numberless, in short, are the ways, and sometimes imperceptible, in which the affections color and infect the understanding.

—Francis Bacon, 1620

Postmodern Antiscience

We now know the historical context of science and democracy, and how the relationship has become fraught. To understand how to win the war on science, we next need to turn to the three major fronts of the battle; and to understand science's attackers, their motivations, and their tactics.

The first major front in the war on science is the identity-politics or postmodernist front, waged by academics and the press. On this front, science is subordinated to the "science studies" of humanities scholars and the journalistic denial of objectivity. Such academics and reporters insist that all truth is subjective or derivative of one's political identity group, and they confuse the process of science with the culture of scientists, thereby falsely equating knowledge with opinion. After all, they argue, ancient truths have proved to be false, so how can we know for sure that knowledge from science is true? Some academics even seriously argue that one cannot know with any degree of certainty that Earth orbits the sun. This front isn't as powerful as the other two major fronts—fundamentalist religion and incumbent industry—but its thinking legitimizes their arguments in the public dialogue and provides them with useful tools, thus multiplying their power.

Postmodernism was born out of an antiauthoritarian perspective that was, ironically, quite similar to that of science. But, unlike science, it began as a philosophical reaction against the modern ideas of rationalism that lay at the foundation of the Enlightenment and its attempt to bottom out arguments in the physical world, as John Locke had counseled. The central idea of postmodernist thinking was that both traditional religion and the Enlightenment had gotten it wrong: there is no such thing as objective truth, and any claim of objectivity is suspect. Objective truth is just a stalking-horse that reigning powers use to make dubious claims of authority, which postmodernists seek to deconstruct.

The only problem with this thinking is that, by and large, postmodernist philosophers didn't really seem to understand science. Even eminent postmodernists like Jacques Derrida, who saw postmodernism as an outgrowth of science, mistook the authority conferred by the theatrics of the white lab coats and the way science was being used by the military-industrial complex for science itself, seeing it as an authority system when, in fact, it is just the opposite. It bears repeating: while scientists may be "authorities" in their fields and able to speak "with authority" on a given topic, what authority they have comes only from the antiauthoritarian exploration of nature. It is not grounded in the scientist, but in the evidence from nature itself. It is the authority of gravity. Ascribing it to the scientist because he warns an apple may fall to Earth is a mistake.

Postmodernism has many dimensions in literature, art, the humanities, science studies, feminist studies, and other socio-political studies, and is known by a number of other names, depending on how it is being applied and who is talking. These include *social constructivism, multiculturalism, deconstructionism, post-structuralism,* and *cultural* or *moral relativism.* We are less concerned with the social or artistic aspects of postmodern culture here than with the academic, political, and science-studies aspects, which sought to circumscribe science and reclaim it as a subset of the humanities. In so doing, postmodernism created the intellectual ammunition being used in the war on science to confuse the public about its role and function.

Deconstructing Truth

The truth of things, postmodernists argued, was not external to the human mind, something waiting "out there" to be discovered. It was something we constructed with our language. The fact that it was "out there" was simply because that's where we put it with our language. If there were such a thing as truth, then, it was to be found not in religion or in science or in any absolutist claims, but in the linguistic context of a claim, and in the cultural identity and individual perspective of the person making the claim.

Any claims of scientific objectivity were simply myths used to preserve or gain power, similar to the claims of religion. You had to figure out what the vested interests of the speaker were to get at what he or she was really communicating. For example, a white man who was talking about the primacy of knowledge and science also enjoyed unexamined white privilege that allowed him to think in those terms. The hidden agenda—often unexamined—of the speaker was contained like a secret code, embedded in the text of what he said, and the job of the postmodernist, like that of the Freudian therapist, was to decode this agenda in a way similar to Freudian analysis. When that was done, the statement would be deconstructed and the true biases of the speaker would be laid bare. "Simplifying to the extreme," French philosopher Jean-François Lyotard wrote, "I define postmodern as incredulity toward metanarratives."

In this way, postmodernists viewed all of science as a sort of public-relations campaign by the elite. Ironically, by opposing and critiquing it, they would lay the groundwork for exactly such a campaign by later antiscience forces. In the beginning, however, this skepticism, or "incredulity" toward the stories the mainstream culture's power elite told about itself—including claims of objectivity or tradition—was similar to the antiauthoritarianism of science itself. After all, science refuses to take any claims on faith and instead says, "Show me the evidence from nature, and the thought process you used to establish it, and I will conclude for myself if your observations are accurate, your thought processes are sound, and, therefore, if what you say is likely to be true." Postmodernists would say something similar, but, instead of seeking evidence from nature and the process used to establish it, they would seek to deconstruct the claims by linguistic criticism and analysis—"the dismantling of conceptual oppositions, the taking apart of hierarchical systems of thought." Their thinking borrowed significantly from Sigmund Freud, as did their process of deconstruction. Like Freud seizing on the seemingly inconsequential presence of a cigar in a patient's dream, to deconstruct a statement or a piece of writing (a "text") postmodernists would, wrote philosopher Christopher Norris, "operate a kind of strategic reversal, seizing on precisely those unregarded details (casual metaphors, footnotes, incidental turns of argument) which are always, and necessarily, passed over by interpreters of a more orthodox persuasion.

For it is here, in the margins of the text—the 'margins,' that is, as defined by a powerful normative consensus—that deconstruction discovers those same unsettling forces at work." Postmodernism was, in this sense, psychoanalysis writ large.

Even the Subject, the thing postmodernists assumed an objective view supposedly studies (but which postmodernists would also argue is the thing that does the studying), can only be understood in terms of differences from other things. Thus *all* approaches to knowledge must be deconstructed to expose hidden biases and assumptions. "What differs? Who differs? What is *différance*?" asked Derrida.

> If we accepted this form of the question, in its meaning and its syntax ("What is?" "Who is?" "What is that?"), we would have to conclude that différance has been derived, has happened, is to be mastered and governed on the basis of the point of a present being which itself could be some thing, a form, a state, a power in the world to which all kinds of names might be given, a what, or at present being as a Subject, a who.

He declared, in his inscrutable way, that his mission was "to make enigmatic . . . the very words with which we designate what is closest to us" by asking Socratic questions about the most basic elements of the language we use to describe reality. From the answers, he sought to expose contradictions that laid bare our implicit biases.

If it sounds paradoxical and confusing, that's because it is, even to many writers, editors, and humanities and journalism professors. As we'll soon see, therein lies the problem.

Thus Spake the Antiscientist

Among its other ironies, postmodernism's philosophical roots lie in the writings of the antiauthoritarian German philosopher Friedrich Nietzsche, who famously proclaimed that God is dead and invented the idea of the Übermensch, or the Overman, in *Thus Spake Zarathustra*: "out of you who have chosen yourselves, shall a chosen people arise: and out of it, the Overman." Contrary to how it sounds, the Overman (or "beyond-man") is not one ruler or a supreme being but a state of human development. In some ways, it is akin to the New Age concept of the higher self, or to psychiatrist Kurt Goldstein's most actualized self.

"Man is a rope connecting animal and Übermensch—a rope over a precipice," Nietzsche wrote, suggesting that there was a spectrum of moral and ethical—or perhaps intellectual and spiritual—development, and that nature was engaged in a process of "becoming," with man being the passage to beyond-man. "What is great in man is that he is a bridge and not a goal," he wrote, "what can be loved in man is that he is a *transition* and a *destruction*."

Nietzsche's sister Elisabeth invited Adolf Hitler to her brother's shrine in Weimar in 1934, some three decades after his death. The führer had never read the philosopher's works, but after the meeting he adopted the quotes that Elisabeth, an anti-Semite, had provided, and he took up the idea of the Übermensch as a symbol of the Aryan master race. Nietzsche himself would likely have been horrified by the Nazi adoption of this idea. He had written that he would have "all anti-Semites shot." He would likely also have opposed the adoption of the Übermensch by screenwriter-turned-philosopher-and-novelist Ayn Rand in her philosophy of objectivism.

Reacting against the Enlightenment (and suffering from paranoia), Nietzsche railed against uniformity and questioned the very idea that there could be objective truth, which he saw as stifling and totalitarian. He argued instead for something he called "perspectivism," which held that truth is a matter of perspective.

Following Nietzsche, a number of Austrian, German, and French philosophers—among them Martin Heidegger (who himself became a Nazi in 1933), Michel Foucault, Jacques Derrida, Jean-François Lyotard, Jacques Lacan, Julia Kristeva, and Bruno Latour—together with a few Americans—including Richard Rorty and Austrian-American Paul Feyerabend—began rejecting the idea that reality and facts existed independently of our thinking about them. Like Nietzsche's, their writing reflected the idea that knowledge claims must be evaluated within the context of class, race, gender, and other group affiliations. In other words, like Nietzsche, they sought to replace the idea of knowledge as a "view from nowhere" with knowledge as a "view from somewhere."

The intellectual descendants of Descartes, these philosophers pointed to examples from quantum mechanics, relativity, and cultural anthropology to illustrate their argument that truth was a construction of the mind of the beholder. Derrida in particular pointed to these scientific roots, and sought to deconstruct claims of objectivity and the Subject alike. But all such deconstructions, even those aimed at the Subject, nevertheless elevate the subjective perspective, or "view from somewhere," as the only honest statement. In particular, the field of cultural anthropology became the postmodernists' tool, and they subjected science to study as if it were a foreign culture (so-called "science stud-ies"). Cultural anthropology was also a justification for their arguments in that it sought to shed cultural bias by observing cultures within their own frames of reference and evaluating them in terms of their authenticity. "Authenticity," then, for a time, became a virtue under postmodernist critique. And what made one authentic? In cultural anthropology, one's primitiveness or lack of influence from the dominant culture. One's integrity. In postmodernism, being conscious of one's agenda and defending it. In both, the dominance of Western white cul-ture, that one great emerging "universal," became the bogeyman, and in post-modernism it became conflated with science.

Similar to cultural anthropology were the ideas of relativity coming out of physics. Albert Einstein showed that measurements were dependent on one's frame of reference in special relativity. German physicist Max Planck demon-strated how the observer affects the event being observed in quantum mechan-ics. By elevating subjectivity, one could dismiss science's claim to objectivity as just the internal values of one of many cultures. Then the troublesome impli-cations of science (that there is an objective reality) were easily dispensed with as the undeconstructed claims of a power elite seeking to maintain that power. Suddenly everything seemed new and mysterious and possible again.

In the humanities, and subsequently in Western politics and education, this came to mean that all systems of thought, or metanarratives (narratives used by societies to claim legitimacy), were simply different linguistic "con-structs" for assembling our experience of reality. As Derrida said, "What dif-fers? Who differs?"

Lyotard captured the postmodernists' criticism of the modern age:

> I will use the term modern to designate any science that legitimates itself with reference to a metadiscourse . . . making an explicit appeal to some grand narrative, such as the dialectics of Spirit, the hermeneutics of meaning, the emancipation of the rational or working Subject, or the creation of wealth.

Lyotard rejected what he called the defining narratives of modern white Western culture, and announced the postmodern age. The postmodern era was to be a time when we could do away with the ideologies upon which modern culture had relied: religious, political, economic, familial, scientific, and others. These ideologies were exposed as so many fables we had created to comfort ourselves or oppress others. Lyotard and other thinkers sought to deconstruct these narratives by critiquing all of modern society's institutions as expressions of power, with the goal of showing that the emperor is, in fact, naked.

In campus humanities departments—and, later, in broader political discourse and society at large—postmodernism came to mean that a middle-aged black male will have a different experience, and so a different truth, from a young white male, who will have a different truth from an older Hispanic female. Their individual perspectives determine "what is true for them," and anything they say on a given subject is their political right—and that right comes only from personal experience as a member of an identity group, and is tempered by the relative implicit power of that group in society. While this may be true in certain realms of politics and social intercourse, it is not true when it comes to objective knowledge from science. Atmospheric CO_2 is the same whether the scientist measuring it is a Somali woman or an Argentine man.

After World War II and the atomic bomb, and during the civil-rights movement, this school of thought expanded exponentially into what it represents today: a value proposition that emphasizes personal authenticity and sensitivity to the different perspectives of others, that rejects modernism as a political narrative and science as an expression of that narrative, and that recasts science as a branch of the humanities. In postmodernist thinking, the nature of knowledge is inseparable from the knower. There is no objective "view from nowhere," only a "view from somewhere." This elevation of subjectivity (as opposed to the Subject) points to science's foundation in the arts, and casts it as a form of language and culture—which it is, but not as postmodernism portrays.

Postmuddyness

To understand what this means and how this became such a major problem—one that ultimately worked against postmodernists' own interests, as well as those of scientists—let's go back and consider the top-wing "rationalist" view of the world born out of al-Haytham's *Optics* and Bacon's *Novum Organum*, and the Enlightenment thinking to which the postmodernists objected:

> There is a world. It is real. It is filled with objects and processes that exist independently of us and our beliefs or language about them.

The goal of science is to create descriptions of reality that are independent of us and our identities, opinions, or beliefs. We call these descriptions knowledge. This knowledge can be expressed in language, mathematics, graphs, images (drawings, paintings, photographs, films), or some combination of these.

To create this knowledge, we use the scientific method, which is a collection of techniques to measure the way things are in nature independent of our perspectives. These techniques include observation, inductive reasoning, hypothesizing, unique prediction, experimentation, recording, critical peer review, and replication. These techniques help cull objective, reliable knowledge out of our subjective perspectives. They have evolved over time and will continue to evolve.

Like our senses, the scientific method is fallible and often leads us astray. But it is the best method we have come up with so far of building a literature of reliable knowledge about nature, and it has proven to be very powerful. In fact, it has proven to be the most powerful tool humanity has ever developed.

The religious right took issue with these claims when they conflicted with dogma or a literal reading of the Bible or the Quran. Postmodernists (for now, let's use this term to mean the academics whose fields end in the word "studies," such as science studies, gender studies, and the like, and their intellectual allies in the philosophy of science, as well as in journalism, politics, and literature) took issue with these claims of science on principle, arguing that they are based on unexamined assumptions of the Western white male-dominated culture that created modern science. In effect, they sought to pull the rug out from under science and objectivity.

Postmodernists also objected, on similar grounds, to the claims of absolute truth made by organized Christianity, which was also largely Western, white, and male. Postmodernism was thus a secular reevaluation of every Western claim of "truth," all of which were suspected of being linguistic means, explicitly or implicitly, of domination by the triumphalist white male culture of the postwar West. Truth was *subjective*, not objective. There were *many* possible legitimate accounts of an event, not just one. Over the course of twenty years, this thinking came to influence all of Western discourse, education, and politics, and, as we will see, it become central to the late-twentieth-century culture wars.

Aside from their confusion of science with the politics surrounding the power that science creates, many postmodernists seemed to mistakenly believe that a scientist must be able to "prove" his or her propositions. Lyotard offered the analogy of Copernicus making the proposition that the planetary orbits are circular:

The referent (the path of the planets) of which Copernicus speaks is supposed to be "expressed" by his statement in conformity with what it actually is. But since what it is can only be known through statements of the same order as that of Copernicus, the rule of adequation becomes problematical. What I say is true because I prove that it is—but what proof is there that my proof is true?

Lyotard perceived the authority or veracity of a scientific statement as coming from its proof:

> Not: I can prove something because reality is the way I say it is. But: as long as I can produce proof, it is permissible to think that reality is the way I say it is.

But that is a rhetorical argument, not a scientific one. It is the approach of an attorney, not a scientist, and it is not what most scientists actually say or how they think. Scientists are concerned with evidence, falsifiability (vulnerability to disproof), and defensible statements. Science philosopher Karl Popper laid this out quite well, which Lyotard himself refers to later but doesn't seem to understand. A scientist never seeks to prove a scientific statement. He or she seeks to draw and defend conclusions supported by observational evidence but testable by anyone. Any prestige or authority a scientist may have, then, comes from being the first to see and record something that others verify. Thus it is about the *process*, not the *claim*. Here's Popper:

> It would of course be easy enough for me . . . to admit not only testable or falsifiable statements among the empirical ones, but also statements which may, in principle, be empirically "verified."
> But I believe that it is better not to amend my original falsifiability criterion.

Not deduction, but induction—the complete reverse of what Lyotard seems to imply, is what powers scientific claims. Mathematicians do proofs. Lawyers seek to prove things beyond a reasonable doubt. But scientists speak in terms of the preponderance of the evidence, and in terms of disproof. They approach questions of fact as statistical propositions, as we have already seen, due to the inductive method and the relative probability that a conclusion is true. This is why science is often poorly portrayed by reporters, who often want to state things in terms of having been proven, instead of being in a state of uncertainty.

Just because something has been shown by repeated experimentation to be reliably true does not mean scientists consider it "proven." They may consider it "settled," but even settled lake beds can be disrupted, and, because of the rigors of inductive reasoning, scientists hold out the possibility that they could be wrong, and so they simply point to what the evidence suggests. Postmodernists did not seem to grasp this.

Nevertheless, postmodernism enriched political dialogue a great deal. It elevated alternative and minority perspectives, and it paved the way for a more diverse, empathetic, and egalitarian society that values its varied participants. This is a fundamental value underlying both science and democracy: the idea that anyone can discover the truth of something for himself or herself because we all have access to nature, the one universal truth. But what the thinking got wrong was the idea that science is authoritarian rather than antiauthoritarian; the view that science is a culture instead of a process; the confusion of science with the power and politics that surround it; and, because subjectivity has a greater claim to truth in certain realms, the assertion that there is no such thing as objectivity.

Academic Turf Wars

Across Western universities, this relativistic thinking merged with new political ideas about affirmative action, becoming widespread and highly political. Postmodernism provided academic support for a secular, progressive, inclusive interpretation of reality that strove to provide new opportunities for blacks, women, indigenous peoples, gays, lesbians, transgender persons, immigrants, and other people who had been disenfranchised by the dominant culture, of which science was seen as a powerful part, an ally of the man, a tool of oppression.

The postmodern view fit well with the ambivalence toward science that developed after the bomb and during the Cold War. Perhaps science didn't provide an objective view of reality after all. Maybe it was just a bill of goods being sold to us by greedy corporations and the government. Scientific "truth" had included a lot of oppressive outcomes and erroneous conclusions: nuclear weapons, chemical pollution, eugenics, phrenology, not to mention the relative exclusion of women and minorities from the ranks of scientists. The list of offenses seemed endless. Perhaps science really was nothing more than a myth to give legitimacy to the white male society from whence the Enlightenment sprang—a sort of ethnocentric rationalization. Perhaps its so-called objectivity was just a smoke screen.

This view was embraced by large swaths of left-leaning academics in the humanities departments of universities (sometimes referred to as the last medieval institutions), who had found themselves—and their budgets—deposed from their thrones by science departments and their denizens in C. P. Snow's battle of the two cultures. These professors found common cause with political activists who championed feminism, environmentalism, black power, the peace movement, animal rights, and other disempowered groups and causes. Science came to be seen as the province of a hawkish, probusiness, right-wing power structure—polluting, uncaring, greedy, mechanistic, sexist, racist, imperialist, homophobic, oppressive, intolerant. A heartless ideology that cared little for the spiritual or holistic wellness of our souls, our bodies, or our Mother Earth.

Kuhnism

In 1962, the growing ambivalence toward science and the rise of postmodernism crystallized with the publication of *The Structure of Scientific Revolutions* by the American philosopher of science Thomas Kuhn. Stunningly, *Structure* became one of the most cited academic books of the twentieth century. It sold about a million copies, an unheard-of figure for a philosophical text. Science was not the gradual and painstaking accumulation of knowledge, Kuhn argued, but a sociological and thus political phenomenon that happens in sudden paradigm shifts. These shifts were akin to religious revelations or quantum leaps in the energy states of electrons, which accumulate energy and then "leap" to higher orbits in discrete, sudden jumps.

Like the postmodernists, Kuhn cast science as an expression of politics and power. Science is a knowledgeable description of nature. That is inherently powerful, and that power makes it inherently political. But while others may use the results of science for power, in and of itself its practice is not an expression of power over others. Its practice is, instead, a search for truth. This is a critical distinction.

The politics that Kuhn ascribed to science resonated closely with prevailing attitudes. Scientists ("the man") resist new (baby-boomer) ideas, clinging to old (Western, white, male) theories even as the evidence to which they are willfully blind (the "truths" of other people) accumulates (discrimination, sexism, racism, and so on) like energy in an electron, until it finally becomes overwhelming (the civil-rights movement). Then, suddenly, in a crystallizing moment (revolution), the ruling order is displaced (comeuppance) and the intellectual understanding of the old (bigoted) paradigm (attitude) shifts to a new, wider (more tolerant and inclusive) paradigm that incorporates (affirmative action) previously discounted outliers (marginalized groups).

Kuhn was striving to describe science as it really is. However, he appears to have been strongly influenced by the politics of his times. He pointed to several past scientific revolutions as examples and argued that they had not been intellectual so much as sociological upheavals. As evidence, he quoted Max Planck, who, along with Einstein, founded the revolutionary field of quantum mechanics. "A new scientific truth does not triumph by convincing its opponents and making them see the light," Planck said, "but rather because its opponents eventually die, and a new generation grows up that is familiar with it."

This speaks to the politics of science and the process-oriented view of science that is the subject of this book. Of course prior views of truth have their defenders. The defenders have, by then, become the vested interests that often push antiscience, and their ranks may at times include scientists themselves, who have become overly invested in the past. But that is a cultural phenomenon within science that has everything to do with politics and little to do with knowledge creation or discovery itself. It does not speak to the fundamental truth or falsity of a new theory. Kuhn's error was one of overextension—to intertwine the politics of science and the discovery of truth and call them one. People have vested interests. Abandoning them to accept a newly or more completely revealed truth is done at some personal, emotional, and often financial and political cost, and that is hard.

That is why the intellectual honesty demanded by science is both so brutal and so nourishing, so feared and so beautiful—all qualities of innovation, of "creative destruction." Thus, as science progresses, "bad" or less complete science—that is, knowledge still colored by what Bacon called the "numberless" ways our assumptions, prejudices, and "affections" impact our understanding, as well as knowledge colored by the limitations of our senses or instruments—is eventually replaced by "good" or more accurate knowledge. Eugenics and phrenology are discredited, for example. But so is Newtonian physics.

But, during a paradigm shift, it is usually only the progenitor of the previous paradigm that hangs on so dearly, not the entire scientific community, as Kuhn implied. If something new is found that better explains things, the scientific community is all over it. After all, that's where the excitement and opportunities lie: at the frontier. Kuhn was writing as a social critic as much as a philosopher.

Long Division

Many scientific ideas are revised upon closer observation, and it doesn't necessarily have anything to do with political biases. It can simply be because we have finally developed the tools to make close enough observations. Locke said that sensory knowledge is the least reliable, and so it is. Things which once appeared real we can now see were misinterpretations based on limited observation. The sun does not orbit Earth. The solar system is not at the center of the universe. The Milky Way is not the universe. And so it goes, as the power of our tools increases.

This process of refinement charts the history of science itself. Modern science began as natural philosophy. Then, with Galileo's telescope, astronomy was carved out of that to become its own separate science, followed by Robert Boyle's 1661 masterpiece *The Sceptical Chymist*, which separated chemistry. Then in 1687 Newton carved out physics with his *Principia*. After Darwin's 1859 publication of *On the Origin of Species*, biology became a separate field, and with the 1875 publication of Wilhelm Wundt's *Principles of Physiological Psychology*, we carved out psychology as a unique area of scientific study. In the twentieth century, we finally separated neuroscience, the science of the nervous system and the brain, and, ultimately, the physical basis of the mind—the subject from which much of natural philosophy originally sprang.

Why did neuroscience come so late? Because much of the field became possible only with the development of electronics, computer technology, and the imaging systems required to study, measure, and stimulate brain function. These tools were then combined with new understandings of the interaction of software and hardware, and the biology of how chemical and electrical signals give rise to one another. We are still asking the same questions we were in the days of natural philosophy, but now we can ask them in a more scientific context. With the increasing power of our tools of observation and measurement, we can see that what once appeared as an oasis is really a mirage.

Concept Collapse

Neurophilosopher Patricia Churchland calls this process concept collapse. As examples, she offers the concepts of *impetus* in physics and *vitalism* in biology. For centuries, impetus was thought of as an inner force that kept things moving, but Newton revealed that to be an illusion. "We had something that had seemed to be observable, and it turned out to not be a real thing at all," Churchland says.

It was similar story with vitalism, the life force that was thought to distinguish living things from nonliving things; what makes a rock different from a living being and a living body different from a dead one. "In 1900, we used to think it was one thing. Some vital force or spirit," says Churchland. "In fact, we now know that it is many things. With the discoveries of ATP [adenosine triphosphate] and the basic chemical building blocks of life, the understanding of the role of mitochondria, of ribosomes, of cell biology, the chemical nature of DNA, the folding and unfolding of proteins, and so on, that concept too has been revealed as illusory."

Another example is our age-old idea of fire. What is fire, really? It's the burning of wood. It's the fire of the sun. It's the fire of lightning. It's the magical light in a firefly's tail. But when we applied closer observation, this one concept collapsed into four very different things, none of which has very much to do with any other. The burning of wood, we learned, is oxidation, more akin to rusting than it is to the fusion occurring in the sun. And the fusing of atoms is altogether different from the incandescence of lightning, which turns out to have nothing to do with the phosphorescence produced by the chemicals in a firefly's tail. The only thing they have in common is that they appear bright to our senses; otherwise, they are not similar at all.

It's All Relative

If, as we made closer observations, fire, vitalism, and impetus were revealed not to be as we thought they were, does it mean there is no objective truth, but just an endless regression, or worse, simply a shifting perspective of subjective ideas? Kuhn suggested the answer was yes. As evidence, he offered the work of Einstein:

> I do not doubt, for example, that Newton's mechanics improves on Aristotle's and that Einstein's improves on Newton's as instruments for puzzle-solving. But I can see in their succession no coherent direction of ontological development. On the contrary, in some important respects, though by no means in all, Einstein's general theory of relativity is closer to Aristotle's than either of them is to Newton's.

This suggested that there was no real progress in science. It was simply a grand circle, or an endless regression, or random interpretations whose favor was informed by the political ruling order. But Kuhn's theory, while dramatic and captivating, was incorrect. Individual scientists, like Harlow Shapley, may fall off track and become overly invested in their egos and thus become blind to observational evidence. But the process of overall scientific progress is real. It has political implications because new knowledge gives new power, but it is not merely a social exercise that relies on outliers changing the minds of the curmudgeonly majority or waiting for them to die off. It matters little what the curmudgeonly think. That is the point of science. There is an observable reality on which empirical science is based and from which knowledge is derived. That is why it has power: it creatively disrupts prior views, both scientific and ideological, enabling us to affect the physical world in ways we couldn't before. The observations and knowledge extend incrementally, by the contributions, risks, and suffering of many. They do not extend in sudden and dramatic paradigm shifts, and they didn't in Einstein's day, either.

In fact, many of the ideas Einstein developed were done collaboratively, with considerable debate, a prime example being the cosmological constant. His early papers were extensions of the work of Max Planck, the Austrian physicist Ludwig Boltzmann, and others, and his revolutionary findings on Brownian motion were independently discovered by Polish physicist Marian von Smoluchowski, who was also building on Boltzmann's work. Hubble's discovery of the expansion of the universe also extended from ideas that were talked about for years. Redshift was first noted by American astronomer Vesto Slipher in 1912—nearly two decades before Hubble's discovery. Galileo's revolution was an extension of Copernicus's writings from some seventy years before, which were widely discussed. The discovery of the double-helix structure of DNA was also an extension, building on the work of biochemist Erwin Chargaff, and many of the ideas of Bacon that lie at the foundation of Western science were previously expressed by al-Haytham.

It is true that science does not proceed linearly; it advances more like a pack of dogs sniffing out a fox. But that is because of its trial-and-error, observational approach, which adopts whatever new tools become available, applies metaphor, builds on the latest recorded knowledge ("the literature," as scientists call it), and makes and tests bold predictions. Science is our very best tool against prejudices and unexamined attitudes, not the cultural expression of them.

The Age of Equality

At the time, Kuhn's work seemed to offer a resounding refutation of claims of the truth and power of "mechanistic" science, and it served as a catalyst, transforming modern culture's entire relationship to science. If science was simply one way of knowing about the world, other, previously discounted ways of knowing were equally valid. This seemed obvious. The world wasn't mechanistic after all. There was still room for spirit and mystery and miracles and higher levels of consciousness. We could throw out the past and take up whatever view of reality best suited us. In an age of imminent nuclear holocaust, we could find a new hope. It was the dawning of the Age of Aquarius.

In politics, this thinking became entwined with the goals of the civil-rights movement and postmodernism: examination and critique of power structures, discovery of voices not valued by history, cultural tolerance, acceptance of diverse viewpoints, suspension of judgment, affirmative action, mindfulness of the biases of the speaker, a pullback from American exceptionalism and white supremacy, and the celebration of the self-evident truth that lies at the foundation of democracy—that all people are created equal. If science was the voice of white Western culture, then it was not the voice of these other, discounted cultures.

In a sort of intellectual affirmative action, academics, writers, politicians, and teachers took this idea to its logical conclusion: from all *people* being created equal to all *cultures* being created equal, and, from that, to all *ideas* being created equal. Suddenly, truth was a matter of one's perspective. There was no objective truth; there was feminist truth, indigenous truth, African-American truth, Latino truth, LGBTQ truth, Muslim truth, working-class truth, and so on, all of which had to be equally respected as long as they were authentic expressions. Thus if someone from a disempowered political group did something morally reprehensible, he or she had to be given extra understanding, because it was probably partly due to disenfranchisement.

Cultural conservatives objected to this on a rationalist basis and were crowded into the bottom-right political quadrant (authoritarian conservative) with scientists, who didn't belong there. But suddenly rationalism and modernism seemed like old, conservative ideas—like expressions of authority. The political left lost many brilliant and otherwise liberal thinkers, such as E. O. Wilson, who could not stomach elevating a political goal over the ideas of reason and the Enlightenment. It was antithetical to the whole egalitarian view of modernity, because there was no common authority of evidence. Their opposition was often judged not on its intellectual merits but through a political lens, and considered to be ignorant, racist, sexist, supremacist, ill-read, or right-wing, often in vast, sweeping indictments of everyone who did not embrace the new politics, whether they actually were racists or simply rationalists. In fact, the postmodernists argued, rationalism itself was part of the problem. What rationalists were unwilling to admit, they argued, was that irrational processes, not rational ones, lay at the core of Kuhn's scientific revolutions. Rationalism itself was simply a thinly veneered tool of domination.

Wolves in Sheep's Clothing

Building on the work of Kuhn, whose manuscript of *The Structure of Scientific Revolutions* he had critiqued in 1960, postmodernist Paul Feyerabend succinctly summed up the thinking:

> The world, including the world of science, is a complex and scattered entity that cannot be captured by theories and simple rules. . . . There is not one common sense, there are many. . . . Nor is there one way of knowing, science; there are many such ways, and before they were ruined by Western civilization they were effective in the sense that they kept people alive and made their existence comprehensible. . . . The material benefits of science are not at all obvious.

Feyerabend has been widely quoted by antiscience postmodernist authoritarians on the left, but also by antiscience religious authoritarians on the right. In 1990, archconservative German cardinal Joseph Ratzinger was in charge of Roman Catholic doctrine for the Vatican. Ratzinger was a strict authoritarian who would eventually become Pope Benedict XVI, steering the church back into the political bottom wing. He gave a major speech in which he condoned the 1633 trial and conviction of Galileo for heresy, using a quote from Feyerabend to make his argument.

> The church at the time of Galileo was much more faithful to reason than Galileo himself, and also took into consideration the ethical and social consequences of Galileo's doctrine. Its verdict against Galileo was rational and just, and revisionism can be legitimized solely for motives of political opportunism.

Seeking authority in the cultural dialogue, conservative Catholic scholars make similarly revisionist, neo-postmodern statements today, arguing that anyone who doesn't agree with them is practicing "scientism"—that scientists think they are superior and look down on those who don't "believe" in science, as if it were a belief system instead of a process of measuring nature with our hands and eyes and tools. It's a brilliant tactic, and an excellent example of the intellectual handicap postmodernism has created for itself and science: as long as everyone has their own truth, no one can presume to question anyone else's.

Australian science historian Peter Harrison makes this sort of revisionist argument, saying, "When the Catholic Church responded to Galileo, they were in fact endorsing what the majority position was amongst the scientific community. So if we ask is this a conflict between science and religion, Catholicism was endorsing, rightly or wrongly, what the majority viewpoint was." Harrison's attitude suggests that the church was simply endorsing the prevailing, peer-reviewed, well-considered, evidence-based consensus of the time as we think of such a thing today, but that was not the case. The church did not make a scientific or evidence-based argument at all, as one can see if one reads its indictment of Galileo. Instead, the church made a scriptural argument, indicting him for holding "that an opinion can be held and supported as probable, after it has been declared and finally decreed contrary to the Holy Scripture." That is the essence of the intellectual conflict between the scriptural, authoritarian approach of the church and the observational evidence-based, antiauthoritarian approach of the era's emerging science.

Why did the church do that? If Galileo was not a threat—if he was just a crank whose views were not supported by the evidence, as Harrison and others claim—and religious authorities had the majority of the scientific community on their side, why indict him at all? And why sequester him under a house arrest, limiting his influence? It's because his evidence-based arguments—*look through my telescope and see for yourself*—were compelling, especially to his adoring, young audiences. Knowledge is power, and that power is political.

The End of Objectivity

"While such views remained contained within relatively limited intellectual and political groups, little attention was paid to them by the mainstream scientific community," reflected the editors of the science journal *Nature* in a 1997 opinion piece. "But, over the past few years, their influence has appeared to flourish not only in the academic world—including school-teaching—but also in the wider community."

The entire postmodern movement was tinkering with the foundations of democracy in ways few understood. By painting objectivity as supremacist, the subjectivism and authoritarianism from which America's founders had sought to free the country (but had partially failed to do, by excepting slaves and women from those created "equal") was restored to the throne. Much of Western education and thought after the 1970s lost its grip on reality and became embroiled in "but faith, or opinion, but not knowledge," in the words of John Locke.

It was an unfair and vastly oversimplified criticism of science to say, because it was a field predominantly populated by white men, and because it was being used by governments and industry for nationalist and capitalist purposes, that it was simply another subculture trying to retain its seat of power. This was a case of throwing out the baby with the bathwater. At that time, many professional endeavors were the field of white men. Postmodern thinking mistakenly focused on scientists as a group of some particular background rather than on science as a process of ideas, as something anyone can do regardless of their group.

In fact, it was science itself that laid the very foundation for the values of tolerance and diversity that were later used to sell postmodernism. Science originally advanced the idea that, based on our equal opportunities to observe nature and derive knowledge, all men and women are created equal. It's true that equality in practice was not immediately achieved, but to focus on that is to make the perfect the enemy of the good. No progress would have happened at all without the conceptual foundation of early Enlightenment thinking.

As a foundation of democracy, science itself is a font of tolerance. It is also a great beneficiary of diversity because it thrives on challenges from differing viewpoints to find unexpected breakthroughs and to make its conclusions stronger. All these cherished values of postmodern society, ironically, have their foundation in the thinking and practice of science. In the sense that it is now international, with scientists from around the world collaborating on research projects over the Internet using the language of science, the global science enterprise is the most diverse and yet universal undertaking in human history.

When viewed from this perspective, it is clear that, if taken to the academic and political extreme of classifying science as merely another subjective way of knowing, postmodernism can embroil society in the same sort of culture wars that roiled Europe at the beginning of the Enlightenment by casting us back to the time when every sect had competing claims on the truth. These are the very conflicts that led to the English Civil Wars and motivated Locke to seek a way to define how knowledge is universal and separate from denominational identity. His thinking, in turn, inspired Thomas Jefferson, Benjamin Franklin, and others to found the United States and create modern democracy on that bedrock.

As Islamic scholar Bassam Tibi writes, the authoritarianism of postmodernist politics is not so different from radical Islam. While the most fundamentalist Muslims forbid Western science, "Westerners have no reason for self-congratulation. Western scholarship may not be characterized by 'flat-earthism,' but in the United States, as well as in European academia, some beliefs—such as postmodernism—seem today to replace scholarship. It is no longer possible to ask 'certain questions' and remain unscathed. The sanctions against the spirit of free inquiry are not only tough, but at times also highly primitive." In other words, authoritarian.

Science Class without Objective Truth

The 1960s and 1970s were a time of momentous social change, particularly related to civil rights. For the first time, the United States was making a serious effort to educate African-American children to the same standard as white students. One of the primary methods employed was school desegregation. This posed complex challenges for teachers, who were tasked with educating more diverse classrooms. Black students whose communities had been uprooted— first by highway construction, then by busing—found themselves thrown into the mix with more advantaged white students. It seemed unrealistic to demand equal performance from students who did not have the same level of socio-economic support or shared cultural references.

Similar conclusions were coming from science educators involved in efforts to transfer scientific knowledge and educational methods to developing countries. "Why should we suppose that a program of instruction in botany, say, which is well designed for British children, familiar with an English country-side and English ways of thinking and writing, will prove equally effective for boys and girls in a Malayan village?" researchers Francis Dart and Panna Lal Pradhan asked in 1967. "It is not merely that the plants and their ecology are different in Malaya; more important is the fact that the *children* and *their* ecology are also different."

Beginning in the 1970s, science educators began to express "a growing awareness that, for science education to be effective, it must take much more explicit account of the cultural context of the society which provides its setting, and whose needs it exists to serve," as British science education researcher Bryan Wilson wrote in 1981. While white teachers once taught white students using white cultural references, now all teachers had to develop strategies to reach diverse classrooms. And while this first began to happen in the United States, it soon spread throughout Europe and other developed countries, with more remote developed countries that were far away from immigrant influxes, like Finland, rising to the top in international science and math rankings. Less diversity, as it turns out, equals, on average, less classroom management and easier teaching. But in places like the United States, followed closely by England, France, Germany, and other leading developed countries, diversity became both an advantage and a cost center. Both ethics and economics required finding new ways to reach diverse classrooms.

These new strategies were built on the postmodern assumptions of social constructivism, wherein learning is regarded as a social process. "Educators have long viewed science as either a culture in its own right or as transcending culture. More recently many educators have come to see science as one of several aspects *of* culture," wrote science-education professor Bill Cobern, referring to the idea that science is the cultural expression of Western white men.

But if the postmodern view of science is true, wouldn't teaching science be a form of cultural supremacy or, worse, cultural genocide? Certainly, if one accepted the precepts of science studies, that was the case, and such an action was politically and morally indefensible. The only answer to this logical contradiction lay in redefining what is meant by "truth." And that is where the postmodernist academics and science educators, seeking a way out of the dilemma, went wrong.

Social constructivist thinking became the mainstream paradigm in Western teacher education in the 1970s and 1980s, eventually influencing the educations of tens of millions of Western students. Parts of this transformation have been very positive for science—the emphases on hands-on learning and on process over product have, in the limited number of classrooms where they are practiced, had very good outcomes. After all, at its best, science is, like art, a hands-on activity. It is emphatically not about producing the previously known correct answer to an experiment or equation; it is about exploration and (faithful) representation of nature, and the testing of previous knowledge. So these have been some of the good outcomes of a more postmodernist approach.

But constructivist thinking in education also came to hold the well-intentioned but incorrect belief that "there is no representation of reality that is privileged, or 'correct,'" the education professor's version of the journalism professor's "there is no such thing as objectivity." One education professor described it this way:

> Because reality is in part culture dependent, it changes over time, as cultures do, and varies from community to community. Knowledge is neither eternal nor universal. . . . We must think increasingly in terms of "teachers and students learning together," rather than the one telling the other how to live in a "top-down" manner. This is necessary both so that the values and interests of students are taken into account, and so that the wealth of their everyday experience is made available to fellow students and to the teacher.

This confusion of reality with culture devalues actual knowledge and presumes that students have some "wealth of . . . everyday experience" of equal value to the accumulated knowledge from hundreds of years of experimentation, risk, and sacrifice by intrepid scientists that is informing the lesson at hand. That is to say, despite the teacher's extensive training, the student has access to a "wealth of . . . everyday experience" that the teacher does not.

What could that experience be? The only possibility is the student's membership in a political identity group—racial, gender, sexual orientation, disability—that is different from the teacher's.

And what, in turn, does this teach the student? That there is no commonly shared knowledge "out there" that we all strive to attain. Instead, we each construct our own reality, using our own language, and the perspective *already held* is of equal value to *anything to be learned*. This emotional and political goal may make the student and the teacher feel good, but science has shown that it is simply not true. Scientists argue that the purpose of education had shifted from teaching knowledge and skills to providing a learning environment in which students construct their own knowledge, which, in the case of teaching science, disregards the accumulated knowledge of more than five hundred years. Is it surprising that students in diverse postmodernist classrooms, such as those across the United States or the United Kingdom, perform poorly when compared to their peers in China, Korea, Poland, Finland, and Japan, which rank among the least diverse countries and also among the highest performers in OECD educational rankings in science and math?

While inclusiveness is important in closing the education gap, high and positive expectations are also important, as are socioeconomic factors, familial support, focusing on objective knowledge, and teaching students how to tell knowledge from opinion. Overemphasizing political identity undermines the knowledge teachers are trying to impart, thereby shortchanging students. There is a difference between being inclusive and elevating all ideas to the same level. There is also a difference between not forcing assimilation and cheating students of objective knowledge and the tools to find it. Such classrooms deprive students of the very tools they need to access real power—and that is itself a discriminatory act. Knowledge knows no color, no religion, no cultural background. It has no sexual orientation, no political party; it is of higher value than these because it is tied to nature. What defines it as knowledge is its separability from the individual; when it is not separable, it is opinion.

The confusion this causes makes some scientists want to throw their hands up. Consider the following introduction to a 1998 science education paper presented at a national conference:

> As Richard Dawkins likes to put it, there are no epistemological relativists at 30,000 feet. But today some will say, "Not so fast!" Dawkins offers a brute definition of universality completely devoid of any nuance of understanding and equally devoid of relevance to the question at hand. No one disputes that without an airplane of fairly conventional description, a person at 30,000 feet is in serious trouble. The question of universality does not arise over the phenomena of falling. The question of universality arises over the fashion of the propositions given to account for the phenomena of falling, the fashion of the discourse through which we communicate our thoughts about the phenomena, and the values we attach to the phenomena itself and the various ways we have of understanding and accounting for the phenomena—including the account offered by a standard scientific description.
>
> In today's schools, there are often competing accounts of natural phenomena especially where schools are located in multicultural communities. There are also competing claims about what counts as science.

The teaching that there is no objective reality, but rather many subjective realities—or, in this case, that subjective realities are on par with objective reality—degrades students' views of the primacy of knowledge and increases the education gap rather than closing it. It is no wonder that there is so much antiscience in Western culture—we've been teaching it for forty years.

And now the fruits of the postmodern vine are in our legislatures, our city councils, our political candidates, and our appointees to our high courts. This sort of education teaches that politics is what matters. Thus history is no longer the search for what really happened, but rather the victor's interpretation as seen through the lens of power and oppression, and it bears a cultural and political focus. Literature is no longer a study of what the author meant, but of the feelings it arouses in the reader because of his or her own cultural perspective. Science is no longer an account of objective reality but a meta-narrative of the dominant culture. Reading the classics is no longer required because they are sexist and racist and not germane to today's political realities. Truth and the right to say something must be evaluated in the context of the speaker's cultural and political identity. Teachers, when they do not have the same cultural and political identity as their students, cannot presume to teach them, but must be guides at the side.

The Unexpected Closed-Mindedness of Openness

By the late 1980s, classics professor Allan Bloom lamented the effects of post-modernist education on the thinking of the students in his classroom at the University of Chicago. His concern spawned a nationwide discussion when he wrote about it in *The Closing of the American Mind*.

> The relativity of truth is not a theoretical insight but a moral postulate, the condition of a free society, or so [the students] see it. They have all been equipped with this framework early on, and it is the modern replacement for the inalienable natural rights that used to be the traditional American grounds for a free society. That it is a moral issue for students is revealed by the character of their response when challenged—a combination of disbelief and indignation: "Are you an absolutist?," the only alternative they know, uttered in the same tone as "Are you a monarchist?" or "Do you really believe in witches?"

Absolutism is considered morally objectionable because it leads to intolerance, but that is only true when it is applied to a matter of "faith, or opinion, but not knowledge." In the case of science, precisely the opposite has proven to be true. By acknowledging that there *is* an objective reality, and that we can form knowledge about that reality by using science and observation, we remove questions of fact from the authoritarian argument. This is the great insight that the United States, the world's oldest democracy, was founded upon. The importance is in the knowledge, not in the speaker—just the opposite of the postmodern perspective. Bloom was criticized as a conservative, a sexist, and a racist, all of which he denied. His arguments show he was simply defending knowledge:

> Openness—and the relativism that makes it the only plausible stance in the face of various claims to truth and various ways of life and kinds of human beings—is the great insight of our times. The true believer is the real danger. The study of history and of culture teaches that all the world was mad in the past; men always thought they were right, and that led to wars, persecutions, slavery, xenophobia, racism and chauvinism. The point is not to correct the mistakes and really be right; rather it is not to think you are right at all.
>
> The students, of course, cannot defend their opinion. It is something with which they have been indoctrinated. The best they can do is point out all the opinions and cultures there are and have been. What right, they ask, do I or anyone else have to say one is better than the others? If I pose the routine questions designed to confute them and make them think, such as, 'If you had been a British administrator in India, would you have let the natives under your governance burn the widow at the funeral of a man who had died?,' they either remain silent or reply that the British should never have been there in the first place. It is not that they know very much about other nations, or about their own. The purpose of their education is not to make them scholars but to provide them with a moral virtue—openness.

Collapsing Hermeneutic Isolation: The Quantum-Studies Approach to Politics

By the 1990s, a full generation's worth of university departments had grown up around "cultural studies"—women's studies; gay, lesbian, bisexual, and transgender studies; science studies, and so on—all purporting to be among the social sciences, but arguing the a priori political notion, misappropriated from quantum mechanics, that objective reality is subject to the observer.

Academics would often borrow the language of science in this way, though only when it supported their views. Following the inscrutability of leading philosophers like Derrida, they would use jargon to make sometimes outlandish political statements sound highbrow.

In 1994, Rutgers University mathematician Norman Levitt and University of Virginia biologist Paul Gross published a polemic attacking this appropriation of scientific terminology called *Higher Superstition: The Academic Left and Its Quarrels with Science.* Like Kuhn's *The Structure of Scientific Revolutions* and Bloom's *The Closing of the American Mind, Higher Superstition* charted a new course in the culture wars and became a bestseller. Gross and Levitt characterized the conflict as a clash between the academic left and the scientific right, and academics called the book right-wing. But there were plenty of scientists on the left, and others who were ideologically unaffiliated, who were tired of the arguments over the cultural nature of reality that, by then, was being called the "science wars." Among the most notable was eminent Harvard University entomologist E. O. Wilson, who declared in a New Orleans speech that "multiculturalism equals relativism equals no supercollider equals communism." Construction of the superconducting supercollider, located in Texas and more than three times the size and power of CERN's Large Hadron Collider, was cancelled by Congress in 1993.

Then, in 1996, humanities scholar Andrew Ross, the editor of the leading postmodernist journal *Social Text*, made a fateful decision. He devoted an issue to discussion of the science wars, and he accepted for publication a paper by Alan Sokal, a New York University physicist and a self-described leftist who had been inspired by *Higher Superstition* to submit a paper titled "Transgressing the Boundaries: Toward a Transformative Hermeneutics of Quantum Gravity." It was a perfect highfalutin postmodern title whose meaning was both chinstroking and inscrutable. The only problem was that the paper was a hoax— a parody mash-up of the most ridiculous postmodernist writing Sokal could find that appropriated the language of science to argue that there was no reality. It was the kind of politically correct, heavily jargonized, intellectually vapid nonsense Woody Allen's Alvy Singer had famously called "mental masturbation" in the 1977 film *Annie Hall.* It was tailor-made to please its intended audience, complete with the popularly employed quotation marks around certain words to imply their inferior status:

> Deep conceptual shifts within twentieth-century science have undermined this Cartesian-Newtonian metaphysics; revisionist studies

in the history and philosophy of science have cast further doubt on its credibility; and, most recently, feminist and poststructuralist critiques have demystified the substantive content of mainstream Western scientific practice, revealing the ideology of domination concealed behind the façade of "objectivity." It has thus become increasingly apparent that physical "reality," no less than social "reality," is at bottom a social and linguistic construct; that scientific "knowledge," far from being objective, reflects and encodes the dominant ideologies and power relations of the culture that produced it; that the truth claims of science are inherently theory-laden and self-referential; and consequently, that the discourse of the scientific community, for all its undeniable value, cannot assert a privileged epistemological status with respect to counter-hegemonic narratives emanating from dissident or marginalized communities.

Shortly after that, Sokal published an article in *Lingua Franca*, a sort of *People* magazine of the academic world, that described the hoax and how it showed that postmodernists were incapable of distinguishing between a real argument and nonsense. "What concerns me," he wrote—and remember, he's a self-identified leftist as well as a scientist—"is the proliferation, not just of nonsense and sloppy thinking per se, but of a particular kind of nonsense and sloppy thinking: one that denies the existence of objective realities."

The media loved the story because of its emperor's-new-clothes aspect, which, ironically, is precisely what postmodernist deconstruction sought to expose: the myths perpetuated by elite power structures. Pompous college professors being revealed as vacuous dupes perpetuating the very myths they decry is a narrative with considerable popular appeal. The right, in particular, which had been railing for years against political correctness and identity politics, saw it as a long-awaited skewering of the "effete, elitist academic who's 'above it all,'" as Rush Limbaugh said of Stanley Fish, *Social Text*'s publisher at the time—a skewering that dismayed progressives, since Fish is an eloquent and widely respected writer on the left, a political perspective Sokal claimed to share.

"He [Sokal] says we're epistemic relativists," complained Stanley Aronowitz, a cofounder of *Social Text*. "We're not. He got it wrong. One of the reasons he got it wrong is he's ill-read and half-educated." But this snobbish nonresponse was an ad hominem attack (attacking the speaker and not the argument, a classical logical flaw), and so it only bolstered Sokal's—and cultural conservatives'— criticisms. "Conservatives have argued that there is truth, or at least an approach to truth, and that scholars have a responsibility to pursue it," wrote Janny Scott in a *New York Times* article about the hoax, news of which spread worldwide. "They have accused the academic left of debasing scholarship for political ends."

But it wasn't just conservatives. The hoax also reinforced the view of modernists on the left (including several prominent feminists) who were alarmed by the rigor that was being jettisoned in Western intellectual culture—and suspected that fields like women's studies and science studies were nonsense. Prominent gender historian Ruth Rosen wrote in the *Los Angeles Times*,

> It took a New York University physicist named Alan Sokal to expose the unearned prestige that the Academic Emperors have heaped upon themselves. A self-described progressive and feminist (to which I can attest; I helped with his exposé), Sokal became fed up with certain trendy academic theorists who have created a mystique around the (hardly new) idea that truth is subjective and that objective reality is fundamentally unknowable. To Sokal, the denial of known reality seemed destructive of progressive goals.

Michael Bérubé, a Pennsylvania State University literature professor and one of Sokal's most prominent critics, later changed his mind. In 2011, he wrote a brilliant mea culpa describing how the hoax gave ammunition to the right, but he also said it empowered a previously silent group on the left "who believed that class oppression was the most important game in town, and that all this faddish talk of gender and race and sexuality was a distraction from the real struggle, which had to do with capital and labor."

Most important, the paper that "punk'd" *Social Text*, as Bérubé called it, exposed the erosion of reason and objective knowledge in postmodern classrooms, where the next generation of leaders was being educated. The result of this erosion, as the feminist essayist Katha Pollitt wrote in the *Nation*, was "a pseudo-politics, in which everything is claimed in the name of revolution and democracy and equality and anti-authoritarianism, and nothing is risked."

A Marriage Made in Hermeneutics:
Postmodernist Journalists and Antiscience Politicians

Most of today's journalists and policymakers did not come up through the sciences. In high school and college, they were exposed to the political correctness of the postmodernists who populated humanities, education, English, journalism, and political-science departments. Conservative students, and those who would eventually become conservative, did not forget the arguments they learned in those classrooms—that science had no real intrinsic authority, that it was just another way of knowing, and that people in positions of authority could pass their own political agendas off as truth as long as their arguments had complex presentations and included cherry-picked bits and pieces of science.

This doublespeak seemed to confirm the idea that was at the very heart of what would become the neoconservative movement: to paraphrase Donald Rumsfeld, the winners write the history books. The postmodernists of the ascendant left academic culture were the living proof. The neoconservatives could even, as their professors had for close to a generation, attack the credibility of science and reason itself, on which the secular world order was based, by casting it as just another worldview.

It is perhaps no coincidence that, as these students became journalists and policymakers, democratic society was swept into an era of endless rhetorical debate over faith and opinion and national discussions dominated by identity politics and partisanship—an intellectual morass justified as the "marketplace of ideas."

Major media outlets can thus give equal platforms to scientific outliers, celebrities, and political whack jobs on important issues ranging from climate disruption to vaccines. It's not their role to establish the objective truth of the matter, leading members of the press argue, any more than it was to question the lack of substantive rationale for invading Iraq.

Science has shown, of course, that this is nonsense. There is objective truth to be learned by observation, and the knowledge gained grants us power that other "ways of knowing" quite simply do not. But the more dangerous problem with postmodernist thinking is its a priori nature. Not truth, but a political goal must be served in this case, the goal of openness, or tolerance without judgment. *Your truth is your truth, and who am I to judge?* Or: *Different strokes for different folks.* But, without objective knowledge, all arguments become "but faith, or opinion," and can go on forever. We are either paralyzed by it or we must resort to authority instead of objectivity to make decisions. This casts us all the way back to Hobbes's predemocracy, pre-Enlightenment, pre-Locke war of every man against every man, where not evidence but raw power determines the outcome of a dispute.

Thus, when taken to an extreme, postmodernist thinking can lead to the very brutality and oppression of minorities it sought to avoid, by developing the intellectual arguments used to deny the only thing that stands between these disempowered groups and power: the leveling evidence from science. This was becoming increasingly evident in political arguments as the later baby boomers took charge of the culture. As the first children taught under postmodernist philosophy, they were generally unable to articulate positions based on shared knowledge, and so were left with "but faith, or opinion" and the belief that what matters most is finding the nuggets that match one's own argument—and in never, ever compromising, because the winner writes the history books.

Today, in the United States, Canada, Australia, and other countries, serious candidates for political office can openly state views that run counter to all known science and history on a variety of topics, and many journalists don't feel it is their role to point out that these emperors have no clothes. Further, it is journalists, more than any other profession, whose confusion about the nature of objectivity has both enabled and—through journalists' development of the field of public relations—directly caused much of the assault on science by industry.

The dissociation from history and the hard-won knowledge of science has thus led to a generation of leaders who are at once arrogant and ignorant, and thus likely unable to lead the world out of the morass. We embrace the forms of tradition but not the substance, focused only on winning, unable to discern between what feels good and what is true. It is a condition that threatens to leave the world permanently damaged.

A Brutal New Age

From postmodernism emerged its cousin the New Age, a pop-culture spiritual movement built on the idea that truth is subjective. Traditional religions rejected this notion, arguing that truth is objective, and so in many ways the New Age became the religious aspect of the secular postmodern movement. Much of its early formation can be traced to the writings of novelist and poet Jane Roberts, who, in 1963, the year after Kuhn's *Structure of Scientific Revolutions* became a massive bestseller, began "channeling" the words of a disembodied spirit named Seth. In 1970, Roberts published *The Seth Material*, followed by *Seth Speaks: The Eternal Validity of the Soul* in 1972. The books became bestsellers, and are similar to Nietzsche's *Thus Spake Zarathustra* in that Seth argues there is a higher self, a sort of driver of the vehicle that is one's mind and body.

Over the course of the 1970s, the New Age exploded as a spiritual movement, gentrifying a collection of ideas and practices from pre-Christian, non-Western, and indigenous religions. New Age retreats, which were attended almost exclusively by white, middle-class people, featured psychic readings by clairvoyants, spiritual lectures by trance mediums like Roberts, aura readings, the laying on of hands, homeopathy, extrasensory perception, Reiki healing, hypnosis and past-life regression, transcendental meditation, spoon bending, telekinesis, psychokinesis, remote viewing, astral projection, sweat lodges, vision quests, psychic channeling, ESP testing cards, classes on the *I Ching*, the *Kama Sutra*, the *Tao Te Ching*, Bach flower remedies, Jungian dream analysis, astrology, tarot readings, chakra adjustment, crystal healing, automatic writing, spirit guides, Ouija boards, biofeedback, consciousness raising, yoga, and the chanting of the word "om," which is said to contain all the sounds in the universe.

In short order, the movement became big business. The retreats were like carnivals. They were held around the country and included workshops put on by traveling New Age teachers and authors who were revered like holy gurus. A flood of bestsellers followed, creating the New Age literary genre: *Jonathan Livingston Seagull, Illusions, Be Here Now, The Dancing Wu Li Masters, The Teachings of Don Juan, Handbook to Higher Consciousness, The Hundredth Monkey, Love Is Letting Go of Fear, The Crack in the Cosmic Egg, A Course in Miracles, Your Erroneous Zones, The Aquarian Conspiracy*, and many others. Each purported to peel away the gauze and reveal the true, underlying, magical nature of a reality that one could create simply by believing it was so—the objective as a creation of the subjective.

If all "ways of knowing" are equal, then homeopathy—in which a substance that would create a patient's symptoms in a healthy person is diluted until none of its molecules remain, with this dilution then administered as a cure—should be as efficacious as conventional pharmaceutical treatment for anyone who values "open-mindedness." Anyone who challenges this idea is just a skeptic (the New Age version of a spiritual bigot) who is trying to oppress "holistic," "higher" thinking because of his or her paradigm of Western materialistic white supremacy.

Theodore Schick Jr., a professor of philosophy, and Lewis Vaughn, the former managing editor of *Prevention* magazine, have written extensively and critically about this spiritual version of postmodern antiscience. They summarized how New Age thinking plays out through popular culture in their book *How to Think about Weird Things: Critical Thinking for a New Age.*

There's no such thing as objective truth. We make our own truth.
There's no such thing as objective reality. We make our own reality.

There are spiritual, mystical, or inner ways of knowing that are superior to our ordinary ways of knowing. If an experience seems real, it is real. If an idea feels right to you, it is right. We are incapable of acquiring knowledge of the true nature of reality. Science itself is irrational or mystical. It's just another faith or belief system or myth, with no more justification than any other. It doesn't matter whether beliefs are true or not, as long as they're meaningful to you.

Thus, Reiki healing may offer as much hope as chemotherapy. If it fails, it is because the patient is "blocked"—emotionally or spiritually unable to accept the healing energy, or unable to let go of anger from a past life, for example—and so the fatal illness progresses. It's the patient's fault. The reasons for failure are emotional and spiritual and have nothing to do with the practitioner, the method, or the physical world.

But like other a priori approaches, such as religion and postmodernism, the New Age lacks a method of establishing knowledge independent of the authority of the practitioner. It thus requires faith in the guru, faith in the invisible, faith in the energy. This road is appealing because it offers ready access to mystery and hope—exactly the fare that science once sold to the public. It is nonjudgmental and welcoming precisely because all judgment must be suspended. But in the end, without objective standards, it must fall back on authority as the arbiter of what is true, and so it, too, leads back to brutality and authoritarianism.

This brutal, blame-the-victim aspect of New Age thinking is on ready display when the patient fails to be healed. This is a common scenario in New Age healing classes, since they often draw desperate people with illnesses for which medical science has not yet been able to find cures. For admission, patients might pay hundreds or thousands of dollars. The healer, who claims to be tuned into the energy of the universe, lays on hands, and may get other students in the group to channel their energy as well. Only the poor patient is "blocked." The rest—these very special students—can see it clearly by clairvoyantly looking at the blackness in the patient's aura and feeling the coldness in the parts of the aura affected by the disease. This block is likely the fault of some anger or other negativity. Because we manifest our own reality, physical maladies are usually merely symptoms of deeper spiritual causes. In a Freudian turn, it may even be some emotional trauma of which the patient is not aware—not from childhood, but from a past life, or from the interregnum between lives. This is why he is sick. If only he could see the truth of this, he could stop refusing the healers' loving energy. But alas, he cannot. The students and healer can see it. Why can't he? And so, sadly, his disease progresses and he eventually dies.

It's not that the cruel judgment of this thinking is intentional; many New Agers have the best of intentions. After all, they embrace tolerance and openness. But when confuted by reality—for example, by the deterioration and death of a well-adjusted and otherwise spiritually innocent patient—they have nothing to fall back on. That is because none of this is based in objective, testable knowledge about reality. And so New Agers are forced to either reject New Age principles or become intellectually dishonest. Rejecting both a belief system and a community is a hard thing to do—just ask Thomas Kuhn. It is what he argued scientists are unable to do. A paradigm change comes at great cost to one's personal vested interests, be they social, emotional, financial, political, or spiritual.

But in refusing to accept objective truth and the possible pain of intellectual honesty, we are left with the heartlessness of hubris, false hope, and the cruelty of blaming the victim. If we are unwilling to let go of our a priori principles and muster the courage to look at the situation as it is—that good people sometimes become sick and die due to reasons that are no fault of their own—rather than as we wish it would be, if we insist on harboring "sciences as one would," as Bacon called it, colored by our feelings, then it must be the *other* that is wrong. Thus Shapley wipes Humason's photographic plate clean, disease victims die due to their own repressed anger, and the emperor's tailor is paid a fortune, as many New Age gurus are. This is a retreat into superstition and darkness with heartbreaking human consequences—and even more heartbreaking political ones—rendered under the auspices of openness, tolerance, and love.

In the end, the fundamental idea that objectivity doesn't exist, and that science is little more than a metanarrative of the dominant culture, whose elite use it to hang on to power, was perhaps postmodernism's most corrosive contribution—one that informed and enabled a much more massive attack on science by religious ideologues and powerful industrial interests in the years to come. As Bérubé, the once-vocal critic of Sokal and his hoax, put it:

> These days, when I talk to my scientist friends, I offer them a deal.
> I say: I'll admit that you were right about the potential for science
> studies to go horribly wrong and give fuel to deeply ignorant and/or

Chapter 9

THE IDEOLOGICAL WAR ON SCIENCE

Enlightenment is man's emergence from his self-imposed immaturity.
Immaturity is the inability to use one's understanding without guidance
from another. . . . It is so easy to be immature. If I have a book to serve
as my understanding, a pastor to serve as my conscience, a physician
to determine my diet for me, and so on, I need not exert myself at all. I
need not think, if only I can pay: others will readily undertake the irk-
some work for me.

—Immanuel Kant, 1784

God Help Us

At the same time that government and industry scientists were turning away
from the public dialogue, postmodernist academics were undermining science's
intellectual standing, and journalists were becoming confused about the nature
of objectivity, churches, especially Protestant churches, were reaching out aggressively. Unlike scientists, churches still depended upon engaging the public for
financial support, and they were alarmed by baby boomers' deep skepticism of
religion in the 1970s. Environmental disasters and the bomb were rendering life
increasingly hopeless, but organized religion seemed staid, out of touch, and
unable to help parse the new moral and ethical challenges. New Age mysticism,
drugs, sex, and materialistic hedonism became mainstream alternatives for cop-
ing, along with an explosion in television watching—the average American adult
and child watched four hours of the "boob tube" every day. With membership
and collections falling, many emerging Protestant leaders believed the answer
lay in reaching out in new, more passionate ways, including going where the eye-
balls were: television. One by one, charismatic preachers began to take to the airwaves. Surprised by this new energy, the public responded. Suddenly, evangelism
was relevant again, and this time its leading figure was Billy Graham.

Graham's apparently overnight success as a television evangelist was, in fact, two decades in the making, driven by a twenty-year career railing against science. On September 25, 1949, just a month after the Soviet Union shook the world with its successful nuclear-bomb test, the thirty-year-old Baptist preacher stepped onto the stage inside a giant tent he had erected on Washington Boulevard in Los Angeles. He billed this as the Greater Los Angeles Billy Graham Crusade at the Canvas Cathedral with the Steeple of Light. Backed by hundreds of Christian leaders from across Southern California, Graham drew an average of fifty-four hundred people every night, with thousands more standing out in the twilight, straining to hear or listening over their car radios. The spectacle went on for sixty-five sermons over eight weeks, drawing three hundred and fifty thousand people. Graham's evangelical sermons were the pitch-perfect blend of Northern progressivism and Southern conservatism, building a mainstream, Billy Sunday–style brand that captivated Californians and, later, the rest of the United States. It was a formula that others would copy.

Graham quoted scripture and railed against science as he talked about his experiences as a traveling preacher in the days after World War II. "All across Europe, people know that time is running out," he told his worried audience. "Now that Russia has the atomic bomb, the world is in an armament race driving us to destruction." The time to accept Jesus was now or never.

> First of all, I want you to see the need in the philosophical realm. We have just come through an era of materialism, an era of patronism, and humanism in the educational circles of this country. We have been deifying man. We have been humanizing God. And all over the religious world there is a stark unbelief in the supernatural. All through this country of ours we have denied the supernatural, outlawed the supernatural, and said that miracles are not possible now. And we have taken up with things, rather than the spirit of God. . . . Because of the goodness and grace of God I can say tonight that I am not ashamed of the Gospel of Jesus Christ, for it is the power of God and the salvation to everyone that believes. I do not believe that any man, that any man, can solve the problems of life without Jesus Christ. There are tremendous marital problems. There are physical problems. There are financial problems. There are problems of sin and habit that cannot be solved outside the person of our Lord Jesus Christ. Have you trust in Christ Jesus as savior?

Tonight, I'm glad to tell you as we close that the Lord Jesus Christ can be received, your sins forgiven, your burdens lifted, your problems solved by turning your life over to him, repenting of your sins, and turning to Jesus Christ as savior. Shall we pray?

Graham was offering a message of hope and wonder in the face of fear. It was a message that, for the past forty years, had been offered by scientists like Edwin Hubble. But now Graham was inviting people back into the world of miracles and belief, a world characterized, in the words of Immanuel Kant, as the "cowardice . . . of lifelong immaturity."

In 1957, the year of *Sputnik 1*, Graham continued to use public anxiety over the H-bomb to sound an antiscience, anti-intellectual theme. "When Sir Walter Raleigh had laid his head on the executioner's block," he preached, "and the officer asked if his head lay right, Sir Walter Raleigh said 'It matters little, my friend, how the head lies, provided the heart is right.' The heart has come to stand for the center of the moral, intellectual, and spiritual life of a man." The head, Graham argued—and with it, science—should not be considered so important. Look at the trouble it got us into: marriages falling apart, society crumbling, and the end of days at hand due to the bomb and the rest of science, which was associated with man's turn away from God.

Those who would dismiss Graham's impact should note that his work reached an estimated 2.2 billion people. According to Gallup polls, he ranked among the ten most admired men for half a century. They would do well to understand the chord Graham struck—an intensely personal, emotional, and spiritual one. The desire to create knowledge that motivates science ultimately shares some of the same drives as that of its progenitor, religion. Playing to these drives is one way science can reach the masses, by helping them to understand the mystery and wonder of the world and our place in it, to find meaning and hope, and to make life better. These are courageous aspirations in the face of fear, which scientists would do well to trumpet—along with science's track record of actually achieving them.

The Civil War of Values

By the 1970s, the evangelical movement, which Graham had played a large part in reviving, had grown exponentially—thanks, ironically, to technology. Much as the Nazis did with radio a generation earlier, pioneering televangelists like Oral Roberts, Robert Schuller, Pat Robertson, Jerry Falwell, James Dobson, Jim and Tammy Faye Bakker, and Graham produced gospel shows that used mass communication to instill ideology in viewers. Together, according to Arbitron, these preachers reached more than twenty-two million viewers per week. Their common and primary aim was conversion, as dictated by Matthew 28:18–20 and Mark 16:16:

> And Jesus came up and spoke to them, saying, "All authority has been given to Me in heaven and on earth. Go therefore and make disciples of all the nations, baptizing them in the name of the Father and the Son and the Holy Spirit, teaching them to observe all that I commanded you. . . . He who has believed and has been baptized shall be saved; but he who has disbelieved shall be condemned."

Conversion demanded the delivery of constant, strident, emotional, and inspiring sales pitches to any and all who would listen. This was just the opposite of what was going on with science, which, as it enjoyed the fruits of Vannevar Bush's ability to secure government funding for basic science, and corporate funding for applied research, was cloistered in its own abbeys: laboratories— and there it stayed, almost totally silent.

A logical outgrowth of making "disciples of all the nations" was the coopting of the democratic process, and evangelicals moved into the political sphere. Graham was a registered Democrat who publicly opposed intolerance and said that religion should not choose political sides, but he became a minority. Sara Diamond, a sociologist who follows the growth of the Christian right, describes it this way:

> It is a political movement rooted in a rich evangelical subculture, one that offers participants both the means and the motivation to try to take dominion over secular society. The means include a phenomenal number of religious broadcast stations, publishing houses, churches, and grassroots lobbies. The motivation is to preach the Gospel and to save souls, but also, with equal urgency, to remake contemporary moral culture in the image of Christian Scripture. On the front lines of our persisting battle over what kind of society we are and will become, the Christian Right wages political conflict not just through the ballot box but also through the movement's very own cultural institutions.

While the voice of science, the very foundation of secular democratic society, had gone silent, the voice of Protestantism had grown evangelical, angry, antiscience, and intensely political, engaging in a "civil war of values," as James Dobson's *Focus on the Family* radio ministry put it, declaring that the 1990s would be "the civil-war decade." It was to be a cultural revolution with the goal of remaking America in the image of Christian Scripture.

Few scientists saw any connection between the rise of right-wing religion and the absence of science in the public debate. A survey of *Science* magazine, a leading publication of the global science enterprise, shows no mention of the phrase "religious right" until a November 1989 article about attempts to teach creationism in science classes—two months before Dobson's proclamation. Prior to this date, references were chiefly to "fundamentalists," the dogged but easily dismissed foes of evolution whose periodic school-board flare-ups had been chronicled in the magazine since the days of creationists George McCready Price and William Jennings Bryan in the early 1920s. With few exceptions, the fact that the religious right had become a national political force, and that the voice of science in the public dialogue was weakening, appears to have been largely ignored in the professional conversation among scientists.

Dobson's traditionalist point of view is part of a strain of American thought going back to the Second Great Awakening in the early nineteenth century. To be sure, religion has a role to play in a democracy's national dialogue. But, to arrive at balanced public policy, democracy relies on a plurality of voices— not only religious but also economic, scientific, and others. With science gone silent, academics gone postmodern antiscience, and religion gone vehement, American democracy no longer had that plurality. The world's oldest democracy's policies and politics became unbalanced, and a generation grew up regarding science as irrelevant in shaping the public dialogue—even as it was impacting their lives more and more powerfully. The absurd outcome of all this was that, as national and global policy challenges came to revolve around science, their proposed solutions came to revolve around faith.

Don't Let Satan Fool You

The ideological front of the war on science is being waged by religious conservatives in three major battle zones, all of which deal with origins: the nature and age of Earth and the universe, the theory of evolution, and the origin and nature of life and reproduction. Our answers to these three questions lie at the center of physical science, biology, and the health sciences, and of our capacity to make effective policy decisions in education, economic competitiveness, and public health.

The issue that, in many ways, is still the most contentious is the teaching of evolution. Fierce court battles fought in numerous states since the 1925 Scopes Trial have made contestants on both sides famous, with some cases going to the US Supreme Court. Each time evolution wins, authoritarian religious conservatives adjust their language and redouble their efforts.

Classroom teachers often choose to simply skip the subject altogether rather than fight with creationist parents. One science-museum director described overhearing a group of homeschooled children about to enter the paleontology exhibit being pulled aside by their parents and told, "Now, remember, those bones were put there by Satan to fool you."

The question of whether evolution is real is dynamic because both views are uncompromising. One is based on knowledge, which is antiauthoritarian; the other on a literal reading of the Bible, which is authoritarian. They are thus opposites on the vertical political axis.

Ben Carson, a retired neurosurgeon and 2016 US presidential candidate, provides an example of someone who is obviously intelligent, but whose conservative Seventh-Day Adventist faith doesn't allow him to look at a problem from an evidentiary standpoint. Instead, he seeks to find evidence that supports his faith, like a lawyer. This is called "confirmation bias" or "motivated reasoning." Carson has said he believes that Satan encouraged Darwin to come up with evolution. "I personally believe that this theory that Darwin came up with was something that was encouraged by the Adversary, and it has become what is scientifically, politically correct. Amazingly, there are a significant number of scientists who do not believe it but they're afraid to say anything," he added, saying he would write a book, *The Organ of Species*, that would show how the organs of the body refute evolution.

Carson also denies the astronomic and cosmological science that led to the big bang as a strongly supported theory of the origin of the universe:

> I find the big bang really quite fascinating. I mean, here you have all these highfalutin scientists and they're saying it was this gigantic explosion and everything came into perfect order. Now these are the same scientists that go around touting the second law of thermodynamics, which is entropy, which says that things move toward a state of disorganization. So now you're gonna have this big

explosion and everything becomes perfectly organized and when you ask them about it they say, "Well we can explain this, based on probability theory because if there's enough big explosions, over a long period of time, billions and billions of years, one of them will be the perfect explosion." So I say what you're telling me is if I blow a hurricane through a junkyard enough times over billions and billions of years, eventually after one of those hurricanes there will be a 747 fully loaded and ready to fly. Well, I mean, it's even more ridiculous than that 'cause our solar system, not to mention the universe outside of that, is extraordinarily well organized, to the point where we can predict 70 years away when a comet is coming. Now that type of organization to just come out of an explosion? I mean, you want to talk about fairy tales, that is amazing.

Here we find another argument popular with creationist policymakers, which cosmologists call the anthropic principle. It is related to the problems we have estimating probabilities and statistics in such things as gambling, life, and economic situations that involve insurance, investment, and risk. Imagine, for instance, the feelings a blade of grass might have when the great holy golf ball landed upon it. "What are the chances," the blade might wonder, "out of all the millions and billions of blades of grass, that the great holy golf ball should choose *me*? I am special." And yet, with every stroke, the golf ball must land somewhere.

Adopting the Skepticism Argument

Carson frequently uses ridiculous-sounding analogies like the self-assembling 747 to refute scientific concepts about origins. A 2006 panel discussion provides a similar example. Carson asked the audience to imagine aliens comparing a Rolls-Royce to a Volkswagen and concluding that the Rolls-Royce, being superior, had evolved from the VW. Based on the ridiculous-sounding possibility of cars evolving or planes self-assembling, he then flips the argument on its head, saying, "I simply don't have enough faith to believe that something as complex as our ability to rationalize, think, and plan, and have a moral sense of what's right and wrong, just appeared." This implies that it is he who is the skeptic, and scientists (tens of thousands of them working over the last 150 years, in the case of biology and evolution) are the gullible ones.

But science doesn't suggest order "just appeared" out of nothing, or that cars, which are designed machines, are in any way comparable to the processes of evolution going on in life. By starting with a false premise, Carson can claim anything he wants. And it's not an act of faith to "believe" in evolution in the face of the tremendous number of independent lines of evidence from a wide variety of scientific disciplines that all point to the same conclusion.

Carson frequently uses this argument-flipping technique to make himself appear to be the skeptical freethinker and—using the linguistic tricks of postmodernism—scientists the intolerant authoritarians:

> The politically correct crowd in academia is extremely domineering and intolerant of anybody who believes anything different than they do. And if you're a young assistant professor at a major university, and you're trying to be promoted, and you indicate you believe in creationism you're in big trouble. There's a very strong chance that you will not be promoted and in fact you may well be driven out of the university. Extraordinarily intolerant. And you know that's one of the things about PC and one of the reasons I rail about political correctness so much.

The Scientific Definition of Life

The conflicting claims of truth made by postmodernists, ideologues, industry, and science are at the heart of most of today's major political disputes. Evolution has nothing to do with belief or political correctness; it has to do with evidence from observing nature, the truth of which Carson is seeking to deny. Modern medicine and biology are based on evolution—biology is essentially applied chemistry and physics in the context of evolution. Evolution is the most fundamental principle in biology, the one that unified it into an organized science. Without the theory of evolution, there would be no biology and no modern medicine. It connects and provides a framework for understanding all the disciplines within the life sciences, from genetics to virology to oncology to organic chemistry.

In fact, evolution is central to our current understanding of what life is. At its most basic, life is made up of polymers—molecules called monomers that are chained together like pearls on a string. You may recognize the word "polymer" because we've been able to make polymers that are not normally found in nature, like plastic. All life we know of is based on a two-polymer system of a protein polymer and a nucleic-acid polymer. Monomers called nucleotides make up the nucleic acid polymers RNA and DNA, which, based on the order in which the pearls are strung, carry an organism's genetic code, thereby storing the information and passing it on to descendants. A mutation in these polymers is the basis for evolution. Nucleic acids also direct the formation of protein. Proteins, which are made up of amino acids, are essential for the growth and repair of tissue. Proteins are the workhorses of the cell, acting as enzymes, hormones, and antibodies. In combination, these two polymers—nucleic acid and protein—are present in all life, and are able to build structures of life that are resilient and responsive to the environment by evolving flexibly to fit their circumstances. Because of this, a 1992 NASA workshop defined life as "a self-sustaining chemical system capable of Darwinian evolution."

Without the theory of evolution, there would be no modern medicine, antibiotics, virology, or pharmaceuticals. The medicines Ben Carson prescribed to keep his patients from dying of infections after he operated on them would not exist.

Superbugs

The knowledge gained from the theory of evolution helps us to be better farmers, better environmental stewards, better computer programmers, and better doctors, improving health and productivity, and saving trillions of dollars and millions of lives. Antibiotics provide an example. We have many times the number of bacteria in our bodies as we do human cells, most of them working away in symbiotic relationships with us. Without them, we would not be human. We need them to help us digest our food, moderate our moods and our immune responses and even our body weight, to fight off nastier bacteria, and to stay healthy.

But every year, pathogens try to attack us and take us down. Diseases like strep throat, meningitis, mononucleosis, gonorrhea, cholera, tuberculosis, Lyme disease, syphilis, tetanus, diphtheria, typhus, and pertussis are all bacterial diseases, and they're just the beginning. It used to be that a course of antibiotics would cure most any of them, but over the last two decades these bacteria, which reproduce very quickly, have started to evolve into antibiotic-resistant forms that we can't kill in the same old ways.

The theory of evolution is helping microbiologists understand our complex relationship with bacteria, and devise new strategies to slow the evolution of superbugs before these diseases get out of control again. This situation is both likely and disturbing, as the superbugs could kill hundreds of millions of people.

For example, scientists know that antibiotics don't affect viruses, so they're working to get doctors to stop prescribing them across the board. Over 50 percent of antibiotics prescribed for people are either not needed or not optimally effective as prescribed. Often, doctors give patients antibiotics to make them feel like they did something for them. But taking antibiotics when one has a virus does nothing to help beat the virus. Instead, it may disrupt a body's fragile biotic balance, and it most likely will encourage bad bugs to develop antibiotic resistance. In the United States, more than two million people develop serious infections from bacteria-resistant bugs each year, and the number is climbing rapidly. When doctors prescribe antibiotics, scientists now emphasize that it should only be for appropriate uses—to fight bacterial diseases, not viruses. They also want doctors to avoid giving low doses over long periods of time, because they're not high enough to kill all the bacteria, and the ones that remain evolve within our bodies into resistant strains. This is also one of the reasons microbiologists generally don't favor the broad use of antibacterial soaps.

Scientists recommend doctors counsel patients to take all their pills, even after they start to feel better. Otherwise, the more resistant remnants of the bacterial army will regroup and evolve into resistant strains that can resurge with a vengeance. The antibiotics need to kill them all, not leave the toughest ones around to reproduce. An interagency task force led by the CDC concluded that, if the problem of antibiotic resistance is not addressed, "drug choices for the treatment of common infections will become increasingly limited and expensive—and, in some cases, nonexistent." This will change life as we know it. Medical implants? Joint replacements? Brain surgery? Forget it. Before antibiotics, one out of every two hundred women who gave birth died from infections. Tattoos? Botox? Hair transplants? No way. One out of every nine people who got a simple skin infection from a scrape or an insect bite died.

Margaret Chan, director-general of the World Health Organization, described the evolving situation bleakly:

> This will be a post-antibiotic era. In terms of new replacement antibiotics, the pipeline is virtually dry, especially for gram-negative bacteria. . . . Prospects for turning this situation around look dim. The pharmaceutical industry lacks incentives to bring new antimicrobials to market for many reasons, some of which fall on the shoulders of the medical and public health professions. Namely, our inability to combat the gross misuse of these medicines. . . . A post-antibiotic era means, in effect, an end to modern medicine as

we know it. Things as common as strep throat or a child's scratched knee could once again kill. Some sophisticated interventions, like hip replacements, organ transplants, cancer chemotherapy, and care of preterm infants, would become far more difficult or even too dangerous to undertake. At a time of multiple calamities in the world, we cannot allow the loss of essential antimicrobials, essential cures for many millions of people, to become the next global crisis.

Much of this problem is being driven by misuse in agriculture by the very Bible Belt farmers and agribusiness executives that deny evolution exisits. Scientists are now urging farmers and agribusinesses to stop the prophylactic use of antibiotics in animal feed, which is common because it can accelerate livestock growth, thereby improving the bottom line. But its widespread use, which dwarfs use in humans, sprays low levels of antibiotics throughout the environment, helping resistant strains get a much broader and tougher foothold while keeping nonresistant, weaker strains out of the gene pool—helping, again, to create superbugs.

This effect has long been known. Alexander Fleming described it in his 1945 Nobel Prize lecture upon receiving the prize for the discovery of penicillin. "Then there is the danger that the ignorant man may easily underdose himself and by exposing his microbes to non-lethal quantities of the drug make them resistant," Fleming warned.

Data on the use of antibiotics in agriculture are not systematically collected, but a Union of Concerned Scientists report found that as much as 84 percent of the total use of antimicrobials in the United States was for non-therapeutic purposes in agriculture, with 70 percent in livestock. The use of antibiotics as growth promoters was banned in the European Union in 2006. Antibiotics can only be administered to animals under a veterinary prescription and the vet is legally obligated to record use. But data collection varies by country, so it is difficult to say whether or how well the ban is being followed. In Canada, a blue-ribbon panel concluded that "the lack of progress on this recommendation was regarded as 'a continuing international and national embarrassment.' It 'has been studied for 15 years and still no effective regulatory solution has come to light.'"

In economic terms, this is essentially a taking from the public, as farmers, agribusinesses, and feed manufacturers get incrementally richer at the expense of public health. We'll get into this more later, but this is the sort of thing that sin taxes are designed to counter. In this case, there is a strong scientific case to be made for taxing agricultural operations heavily on the excessive use of antibiotics, thereby both discouraging the practice and compensating the public from the increased costs of antibiotic-resistant microbes. Pathogens created in this way are especially dangerous, because they have a direct vector into the human population through our grocery stores.

It's healthier for everyone if we treat the disease overwhelmingly and when it occurs, rather than dribbling treatment around everywhere and encouraging the bacterial insurgency to adapt and spread. The theory of evolution is the only reason we understand any of this.

Teaching the "Controversy"

Despite the life-saving benefits of the theory of evolution, religiously conservative politicians often suggest we should "teach the controversy," a framing mechanism that suggests there is a legitimate scientific controversy over whether evolution is true. For the record: There simply isn't. Evolution is the most well-supported knowledge in science. But what most critics of creationism don't understand is that this doesn't really matter, since the conflict is only tangentially related to the facts of science.

Since the time of Aimee Semple McPherson and the Scopes Monkey Trial, fundamentalist Christians have painted evolution as the cause of most social maladies, because it undermines the idea of humans as God's special creation and distinct from other animals. Consider this cartoon from the creationist group Answers in Genesis: Two castles, each on a tiny island, fire at one another with cannons. The castle on the left is on the island of "Evolution (Satan)" and flies the flag of Humanism, while the one on the right rests atop the island of "Creation (Christ)" and flies the flag of Christianity. In the balloons above the Evolution castle, one can see all the social ills creationists view as being caused by the theory of evolution: euthanasia, homosexuality, abortion, racism, divorce, and pornography. The misguided priests on the island of Christ are stupidly firing at these mere symptoms, or at each other. Meanwhile, on the other side, the lone grim figure, a sort of scientist-pirate-Satanist, is blasting away at the foundation of the creationists' castle.

With this type of portrayal being exposed to children from a young age (and this cartoon has been used in public school science classes, not just in religious publications), one can begin to understand why some on the religious right oppose evolution, just as William Jennings Bryan and Aimee Semple McPherson did in the 1920s. Former US congresswoman and presidential candidate Michele Bachmann provides an example. Bachmann served in Congress for eight years, beginning in 2006, and quickly became a standard-bearer for evangelical conservatism. Like Ben Carson, Bachmann and other creationist politicians use the argument-flipping technique to accuse scientists of being intolerant, claiming that if scientists were not they would be willing to debate the controversy. But most scientists won't debate creationists because it is another example of false balance—equating knowledge from science with a charismatic someone's opinion. As Bachmann said,

> I have no problem with teaching the various theories . . . of origins of life. . . . But, I think there's one . . . philosophy . . . that says only one could be taught and that one would be evolution. And because the scientific community has found that there are flaws in abiding by that dogma, I think it's important to teach that controversy. [Lawmakers and educators should not] censor information out of discussion because it doesn't meet within someone's dogmatic beliefs. Something that I think sometimes people don't like to hear is that secular people can be sometimes even more dogmatic in beliefs than people who are not secular. . . . In some ways, to believe in evolution is almost like a following; a cult following if you don't believe in evolution, you're considered completely backward. That seems to me very indicative of bias as well.

This thinking is similar to many other antiscience arguments, from creationism to climate-change "skepticism": the suggestion that there are "flaws" in the science and that scientists are ruled by "dogma." It shows a postmodernist antiscience educational background, appropriated into conservative Christianity, with science simply another "way of knowing" on equal footing with faith. As Bachmann put it, because there are "eminent, reasonable minds" within the scientific community that disagree with the theory of evolution, "I would expect that teachers would disagree, and students would disagree, and the public would, certainly."

A Scientific Theory Is the Next Closest Thing to a Fact

Scientists will tell us that much of the conceptual problem lies in the public's misunderstanding of the word "theory." A scientific theory is not a hypothesis or guess, as the word commonly means when used in casual conversation. A scientific theory is the one explanation that is confirmed by all the known and validated experiments performed to date. Experiments involving evolution have numbered in the hundreds of thousands over the past 150 years. A theory is thus among the most certain forms of scientific knowledge, with evolution among the most certain of theories. But because science is inductive, scientists recognize that there is still a chance that it could be wrong.

In the case of evolution, the chance that it is wrong is somewhat smaller than that of, say, Earth being destroyed by a meteor within the next five minutes. We can see it working with our own eyes by watching viruses and bacteria evolve under the microscope. When one does, it becomes difficult to see how anyone could construe it as a matter of belief. It's like saying, "I don't believe in gravity."

Dogs of Intelligent Design

If you were a scientist, and you wanted to test whether the theory of evolution was in fact true, what would you do? You'd do an experiment. You'd apply the selective forces applied by nature in a controlled way and see what happens. And, as it turns out, that's precisely what humans have done for thousands of years—by selectively breeding dogs, for example, or by exerting selective forces on other species of plants and animals, causing them to evolve. The wide use of the herbicide glyphosate is an example. With its broad use, weeds are now adapting by becoming glyphosate-resistant in the same way superbugs are becoming antibiotic-resistant.

Ben Carson and many other creationists will admit to microevolution— small evolutionary changes within a species—but reject macroevolution—the gradual emergence of higher species from earlier ones, and thus the common descent of all life. Can we point to any experiment that really captures evolution on a so-called "macro" level? It turns out we can. In the 1930s, a Russian geneticist names Dmitri Belyaev wanted to test just this question, and he carried out an amazing experiment. Belyaev knew that dogs were descended from wolves, and he knew that behavior ultimately arose from biology. But he couldn't figure out the mechanism that could account for the wolf giving rise to so many unique breeds of dog, or how it could be that dogs seemed to be born with an affection for humans. He wondered if it would be possible to test this. He set out on a program of selectively breeding Russian silver foxes, which mature quickly enough that, he hoped, he could perhaps observe evolutionary changes within the span of his lifetime. He captured several wild foxes and put them in cages, and he began a program of selective breeding based only on a single trait: tameness.

Each week, after baby foxes were born, Belyaev and his staff rated their reactions to a researcher. The scientist would attempt to pet and handle the foxes while offering them food, and would note whether they preferred other foxes or were happy to spend time in the company of humans. This continued for a period of six to seven months, until the foxes reached sexual maturity. At that point, the scientists only allowed the foxes with the highest tameness rating to breed, which represented less than twenty percent of each generation—a process akin to evolutionary selection, whereby only those members of a species best fitted to their environmental conditions (the true meaning of "survival of the fittest") survived and produced offspring.

With each subsequent generation they did the same thing, and after forty generations they were able to show something remarkable. By this one evolutionary pressure of selecting for tameness, the wild silver foxes had transformed into animals very different from their wild ancestors. They had lost their "musky fox smell" and were happy to be around humans; in fact, they whimpered to attract attention, and they sniffed and licked their caretakers in much the same way domestic dogs do. They wagged their tails when they were happy or excited. Perhaps most strikingly, they had a whole host of other genetic characteristics we tend to associate with dogs: floppy ears, short curly tails, longer reproductive seasons, black-and-white spotting and other coat changes that are common in dogs, and changes in the shape of their skulls, jaws, and teeth. In addition, they were naturally happy to be around humans, and much more curious about their environment. The experiment is ongoing today, and has identified many of the underlying genetic and biochemical changes that caused the foxes' evolution based on this one trait.

Motivated Ignorance

Part of the problem in the rejection of evolution seems to be educational. Most creationist policymakers, when asked about their views of evolution, display a high level of ignorance about just what it is and how it works. Bachmann, for example:

> Natural selection is not the same thing as evolution. [Note: Yes, it is] No one that I know disagrees with natural selection, that you can take various breeds of dogs . . . breed them, you get different kinds of dogs. . . . It's just a fact of life. . . . Where there's controversy is, where do we say that a cell became a blade of grass, which

became a starfish, which became a cat, which became a donkey, which became a human being? There's a real lack of evidence from change from actual species to a different type of species. That's where it's difficult to prove.

Darwin coined the term "natural selection" to distinguish it from the artificial selection done by breeders. While, as we've just seen, we can use selective breeding to apply evolutionary force and cause certain changes over several generations, no evidence has ever suggested that human beings are descended from donkeys or blades of grass. But a weak grasp of the core concepts doesn't fully explain the issue, which is also caused by strong fundamentalist religious motivation.

Canada's then-minister of science and technology, Gary Goodyear, showed a similar confusion in 2009. A Conservative and a chiropractor by profession, Goodyear was asked by reporters if he "believed in evolution." He said, "I'm not going to answer that question. I am a Christian, and I don't think anybody asking a question about my religion is appropriate." But the question was about science, not Goodyear's religion. Scientists were alarmed, and some in the press called him a creationist. Over the next several days, Goodyear dodged follow-up questions, until he finally stated,

Well, of course, I do, but it's an irrelevant question. . . . We are evolving every year, every decade. . . . That's a fact, whether it's to the intensity of the sun, whether it's to . . . walking on cement versus anything else, whether it's running shoes or high heels, of course we are evolving to our environment, but that's not relevant and that's why I refused to answer the question.

Goodyear's answer showed another fundamental misunderstanding of how evolution works—one commonly held by creationists. The idea of evolution is that genetic variation among organisms results in differences in their "fitness"—a biological term referring to the number of reproducing offspring they have relative to their success in their environment. Organisms that are good fits with their environment will tend to pass on genes at a higher rate than organisms that fit less well. That ultimately leads certain characteristics to become more prevalent among their descendants, as in Belyaev's silver foxes. Life adjusts dynamically to its environment; the types of characteristics that result in more offspring change over time as the environment changes. But it doesn't occur every year or decade, and it has nothing to do with footwear selection.

Creationists often refer to the writings of Michael Behe, a biochemistry professor, creationist, and the author of *Darwin's Black Box*, a book arguing that some structures, such as the human eye, are just too complex to be the result of evolution and thus must be evidence of "intelligent design," a more recent version of creationism. This is undoubtedly where Ben Carson gets some of his ideas about 747s. But the eye is not irreducibly complex; we can show with considerable evidence exactly how the eye evolved, and is continuing to do so. Like Harlow Shapley, Behe and Carson have made the mistake of clinging to an a priori first principle rather than building understanding from observational evidence. Their conclusions are not science; they're what Bacon called "science as one would," full of examples of the "vulgar Induction." In other words, rhetoric.

The other professors in Behe's department of biological sciences at Lehigh University, who felt compelled to publish a statement clarifying their position, also pointed this out:

> While we respect Prof. Behe's right to express his views, they are his alone and are in no way endorsed by the department. It is our collective position that intelligent design has no basis in science, has not been tested experimentally, and should not be regarded as scientific.

Bachmann says that many professors at religious colleges, and some religious professors at secular colleges like Lehigh, teach that "the Earth was created by an intelligent being—God, if you will—and that there are Scripture passages that say that a day is as a thousand years and a thousand years is a day, and that therefore, over time, God could have created all this." Sure. Why not? But it's got nothing to do with knowledge, which flows, as Locke demonstrated, from observation.

This sort of speculation is altogether human and entirely appropriate. But it's not science. Science, by its very foundation on Bacon, Locke, and empiricism, limits itself to probabilistic conclusions that explain observations of the physical world. Science makes no statements about the ultimate reality outside of these limitations, and teaching that it can is a corruption of its clarity of thought. Confusing the matter for our children not only foolishly trades our ideological comfort for their ability to compete in the twenty-first century, but also, by unmooring them from knowledge, dooms them to the political paralysis and inability to problem-solve that plague Western democracies in the first decades of the new century of science.

Meet the Flintstones—and Nessie Too

Spreading largely from the United States, creationism is turning up in classrooms around the world, particularly in Australia, South Korea, Indonesia, Turkey, Pakistan, and several countries in Europe.

In 2012, a commotion occurred in South Korea over the removal of evolution from state-approved science textbooks by the Ministry of Education, Science, and Technology. The move alarmed biologists, who were not consulted. Although only about a third of South Koreans say they don't believe in evolution, antievolution efforts are more successful there than in the United States. The country's leading science institute, the Korea Advanced Institute of Science and Technology, once hosted a creationism display on its campus, and in 2008, the Korea Association of Creation Research sponsored a successful exhibition on creationism at Seoul Land, a popular amusement park. A 2011 survey of teacher trainees found that found that 40 percent of biology teachers agreed with the statement that "much of the scientific community doubts if evolution occurs"; half disagreed that "modern humans are the product of evolutionary processes." Half of South Koreans are religious, and half of those are Christian.

But creationism is not limited to Christian countries. Definitive belief in God or a supreme being is highest in Indonesia (93 percent) followed closely by Turkey (91 percent), according to an Ipsos/Reuters poll. The same poll found that 75 percent of Saudi Arabians, 60 percent of Turks, and 57 percent of Indonesians—three of the world's leading Muslim countries—referred to themselves as "creationists and believe that human beings were in fact created by a spiritual force such as the God they believe in and do not believe that the origin of man came from evolving from other species such as apes."

Pervez Hoodbhoy, an atomic physicist at Quaid-e-Azam University in Pakistan, said that when he gives lectures covering the sweep of cosmological history, from the big bang to the evolution of life on Earth, the audience listens without objection "until the apes stand up." Mentioning human evolution has led to near-riots, and he has had to be escorted out. "That's the one thing that will never be possible to bridge," he said. "Your lineage is what determines your worth." Biology education in Pakistan, which otherwise teaches evolution, leaves humans out. And creationist textbooks by Turkish author Harun Yahya are reportedly used by most of the biology teachers in Indonesia, the world's most populous Muslim country.

Meanwhile, in May 2005, Netherlands minister of education Maria van der Hoeven suggested that debate about intelligent design might encourage discourse between the country's various religious parties. She thought it might also "stimulate an academic debate" on the subject.

That same year, a national survey found that 30 percent of Poles reject evolution, and many of those who accept it regard it as a process guided by God. In 2006, Poland's deputy education minister Mirosław Orzechowski denounced evolution as a lie taught in schools. His boss, Education Minister Roman Giertych, said the theory would still be taught "as long as most scientists in our country say that it is the right theory." Giertych's father Maciej, a member of the European Parliament, opposed the teaching of evolution and claimed that "in every culture there are indications that we remember dinosaurs. The Scots have Nessie; we Poles have the Wawel dragon." In 1993, he sent a videotape entitled *Evolution: Fact or Belief?* to public schools free of charge.

Thousands of copies of a German creationism book with the postmodernist name *Evolution: A Critical Textbook* have made their way into German schools. Many were distributed free of charge, are kept in libraries, and have been taught in state-sponsored schools, particularly in the state of Hessen. Christian Democrat Karin Wolff, the state minister of culture, who is responsible for education, suggested including the biblical creation story in biology classes. This triggered protests from other politicians and biologists, and the book was removed. Still, the authors of one op-ed on the issue stated that "gaps in the theory of evolution are discussed, and the possibility of God's creation is taught" at more than one hundred state-sponsored schools.

The book has continued to be used across Europe, being translated from the original German into Dutch, Finnish, Italian, Portuguese, Russian, and Serbian. In Germany, creationists take advantage of German guilt about the Nazi practice of eugenics, and the textbook draws a "straight connection between the ideas of Darwin and Adolf Hitler."

In 2007, the Parliamentary Assembly of the Council of Europe passed a resolution condemning the dangers of creationism in education, saying "some people call for creationist theories to be taught in European schools alongside or even in place of the theory of evolution," and urging "education authorities in member states to promote scientific knowledge and the teaching of evolution and to oppose firmly any attempts at teaching creationism as a scientific discipline."

In May 2015, although the Scottish parliament didn't embrace creationism, it did reject a bill that would ban the teaching of young Earth creationism in science classes, saying that subject matter decisions should be left to teachers.

Beliefs about Nessie and the Wawel dragon being dinosaurs are similar to ideas many Americans share, as measured by the *Science and Engineering Indicators* reports—issued every other year by the National Science Board (NSB), the body charged with overseeing the National Science Foundation—and in Gallup polls. Both of these sources try to capture respondents' views with short conceptual questions. At the beginning of the millennium, 45 percent of Americans agreed with the statement that "God created human beings pretty much in their present form at one time within the last 10,000 years or so," while 52 percent of Americans were unaware that the last dinosaur died before the first humans arose. That's to be expected. After all, we've seen them together on *The Flintstones*, and there have been fairy tales of dragons for a millennium.

Fifty-four percent knew that Earth orbits the sun, and that it takes a year to do so. Fifty-one percent knew that antibiotics don't kill viruses. And 53 percent knew that "human beings, as we know them today, developed from earlier species of animals."

Onward, Christian Soldiers

But by 2006, just *five years later*, something remarkable had happened in the United States. The last number had fallen significantly—from 53 percent knowing that "human beings, as we know them today, developed from earlier species of animals" to just 43 percent. Why? Did 10 percent of Americans get dumber about evolution during this time? Were evangelical creationists like Michele Bachmann and Michael Behe converting people in droves?

The answer may be suggested by politics. On September 11, 2001, al-Qaeda operatives flew planes into the World Trade Center and the Pentagon, and another hijacked aircraft crashed in Pennsylvania. The United States was shocked and angered, and its lawmakers responded, as traditionally happens in times of uncertainty, by increasing the power of the federal government and curtailing civil liberties—a swing toward authoritarianism. This is when the Bush presidency moved to a full bottom-wing position, and the word "liberal" became an epithet to silence liberal (top-wing antiauthoritarian) conservative Republicans.

Between 2001 and 2006, the White House and both houses of Congress were largely controlled by Republicans, and the Republican Party was largely controlled by authoritarian religious fundamentalists. George W. Bush had become the most profoundly evangelical president in American history and was rewarding his fundamentalist followers. The president's reelection campaign had noticed that evangelical voters hadn't turned out for Bush in 2000, so they worked to target and mobilize evangelicals as a voting bloc in the 2004 election. Bush political advisor Karl Rove famously walked the halls of the West Wing whistling "Onward, Christian Soldiers" while preparing the 2004 Bush-Cheney reelection campaign. Ralph Reed, the executive director of the Christian Coalition of America, was hired as a campaign official and developed a strategy that included campaign activities in evangelical churches, endorsements and voter registration drives by pastors, and mobilization of church congregations.

The "civil-war decade" was bearing fruit, and the same strict moral authority that had stiffened the wills of those venturing into the Wild West, that had emboldened the temperance movement of the post–Civil War 1800s, and that had given courage to people unsettled by evolution, immigration, and materialism in the Jazz Age now granted a sense of security in another time of upheaval.

Viewed in this light, there emerges a possible correlation between the 10 percent of the electorate that had changed its mind about evolution and the swing voters who went overwhelmingly for Bush in the 2004 election. Many surveys suggested that the number of swing voters in the 2004 election was about 13 percent—close to the number of new creationists. Other studies have shown the number of swing voters remained relatively stable—between 9 and 13 percent—in elections dating back to 1972.

In 2004, Bush followed Reed's strategy of church outreach to unify political support in suburban congregations, using those peer groups to capture unaffiliated voters who might swing from one cycle to the next. College-educated married white women, for example, were split almost evenly between the parties in 1992, 1996, and 2000, but broke sharply for Bush in 2004. "You are going to do what we used to call a friends and family program," Reed said, "where you had people take their Rotary Club or church or synagogue, or temple membership, or neighborhood directory, or even a tennis or garden club list, match it up against a voter file, and you will be able to do customized messages just to those people." Members of congregations did indeed send in their church directories and formed "moral action teams" to fight what they increasingly saw as "a spiritual battle."

Reed knew he was pushing churches into an ethical gray area that could cause them to violate the law. But, in December 2001, Rove had told an audience at the American Enterprise Institute that only about fifteen million of nineteen million voters on the Christian right had turned out in 2000. It was critical to Bush's reelection that the remaining four million be motivated to go to the polls. Reed appeared to relish the negative publicity his strategy of church mobilization created. He saw it as an opportunity to further reinforce the campaign's connection to faith and motivate true believers, saying that "Christians should not be treated as second-class citizens."

Activist Judges

At the height of this overheated moment, the school board in Dover, Pennsylvania, voted on October 18, 2004, to require its ninth-grade science teachers to read a statement to their classes questioning the validity of evolution. "Because Darwin's Theory is a theory," the teachers were required to tell students, "it continues to be tested as new evidence is discovered. The Theory is not a fact. Gaps in the Theory exist for which there is no evidence."

The district's science teachers defied the order, countering that it breached their ethical obligations to students "to provide them with scientific knowledge that is supported by recognized scientific proof or theory." Several parents, including lead plaintiff Tammy Kitzmiller, agreed and sued the school district. John Jones, a Republican federal judge appointed by George W. Bush, ruled that teaching intelligent design in public-school science classes was unconstitutional.

After the decision, fundamentalist activist Phyllis Schlafly said that Jones had "stuck the knife in the backs of those who brought him to the dance." In a classic example of the argument-flipping technique, creationists accused Jones of being an "activist judge" and threatened him to the point that he required the protection of US Marshals.

Later, in describing the job of a federal judge, Jones said that, unlike Congress and the presidency, which were created to be responsive to public sentiment, the founders "in their almost infinite wisdom" created the judiciary "to be a bulwark against public will at any given time" and "to be responsible to the Constitution and the laws of the United States."

The Sweet Spot of Politics

As conservative evangelical political candidates campaign in churches across the country, pastors are warned by the IRS against making explicit endorsements of candidates from the pulpit. Instead, "moral values" are often cited as the defining issue of an election. Candidates are invited to speak about these moral values, and millions of voter guides are distributed—in churches, and by congregation members in the broader community—on the moral issues said to be at stake, including abortion, same-sex marriage, faith-based initiatives, stem-cell research, human cloning, HPV vaccinations, abstinence-only sex education, and the teaching of creationism in public-school science classes, all of which certain candidates are on the right side of, and others are not.

A November 19, 2004, Gallup poll noted that "the lowest levels of belief that Darwin's theory is supported by the scientific evidence is found among those with the least education, older Americans, . . . frequent church attendees, conservatives, Protestants, those living in the middle of the country, and Republicans"—all voting blocs that went for Bush, who easily won the 2004 creationist vote, which Gallup then measured at 25 percent of the US population.

Science, by contrast, has no voices speaking on its behalf from the pulpit or almost anywhere else in the political dialogue. So it's not hard to imagine how such a vast swing in public opinion on evolution could happen over a few short years. And it is perhaps not a coincidence that it was during this time of heightened global media coverage of the creationism debate and the resurgence of religious fundamentalism in the United States that several European countries and their education ministers started wondering about evolution too.

Political consultants call this ideologically malleable electorate the "sweet spot" of politics. Clearly, if political speech can influence substantial numbers of Americans' answers to factual science questions—and perhaps those of their counterparts in other countries—something beyond education must be entering into the equation.

When Facts Don't Matter

Scientists will tell us this is just plain wrong. Facts are facts. They're not fungible, and these people are just poorly educated. The problem is that the public doesn't view science as a collection of facts that can't be argued with. In fact, some studies suggest that the brain processes facts and beliefs in essentially the same way. Thus, when a fundamentalist Christian says she believes as a fact that God created people literally as described in the Bible, she means it, in the same way that a scientist believes a factual statement about evolution or geology. This underscores how critical the distinction is between knowledge and opinion, and how hopelessly entangled the conflict between science and politics sometimes seems.

That conflict came to a new head in April 2010, when the US National Science Board deleted the questions and responses about evolution and the big bang (the origins of man and the universe) from its biennial *Science and Engineering Indicators* report: "Human beings, as we know them today, developed from earlier species of animals" and "The universe began with a big explosion."

These two questions had appeared in the report since 1983. The NSB had already released a draft version with the responses, which showed that just 45 percent of Americans had answered "true" to the first question, a percentage much lower than that in Japan (78 percent), Europe (70 percent), China (69 percent), and South Korea (64 percent); and that only 33 percent of Americans had answered "true" to the second question, compared with, for example, 67 percent in South Korea and 63 percent in Japan.

The man responsible for the deletion was the chapter's lead reviewer, John Bruer, a science philosopher focusing on neuroscience, cognitive psychology, and education. The decision took the White House, which had been shown the earlier, unedited draft, by surprise. Scientists were outraged. A staff writer at *Science* magazine, Yudhijit Bhattacharjee, wrote about how stunning a development this was:

> Board members say the answers don't properly reflect what Americans know about science and, thus, are misleading. But the authors of the survey disagree, and those struggling to keep evolution in the classroom say the omission could hurt their efforts. "Discussing American science literacy without mentioning evolution is intellectual malpractice," says Joshua Rosenau of the National Center for Science Education, an Oakland, California–based nonprofit that has fought to keep creationism out of the science classroom. "It downplays the controversy."

But did it? Consider the way the question was asked in a 2004 survey, the year of Bush's reelection, when creation fervor was running high in both the

United States and some European nations: "True or false, *according to the theory of evolution*, human beings as we know them today, developed from earlier species of animals." (Emphasis added.)

When asked this way, fully 74 percent of respondents knew the correct answer, a figure much more in line with the responses in other countries. This suggests that it's not that Americans are ignorant. They just sometimes disagree anyway. Because science is political.

The Antiscience Militia

To more fully understand the political nature of the battle, consider the story of Michael Webber, who stood looking out his office window over Austin, Texas, one day. His business attire—jeans, cowboy boots, and a button-down shirt—was not unusual for Texas, and neither was what he had just done. Webber and a group of other parents had just started a homeschooling cooperative, opting out of the public schools or, in Webber's group's case, augmenting them, with material that they believed was vital for children to know. This was Texas, after all, where religion holds a special place in the public dialogue.

But the cooperative wasn't formed to teach kids about religion—it was formed to teach them about sex. "We're turning that steady march of progress on its head by instead explicitly choosing the widespread indoctrination of stupidity," Webber said in a 2009 opinion piece in Austin's *Statesman*. The cooperative was formed because Texas lawmakers had passed laws cutting sex education from two six-month courses to a single unit of abstinence-only instruction.

According to a 2009 report, 94 percent of Texas schools, which at the time were educating more than 3.7 million students, were giving no sex ed whatsoever beyond abstinence-only, a curriculum that includes emphasizing that birth control doesn't work. "You think I want my daughters learning *that*?" Webber asks. Another 2.3 percent of schools had no sex education at all, and 3.6 percent had "abstinence-plus."

Instead of providing fact-based information about birth control and pregnancy prevention, the programs heavily emphasize the risks—which in reality are small—of sexually transmitted diseases leading to cervical cancer, radical hysterectomy, and death. The programs also underscore Christian morality. The report highlighted one Texas public-school district's handout, entitled "Things to Look for in a Mate":

How they relate to God
A. Is Jesus their first love?
B. Trying to impress people or serve God?

Another public-school district had a series of handouts that referenced the Bible as their basis to promote abstinence from sexual activity:

Question: "What does the Bible say about sex before marriage/premarital sex?"
Answer: Along with all other kinds of sexual immorality, sex before marriage/premarital sex is repeatedly condemned in Scripture (Acts 15:20; Romans 1:29; 1 Corinthians 5:1; 6:13,18; 7:2; 10:8; 2 Corinthians 12:21; Galatians 5:19; Ephesians 5:3; Colossians 3:5; 1 Thessalonians 4:3; Jude 7).

The results? Teen pregnancy in Texas went *up*. It was higher than it had been before the abstinence-only movement, and the 2009 Texas teen birth-rate, at 60.7 live births per one thousand teenagers, was more than 60 percent higher than the national average of 37.9 live births per one thousand teens. The National Vital Statistics Reports data for 2013 showed ratios that were nearly identical, although both the national and Texas birth rates fell. Even more troubling than the dismal failure of the abstinence-only approach, Webber says, was that repeat teen pregnancy also went up. It turns out that Texas kids thought, "If birth control doesn't work, why use it?"

One can't argue with their logic. The conservative Christian proponents of the abstinence-only approach convinced state education and school officials that, contrary to scientific findings, further information about birth control and sex should be taught by parents, according to their morality, in the privacy of their own homes. The thought of teachers talking with their kids about sex made them queasy. But the problem, Webber says, is that those same parents are often too bashful to bring the topic up with their kids at all—and if they do, they often use euphemisms. One girl Webber knew of didn't understand how she had gotten pregnant because she and the boy hadn't actually *slept* together after having sex. It's also extremely tough for teenagers to get contraceptives in Texas. "If you are a kid, even in college, if it's state-funded you have to have parental consent," said Susan Tortolero, director of the Prevention Research Center at the University of Texas, Houston.

Not only was the teen birth rate high in Texas compared to the US average, but the teen birth rate in the United States is by far the highest in the industrialized world. US teens are two and a half times as likely to give birth as teens in Canada, roughly four times as likely as teens in Australia, Germany, or Norway, and almost ten times as likely as teens in Switzerland. Among the more developed countries, Russia has the next-highest teen birth rate after the United States, but an American teenage girl is still around 25 percent more likely to give birth than her Russian counterpart, and 50 percent more likely than her counterpart in the United Kingdom, the next-highest industrialized nation after Russia.

Within the United States itself, the teen birth rate is not even. It is highest in the so-called Bible Belt states like Texas, which skew the national average higher. Teens living in the ten highest teen birthrate states of Alabama, Arkansas, Kentucky, Louisiana, Mississippi, New Mexico, Oklahoma, Tennessee, Texas, and West Virginia—all of them "red" states (with the exception of New Mexico) with Republican elected officials likely to support abstinence-only education—were, on average, more than two and a half times as likely to give birth as teens living in the ten lowest teen birthrate states of Connecticut, Maine, Maryland, Massachusetts, Minnesota, New Hampshire, New Jersey, New York, Rhode Island, and Vermont—all "blue" states. As a result, a Texas teen girl is near the top in the industrialized world for pregnancy risk, with risk equal to that of a girl living in Haiti. And she is more than forty times as likely to give birth as her Slovenian counterpart.

"Abstinence works," said Texas governor Rick Perry during a televised interview with *Texas Tribune* reporter Evan Smith. The audience laughed and Smith pointed out the state's abysmal teen pregnancy rate. "It works," insisted Perry. "Maybe it's the way it's being taught, or the way it's being applied out there, but the fact of the matter is it is the best form of—uh—to teach our children." Smith asked for a statistic to suggest it works, and Perry replied, "I'm just going to tell you from my own personal life, abstinence works." It was the same message Bristol Palin made roughly $1 million repeating as an abstinence spokeswoman after her first out-of-wedlock pregnancy—and before her second.

Webber and his wife chose to protect their daughters by forming a mother-daughter sex-ed cooperative. They brought in experts and armed the girls with what they needed to know. "But not everybody can afford to do that," he says. "In fact, I'll bet even of those who can, most won't. What I really don't understand is why these people are so insistent on instilling stupidity in the next generation. If it's not sex ed, it's climate change or evolution or vaccinations."

The larger question posed by the ironic reversal of parents homeschooling to teach science-based sex ed instead of the religion-based abstinence unconstitutionally mainstreamed in the public schools is, What is the relative value of science versus ideology in policymaking and education? That the question must be asked at all is a testament to the success of a movement largely fought by an armchair army of antiscientists that is loosely organized through evangelical churches and industry-funded astroturf groups. These activists cherry-pick science and misconstrue studies to create rhetorical arguments that support their values. The arguments are a form of amateur propaganda, designed to sound plausible to lawmakers who either don't understand that what is being presented to them is not science or, more likely, don't care, since their natural desire is to grease the squeakiest wheel.

Opponents of evolution, climate disruption, vaccines, birth control, stem-cell research, HPV vaccination, sex education, and other science issues are all using the same methods. Motivated by the sense of identity, belonging, and purpose they receive from these well-funded causes, thousands of laypeople are delving into geology, biology, immunology, paleontology, statistics, climatology, meteorology, geophysics, and oceanography, with the support of churches and industry-funded front groups who, like the Cornwall Alliance, preach a gospel of biblical fundamentalism mixed with a heavy dose of Ayn Rand, free-market economics, science denial, and anti-tax ideology which, when combined, serve the vested interests of wealthy church and business executives. This antiscience militia is aided and abetted by trained scientists and professors like Fred Singer, Willie Soon, David Legates, and Michael Behe, who supply a steady stream of pseudoscience that can be used by foot soldiers to sway the public debate.

Bad Reasoning

These ideologically motivated partisans generally make two arguments that sound plausible to average lawmakers or school-board members because they themselves use rhetorical arguments to navigate their daily lives. They also sound convincing, because they sound like science—but they're not. The arguments are the same whether the subject is climate disruption, evolution, vaccination, tobacco, or sex education.

The first argument is: *Lacking certainty, we should do nothing.*

On the face of it, this sounds prudent. Why take a risk unless one is certain of the need? We cannot be absolutely certain that climate disruption is occurring; therefore, we should do nothing. We cannot be absolutely certain that the theory of evolution is true; therefore we should not teach it. But science has never proposed absolute certainty—only the relative certainty offered by the preponderance of the evidence. That's why scientists use statistics and transparency in order to quantify the relative probability that a statement is true. It is its *lack* of certainty, in favor of rigorous honesty, that has made science so uniquely powerful.

The second argument is: *Since the conclusion is not certain, we should get a balanced perspective from both sides.*

This also sounds reasonable. If we can't say evolution or climate disruption is an absolute fact, then it seems reasonable to hear from people whose views differ, to weigh the pros and cons fairly, and to form a balanced opinion, as would a judge.

This is, once again, applying the rules of rhetorical argument to an issue of knowledge. Imagine that you have been to one hundred doctors. Ninety-seven of them have told you that you have cancer and that it is treatable if they operate now. But that scares you and it's expensive and you'll have to take off from work. Three doctors say there's some doubt. One of them is a homeopath and says he can cure it with water; another says you need more faith in the healing power of Jesus Christ; and the third is a hyperskeptic who says it might be an aberration in the MRI machine, it might be a cyst, it might be benign, it might be malignant but so slow-growing that it's not worth the dangers of surgery, we just don't know, and so on. Do you listen to the ninety-seven and get treatment, or do you listen to the three and wait?

Let's say you're still torn, so you decide to pick one doctor from the ninety-seven and one from the three and, like a judge in robes, you say, "Convince me." This is, of course, the current situation with climate disruption. "For what a man had rather were true he more readily believes," as Bacon pointed out at the beginning of Western science.

Science Is Absolute about What's False, Not What's True

These two arguments capitalize on the fact that, because legitimate scientific conclusions are empirical—that is to say, inductive and based on observation—they will rarely, if ever, make absolute statements without allowing for the possibility of error. Lack of absolute certainty is not a weakness in science—it's a strength. It's the only way one knows it *is* science.

Science philosopher Karl Popper wrote extensively about this distinction. Popper had grown up intrigued by the four great theories of the time in Vienna: Einstein's relativity, Marx's theory of history, Adler's individual psychology, and Freud's psychoanalysis. Each of these theories claimed to be scientific. Each was based on observations, and each was believed by its adherents to be "true." Popper was struck by how a Marxist "could not open a newspaper without finding on every page confirming evidence for his interpretation of history; not only in the news, but also in its presentation—which revealed the class bias of the paper—and especially of course what the paper did not say." Similarly, the Freudian and Adlerian followers "emphasized that their theories were constantly verified by their 'clinical observations.'" Popper could not think of any human behavior which could not be interpreted in terms of either theory. "It was precisely this fact—that they always fitted, that they were always confirmed—which in the eyes of their admirers constituted the strongest argument in favor of these theories. It began to dawn on me that this apparent strength was in fact their weakness." Popper noticed that Einstein's theory alone made predictions that could not be confirmed everywhere one looked; instead it took risks: it described situations in nature that, if contradicted by subsequent observation, would render the theory false.

Such a test of relativity had recently occurred, very dramatically, when Arthur Eddington had confirmed Einstein's prediction that a massive body like the sun could bend space, by observing how the position of stars appeared to shift when nearly behind the sun, during a solar eclipse. If Eddington had not seen light bend as predicted, Einstein's theory would have been falsified.

The distinction seemed critical to Popper. Any adherent of a theory—whether a political theory, a pseudoscience theory like astrology, or a religious theory such as the creationism that was then sweeping the United States—will tend to see confirmations all around, because the adherent is looking at the world through the lens of the converted. What made a scientific theory different, was that it didn't demand conversion; it was testable. It was the boldness or riskiness of its real-world predictions that made it rigorous. And that's what made it science. As Popper put it, "the criterion of *the scientific status of a theory is its falsifiability, or refutability, or testability.*" While science can rarely absolutely say that something is true, it can absolutely say that something is false. One black swan can prove that not all swans are white. It is absolutely false that the world is flat; we can prove it by going around it. All of our knowledge derived using science stands exposed to the same test of falsifiability: it is stated in such a way that an experiment could prove it false. That none has so far gives us high confidence that a given proposition, like the theory of evolution, is probably true, and we quantify that probability repeatedly using statistics. The more often it has been tested and survived, the more likely it is to be a fully objective truth. But this sort of confidence is not an article of faith—just the opposite. It's an article of what survives after the rigors of doubt and scrutiny.

Just the opposite is true of such theories as astrology. Astrology is based on extensive observation of both the stars and the details of a person's history, but its predictions are stated so imprecisely as to take no risks whatsoever, so it can never be rigorously tested. "A theory which is not refutable by any conceivable event is non-scientific," Popper wrote. "Irrefutability is not a virtue of a theory (as people often think) but a vice." This is how we bottom the argument out, as Locke suggested, in natural law and physical reality: by only stating what we know from observation and what we can tie back to observation—but, as Popper noted, with testability, not with certainty. Because we lack absolute certainty, we are required to assume responsibility for ourselves and for the decisions we make as a result. This is sometimes uncomfortable, but it is what both science and democracy are all about.

Espousing antiscience—that the absence of absolute certainty is grounds for inaction or doubt, or that we should then hear out all perspectives equally—is evidence of either a lack of understanding of the most basic tenets of reason, or of motivation by an unreasonable agenda. Such people should not be entrusted to make decisions that have serious impacts upon others.

The Roots of Partisanship

The adoption by schools of abstinence-only sex ed despite its dismal results highlights a central question about American values: Which is more important in education—adherence to our ideological perspectives as parents or the approach the evidence suggests for our children? The same question hangs over the ominous consequences of our decisions about creationism and climate disruption. Let's say that one is a political conservative who sees all around confirmation of the intelligent design of life by God, and who rejects the notion that humans are somehow changing the climate. Let's say one's children will be killed if one is wrong. Is it morally defensible to insist on an ideologically satisfying position without demanding it be testable, at the potential cost of our children's lives? Or is it better to go with what can be tested, and to equip our children with our best knowledge about what nature really says on the subject? Which will likely give them the most competitive edge? The greatest moral and ethical development? The most intellectual rigor? The greatest personal humility? The highest level of honesty? In short, the best chance at success?

These questions are critically important. But this moral conflict can only occur if we accept uncertainty as a good thing, and if we value the rigors of honesty and personal responsibility instead of certainty. Evidence suggests that this is a difficult thing to do if one has become wrapped up in the team-sports mentality of a religious, pseudoscientific, or political ideology. Partisanship thus works against enacting public policy based on sound science. How can we get past it?

First, let's look at what we know about why partisanship is so powerful. There are several factors we're beginning to understand, and the first is called motivated reasoning. When presented with evidence that confirms our beliefs and conclusions, we tend to accept it uncritically. When presented with evidence that contradicts those same conclusions, however, studies show that we tend to subject it to withering scrutiny, ignore it, argue with it, or try to intimidate its proponents, much as the opposing counsel at a trial. Motivated reasoning is fundamentally nonscientific. It's not about finding truth or creating knowledge; it's about winning an argument. It's also how we navigate daily life.

The foundational work in this field is a 1979 paper by Charles Lord, Lee Ross, and Mark Lepper. These Stanford University psychologists recruited a group of subjects, half of whom said they supported the death penalty and half of whom opposed it. They showed each participant one research study that supported the idea that the death penalty deters violent crime and another that contradicted it. The result: each participant identified extensive methodological problems with the evidence that contradicted his or her preexisting opinion, but didn't critically examine the evidence that supported it.

Eugenie Scott, emeritus head of the National Center for Science Education, says she encounters the same sort of motivated reasoning when discussing evolution. "But people are not black or white in their views," she says. "In reality, they distribute along a continuum," with the vast majority of people falling somewhere in the center. It's only when they are forced to choose sides that the continuum collapses into partisanship.

In 2014, Duke University researchers found something similar: when we don't like the politics of the proposed solution to a problem, we are more likely to deny that there is a problem in the first place. They called this motivated denial "solution aversion." We see this sort of motivated reasoning being used by religious conservatives to also deny the problem of climate disruption.

The Partisan Brain

The second factor that influences partisanship has to do with the way our brains process scientific information. Neuroscientist Sam Harris and his colleagues published a study showing that people use the same brain region in the ventromedial prefrontal cortex for belief as they do for ordinary facts. Danish cognitive neuroscientist Uffe Schjødt and his colleagues took this a step further by exploring how the brain responds to authority. Schjødt played recorded prayers to study participants and compared the responses of participants who were charismatic Christians—who believe in speaking in tongues and healing by prayer—to those of nonbelievers. Both groups were asked to listen to three different groups of recordings: prayers read by a non-Christian, prayers read by an ordinary Christian, and prayers read by a Christian known for his healing powers. In actuality, however, all the prayers were read by ordinary Christians.

The researchers used functional magnetic resonance imaging (fMRI) to scan participants' brains as they listened. The scans showed that when the Christian subjects were listening to recordings they thought were made by healers, who have special religious authority, they turned off parts of their medial and dorsolateral prefrontal cortices, which play key roles in critical thinking and skepticism. Nonbelievers, in contrast, did not shut down this brain system.

The authors said this may explain why certain individuals who are perceived to have authority—such as politicians, religious leaders, celebrities, and the news media—can exert influence over others. They suggested that the effect could extend to other interpersonal relationships, such as parents and children, doctors and patients, teachers and students, producers and consumers, and leaders and followers. It is also quite similar to the "field-dependent" style of learning that educators talk about.

Were the Christian subjects idiots or gullible? No. What the fMRIs showed is that the preconceived authority of the speaker, regardless of his or her actual expertise, starts a brain process—a rhetorical frame—that makes us less critical of the speaker's commentary. Thus, partisanship is born. The brain uses it as a strategy for negotiating the environment and living in social groups.

Bad Things in a Just World

Beyond mistaken reasoning, motivated reasoning, solution aversion, field dependence or independence, and a predisposition not to question authority, people are often politically and psychologically motivated by their relative sense of justice. Societies are full of injustices, from wealth disparities to racism and sexism. People react differently to these injustices. Some react with moral outrage and seek to change society. Others react with blame or dismissal of the victims. Still others, like the residents of the antebellum US South or apartheid South Africa, adopt belief systems that justify the situation, in effect changing their intrinsic definition of justice to rationalize society's injustices.

Americans in particular have a high level of what social psychologists call just world belief, perhaps in part because of the country's unresolved legacy of slavery and the gross affront to core American ideals that legacy presents. Not surprisingly, South Africans also score higher than, say, Brits on this scale. Interestingly, Canadians also have high levels, about 66 percent more just world belief than their counterparts in Britain, as one study showed.

All of this helps explain why members of certain countries may hold certain beliefs, like having trouble accepting scientific evidence of environmental problems like climate disruption. The reasons why are understandable at a grade-school level. Many people tend to believe that the world is inherently just: the wicked are eventually punished, the good are rewarded, and problems are corrected. In other words, they believe that people get what they deserve. The relative strength of this belief correlates with differences between those who believe that self-discipline, hard work, and personal responsibility lead to success, versus those who believe that, while those things matter, people are more at the mercy of luck, birth, or other factors beyond their control. The former view is a treasured part of the American ethos, and Americans as a whole—whose country was based in no small part on racial subjugation, first of Native people and then of imported slaves—have a much stronger belief in a just world than, say, Europeans, who tend to be less idealistic, more cynical, and more likely to believe that luck rather than merit plays a significant role in a person's circumstances.

Climate change and other environmental problems seem to violate the just world belief. The idea that despite our best efforts our fate is influenced by luck or the collective actions of others—or that our society is guilty of gross injustice that we are partaking in—is antithetical to the classic American ideals (and, to a certain extent, Canadian ideals) of self-determination and meritocracy. Thus, many Americans and, to a lesser extent, Canadians, tend to react like whites in the antebellum US South, either denying that the problem exists or changing their thinking to justify it.

Research shows that this conflict makes it more difficult for Americans and residents of other countries higher in just world belief to accurately assess personal responsibility. For example, the tendency to blame the victim, which is unusually high in Americans, is an effort, psychologists say, to maintain the just world belief that people get what they deserve. In her elegant and powerful memoir *After Silence: Rape & My Journey Back*, writer Nancy Raine describes how she was raped in her apartment by a stranger who sneaked in while she was taking out the trash. Afterwards, friends and colleagues suggested that she was partly to blame because of her "negative attitude" that might have "attracted" more "negativity," or because she had "bad karma" from a previous life. A close friend told her she had "asked for it" by choosing to live in that particular neighborhood. A crime victim must have been in the wrong place; a sick person must have done something to deserve it; black slaves benefit from white care and protection; as do Native people living in the "domestically dependent sovereign nations" of US Indian reservations, and high poverty levels among, say, Native people have nothing to do with structural issues depriving them of home mortgages or the ability to levy property taxes, but rather have to do with alcoholism and other issues that are their own fault.

If we believe we are individually responsible for our circumstances, and that is a core American ethos, this sort of prejudice makes sense. Although, in America, degrees of just world belief span the political spectrum, they do correlate to some degree with ideology. Strong just world believers tend to be more economically and politically conservative, to see the status quo as desirable, to believe in an active God, to be less cynical, to be less socially and politically active, and, interestingly, to more often be Protestant, possibly because of the Catholic Church's emphasis on serving the poor, which seems to contradict the just world belief of self-determination. This correlation can run headlong into science when science fails to affirm the ethos of unlimited individual freedom and personal responsibility in favor of data suggesting that we have limited resources, that behavior has collective effects, that conditions such as drug addiction may limit control over decision making, that the problems of Native people may be more due to structural legal inequities than they are to inferiorities or alcoholism—and that preaching the self-control of abstinence to teens may make parents feel better, but may not be the most effective way to reduce teen pregnancy.

University of California, Berkeley, psychologists Robb Willer and Matthew Feinberg published a study showing that the just world belief influences how people respond specifically to the threat of global warming. The presentation of "doomsday" scenarios, as conservative Sarah Palin characterized them— the demise of the polar bear, the forced emigration of tens of millions of people, flooded coastal cities, massive wildfires, droughts and floods, devastating hurricanes and tornadoes, death and destruction—have become a common trope in the battle to curb greenhouse gases. They are employed because they are dramatic, which, it is assumed, will get the public's attention and provide motivation.

But does this approach really work? Willer and Feinberg's study found that it might not. Apocalyptic messaging, particularly if it emphasizes that innocent children will suffer from the dire effects of global warming, may actually galvanize *resistance* to the message, increasing skepticism and reducing efforts to minimize carbon impacts, because the idea of climate disruption deeply offends the just world belief in self-determination and control, and because it make parents personally culpable.

Thus we see increased use of SUVs and flagrant energy consumption as an in-your-face rejection of the offending message. "I think the evidence we've seen is pretty clear," says Willer. "The research underscores the importance of keeping in mind that people have a deeply seated belief in the world as just, fair, and stable. Scientists should steer clear of apocalyptic messaging, or if they use it they need to present concrete solutions that suggest ways to control the outcome. Even people high in belief in a just world can handle these extremes if there's a concrete solution. But without that, they find it paralyzing and are motivated to disregard the message."

Using Conservative Messaging to Bridge the Divide

When just world belief is coupled with a high level of patriotism, this effect seems to be multiplied, Willer and Feinberg found in a follow-up study. "Conservatives are on average more patriotic," says Willer. "One thing that sets up is a great deal of cognitive dissonance when it comes to global warming. You think [my country] is great, you know it's a greenhouse-gas emitter, and then you're told that greenhouse gases are bad for the world." They found that if we experimentally increase people's patriotism, their belief in global warming tends to go down.

In other experiments, Feinberg and Willer found that liberals moralize environmental issues and conservatives don't. They wondered, "What if you tried to make conservatives think of global warming as a moral issue? Go to a moral foundation conservatives respect but liberals don't?" They found that conservatives are more likely to talk about moral issues in terms of purity and disgust: they want to keep things pure and sacred, and condemn what disgusts them. For example, Willer says, they believe that "homosexuality is wrong because it's gross."

Michele Bachmann offers an example of this messaging style. In a March 2004 radio interview, Bachmann urged conservatives to attend a rally to ban gay marriage at the Minnesota state capitol. "We will be beseeching the Lord," Bachmann said. "Our state will change forever if gay marriage goes through. Little children will be forced to learn that homosexuality is normal and natural and perhaps they should try it. It will take away the civil rights of little children to be protected in their innocence, but also the rights of parents to control their kids' education, and threaten their deeply held religious beliefs. . . . This is a very serious matter, because it is our children who are the prize for this [gay] community, they are specifically targeting our children. . . . The sex curriculum will essentially be taught by the local gay community."

In 2012, Minnesota was the first state in America to reject an attempt to constitutionally ban gay marriage. Months later, in 2013, it passed a law permitting it. Richard Carlbom, campaign manager for Minnesotans United for All Families, the effort both to fight the amendment and to pass the law, said that the group worked to have as many people as possible come out of the closet, then had them and—especially—their straight friends and family members reach out to voters through thousands of phone calls. This wasn't just a strategy Carlbom pulled out of a hat. Knowing all the prior campaigns had failed in dozens of US states, he turned to psychologist Phyllis Watts. Using social-science research, Watts developed the group's signature technique: asking voters why they got married and then building a common emotive bridge, saying, That's why I (or my daughter, my cousin, my friend, my brother, and so on) want to get married too. The strategy focused on love, empathy, and shared humanity instead of talking about "rights" and framing the debate in terms of militancy and division, techniques which had failed in prior votes. In TV ads, rather than using gay people, the group used straight people talking about their gay friends and relatives, emphasizing family values, and why it was important to them, using the force of familiarity and friendship and empathy instead of fear and purity to win the battle. Scientists would do well to devise a similar bridging strategy based on empathy and our common bonds.

To test what they had observed about purity-versus-disgust messaging,

Willer says, he and Feinberg did something similar. They spoke in conservative language, making a fake environmental advertisement that essentially said, "Now more than ever, it's so important to protect our sacred mountains and our rivers from desecration." The ad also talked about impurities in the environment entering our bodies. It had images of people drinking dirty water. "What really surprised us was that it closed the gap between liberals and conservatives," Willer says.

The Deficit Model:
The Zero-Sum Economics of Science

Faced with this constant stream of emotion-laden messaging that apparently uses neural pathways similar to those used by the science frame and crowds it out, one can't blame scientists for simply wanting to throw up their hands. In the 2006 documentary *Flock of Dodos*, evolutionary biologist and filmmaker Randy Olson asked a group of scientists gathered around a poker table about the creationism-versus-evolution debate. One of them responded by saying, "I think people have to stand up and say, You know, you're an idiot."

An emotional response like that may feel good, but it's likely not going to be a successful long-term tactic. Again, think of the bridging campaign strategy employed by Richard Carlbom in the gay-marriage battle in Minnesota. The scientist's comment also highlights the different ways scientists and the lay public often think about these issues. Scientists are trained to avoid motivated reasoning, the "vulgar Induction" Bacon warned against, and let the chips of reality fall where they may. They prize this intellectual honesty because the stakes for them are very high. They know how value judgments, prejudices, and habits of thought can blind one to the truth—which, in turn, will limit or end one's career as a scientist.

The lay public does just the opposite. They form frames of reference, prejudices, and value judgments as guides for navigating life, and then make rhetorical arguments to get what they need. The idea that the ignorant public just needs to be better educated in order to see the light is called the deficit model—the assumption by scientists that the public thinks the same way they do, and therefore that the public's differences with science are because of a knowledge deficit. If that's the case, it makes perfect sense for scientists to simply try to pour in more knowledge—that is, fill the deficit—to win support and eradicate the willful inculcation of stupidity that Michael Webber bemoans in Texas.

But what if the problem is not that the public lacks knowledge? What if the problem is that the public has the knowledge—or, at least, access to it—but rejects science anyway? A 2008 Pew Research Center survey found that that is exactly what is happening. Thirty-one percent of non-college-educated Republicans accepted the scientific consensus on climate change—a surprisingly low number compared to the public at large. But what was even more surprising was that, among college-educated Republicans, that number *fell*, to 19 percent. The proportion was the opposite for Democrats—an increase from 52 percent to 75 percent. The more educated Republicans were, the less they believed in climate change. They cannot be said to be suffering from a knowledge deficit. As climate disruption has become politicized, Republican support has collapsed.

What could be happening? Most likely, college has given Republicans the permission to question authority and to be skeptical of claims by authority, which they have used to confirm their political beliefs. We are inculcating the attitude of skepticism without teaching the skills of evidence gathering and critical thinking needed to discern what is likely true.

The issue can be seen as a microcosm of the public's relationship with science on a host of looming problems, and it's got nothing to do with lack of knowledge. It has to do with the public using motivated reasoning instead of a scientific approach to evidence—a view encouraged by the subjectivist aspect of postmodernist education, and by scientists not understanding how to bridge the communication gap.

Science Impotence

Towson University psychologist Geoffrey Munro showed in a 2010 study how high the stakes really are in this discussion: he found that people who are pushed into antiscience positions in one area tend to generalize that partisanship to all of science. This may in part be because of the framing neuropathways identified by Uffe Schjødt and Sam Harris. Like Lord, Ross, and Lepper, Munro asked subjects a politically loaded question: "Do you believe that homosexuality is associated with mental illness?" He separated them into one group that said yes and another that said no. He then presented each group with fake scientific studies, half of which presented conclusions that confirmed each group's belief, and half that contradicted it.

Munro then asked the subjects to evaluate the scientific validity of those studies. Regardless of their position, those who were presented with fake studies that confirmed their belief tended to ignore the studies' weaknesses, while those who were presented with fake studies that contradicted their belief were much more adept at finding them. More importantly, they were also more likely to come to the conclusion that science was powerless to settle the question, which Munro called the scientific impotence hypothesis.

Munro next asked the subjects questions about their opinions on other sociopolitical issues on which science could offer definitive information. He found that subjects whose belief had been contradicted by science now felt that science was less likely to be able to solve any issues at all. They had generalized their sense of scientific impotence to all of science. They had essentially been radicalized, set on the path toward an antiscience, antirationalist worldview.

"When a person holds a belief," Munro says, "especially a strong one that is linked to important values (e.g., some sociopolitical beliefs), information threatening that belief creates inconsistency in the cognitive system that threatens one's self-image as a smart person. This produces an unpleasant emotional state."

This has tremendous public-policy and political messaging implications in an age dominated by science. It also implies that, if we wish to improve science education among schoolchildren and college students, we need to understand that some misconceptions about science are the result not of a knowledge deficit but of belief resistance, and to devise ways to short-circuit these processes.

That belief resistance—and this is a critically important point—is largely coming from adults. This is why education is political in the first place, and why the children of scientists are the most likely to become scientists—an effect that is slowly striating society into knowledge haves and have-nots.

Beyond scientist parents, the other major predictor of a child's success in science is having immigrant parents. Fully 70 percent of the finalists in the 2011 Intel Science Talent Search were the children of immigrants—a stunning figure considering that immigrants make up only 12 percent of the US population, and a troublesome one considering the restrictions that have sharply curtailed immigration since 2001. "Our parents brought us up with love of science as a value," said David Kenneth Tang-Quan, one of the finalists.

The Obama administration prioritized placing renewed focus on improving science, technology, engineering, and math education following a somewhat traditional deficit model approach. This is all well and good. But evidence suggested that it would have little effect in transforming the science readiness of children or their ability to compete in a science-driven global economy. That's because it didn't address the larger problems, something I tried to impress on

Obama's Office of Science and Technology Policy to little avail. I said: It's not about the kids, it's about their parents, and it's not just about a knowledge deficit, it's about politics. Transform the parents and you transform the system. This is something that antiscience conservatives know very well—witness sex education in Texas.

By May 2010, this was becoming more apparent. The National Science Teachers Association did a survey that found that 53 percent of American parents didn't feel equipped to encourage their kids in science or to discuss science with them. "Science education has been identified as a national priority, but science teachers can't do the job on their own," said NSTA executive director Francis Eberle. "They need the help and support from key stakeholders, especially parents."

When they were asked what they thought prevented parents from being more involved, 77 percent of the teachers surveyed said they believed that parents don't feel comfortable talking about science with their children. The teachers guessed that part of the problem might be a lack of resources and community involvement, and parents said it would help if communities had science museums. Half of science teachers said that, as it stands now, parents don't have access to materials or community resources that would encourage their children's interest in science. But those solutions are, yet again, taking the responsibility out of parents' hands and putting it in the hands of science museums and other community resources or educational materials. They also ignore the plethora of science information that is freely available on the Internet.

Celebrating Like Maoists

Adding to the emotional factors that drive antiscience partisanship is the celebration of anti-intellectualism, pushed both by the religious right and, more recently, by libertarian climate deniers who begin their sentences with "I'm not a scientist, but . . ."

It is a strategy commonly employed by authoritarians from Hitler on the right to Mao on the left, both of whom demonized intellectuals and had followers who banned and burned books. Mao even went so far as to close most of China's university system during the Cultural Revolution, when every thought and act was to be interpreted in the ideological light of Maoist thought—a history that made the Canadian Harper government's closing of many science libraries and the discarding of much of their contents so alarming to scientists.

In 1968, for example, the national Study Group of Mao Zedong Thought was organized, which denounced many scientific theories, starting with Albert Einstein's theory of relativity. Acupuncture had been banned as unethical in 1822 by the Great Imperial Medical Board of China, but Mao restored it during the Cultural Revolution, giving it equal weight with Western medicine and training peasant "barefoot doctors" to use it in order to disguise the fact that China had so few trained medical professionals. It is interesting to note that today such "traditional" and "alternative" health practices are largely antiscience issues for the political left. Biologist and leading sinologist Joseph Needham wrote a 1978 report for *Nature*, titled "Science Reborn in China: Rise and Fall of the Anti-Intellectual 'Gang,'" in which he described a pathology professor who had been forced to "lecture on carcinogenesis to medical students while they were picking cotton."

People are naturally skeptical, but they are also social. They don't like to think of themselves as antiscience, ignorant, or stupid. But when those qualities are put in conflict with social identity and reinforced by authority, skepticism can be suspended. This bottom-wing authoritarian activity happens more often on the right with religion, but it also happens with identity politics on the right and left.

The right's vocal celebration of anti-intellectualism (Al Gore was "an intellectual" while George W. Bush was "a guy you'd like to have a beer with") recalls the attitudes of the Cultural Revolution and casts the argument in terms of the lone, other, cold intellectual versus us social people having fun together in a group. This shifts it from *thought* to an *emotional and social identity* level that is then reinforced by authority figures, who repeat the message in the media. It is the emotion and the political identity of the messenger or authority figure that causes us to accept or reject the message, even if the message is the solution to our problems.

This shift has been powerful enough to affect US media coverage of events. An example is the October 16, 2006, coverage that NBC News anchor Brian Williams gave to the momentous event of the US population crossing the three hundred million mark. "Tomorrow morning at 7:46 a.m. Eastern time—and don't ask us how they estimate it—the United States population will click over to three hundred million," Williams said. But isn't reporting "how they estimate it" a pretty big part of the story? Williams seemed to simply be pandering to anti-intellectualism. Figuring out "how they estimate it" wasn't really all that hard, as Mark Strassmann showed in CBS's coverage: "Every 11 seconds, America moves one person closer, and number three hundred million could come by birth, by oath as a legal immigrant, or by stealth: someone sneaking into history." Even the United Kingdom's *Guardian* newspaper made it look easy: "Nobody knows the precise second at which the US will cross the 300 million mark, though the time given next Tuesday [9:01 and 48 seconds in the morning] is the literal interpretation of US census projections."

As inventor Dean Kamen says, "We get what we celebrate. If we celebrate actors and celebrity, we get the balloon boy and stupid people acting out to get on reality shows. If we celebrate sports, we get a bunch of kids wearing jerseys, but how many of them will actually become millionaire sports heroes? What if we celebrate science and engineering with that same adoration?"

Kamen's inventions range from the Segway Personal Transporter to robotic prosthetic arms and high-tech portable medical devices. But he regards his biggest accomplishment as forming FIRST (For Inspiration and Recognition of Science and Technology), a nonprofit organization that sponsors an annual competition in which more than 250,000 kids try to build the best robots. Each year, the final competition fills a major football stadium. "When you get successful like we have been in this country," Kamen says, "it's easy to get complacent and lazy and preoccupied with sports and entertainment and nonsense while the rest of the world looks at America enjoying the wealth created by our great-grandparents. The rest of the world is moving even faster than ever before. On a comparative basis, we'd still not be keeping up. Super Bowls are amusement; superconductors can change the energy outlook of the planet. I don't need a president I want to have a beer with. I need a president that takes on the crap and fights for our kids to succeed."

The celebration of anti-intellectualism offers social identity belonging that allows people to let go of their shame of their scientific ignorance, a sort of group bravado in the face of what would otherwise be stigmatized. This sets up a political and social atmosphere in which people can be sold ideas without any grounding in facts or, for that matter, in morals or ethics.

An Alternate Universe

This effect is now moving online with the establishment of sites such as Conservapedia, a "conservative" version of Wikipedia that Andy Schlafly, the son of conservative Catholic activist Phyllis Schlafly, founded because he views the popular Internet encyclopedia as having a liberal bias. Conservapedia seeks to create an alternate intellectual universe by reinterpreting and challenging facts that don't fit its readers' ideology. It also provides rhetorical ammunition to build pseudoscientific arguments. For example, hearkening back to the ideological objections that Nazis and Maoists had to Einstein's theory of relativity, Conservapedia stated, "The theory of relativity is a mathematical system that allows no exceptions. It is heavily promoted by liberals who like its encouragement of relativism and its tendency to mislead people in how they view the world."

The site identified thirty-four "counterexamples" that purported to show that the theory of relativity is incorrect, including "action at a distance by Jesus, described in John 4:46–54"; "despite wasting millions of taxpayer dollars searching for gravity waves predicted by the theory, none have ever been found. Sound like global warming?"; "it is impossible to perform an experiment to determine whether Einstein's theory of relativity is correct, or the older Lorentz aether theory is correct. Believing one over the other is a matter of faith"; and "in Genesis 1:6-8, we are told that one of God's first creations was a firmament in the heavens. This likely refers to the creation of the luminiferous aether."

With alternate reality sites like these, which place political identity and ideological tests over reality in the ultimate right-wing postmodernist approach to subjectivity, it's easy to see how journalists like Brian Williams and Bill O'Reilly become confused and allow narrative to triumph. After all, as both liberal postmodernists and neoliberal conservatives argue, the winner writes the history books.

But the battle between social identity and natural skepticism also shows that people never completely give up their respect for science. Because of this, convincing people to totally adopt antiscience positions requires constructing elaborate explanations to get around it, like postmodernist teachings, climate-science denial, or the alternate right-wing universe of definitions exemplified by Conservapedia—all of them finely constructed rhetorical arguments, but none of them real.

Framing Science

If it's true that our brains process facts and Locke's "but faith, or opinion" in essentially the same way, how can one ever hope to break through? In pondering this question I began to think about times when I've successfully gotten around it, when, say, I was dealing with an antiscience heckler when giving a speech. I have been successful when I have emphasized the scientific process, which changes the frame of reference from an authoritarian assertion to an antiauthoritarian exploration of the senses and intellect: "Look, see it yourself?" Breaking it down, step by step, to the observations, the tests, and the (then) obvious conclusions. Most antiscience arguments are prepackaged and just skip over the surface of science. Digging into just how we know things pulls away the curtain in a way that makes denial much more difficult. But it must be done with respect. Again, this is quite similar to the social psychology approach used so successfully by the Minnesotans United for All Families campaign in that it opens up the conversation and invites people in, to participate with science.

Prominent social scientist Matthew Nisbet and others have built up evidence that supports this. He suggests that people think about science as a "way of knowing" or "worldview" or "frame," and that they think about religion as another "way of knowing" or "worldview" or "frame." This would make sense if, as the neuroscience suggests, we use the same or similar brain centers to process knowledge and matters of opinion/religion. If that's the case, we must choose which realm we will favor at any given time, and that is a framing question. When people are forced to choose between the two frames, the conflict will skew their apparent level of science literacy lower because they tend to choose emotional, religious, or political frames over intellectual ones. Scientists will say this doesn't make sense. The facts are the facts. But people do not consider only facts when making decisions, when investing in the stock market, when choosing a mate, when buying a house—or when voting.

This is an important difference between scientists—who are trained to set aside emotions, assumptions, and ideological predispositions and to adjust their worldview to a careful, detailed consideration of the evidence—and everybody else.

The desire for success in science instills the values of honesty and integrity, which are impossible to fully adhere to when making a rhetorical argument whose purpose is to win by a different standard. While many scientists make observations, think about the implications, and then respond emotionally to those insights about the world, most nonscientists use persuasion to influence others to get what they want. They apply similar strategies to themselves, forming beliefs and principles that help them navigate difficult situations and relieve stress, which are, at their cores, rhetorical strategies. While a scientist seeks power through knowledge, a nonscientist seeks it through persuasion—the "vulgar Induction."

Drawing a Line in the Sand

Facing these challenges head on will inevitably increase stress and raise ire—and, Nisbet argues, it can lead to even further polarization. This in turn alters perceptions and starts a feedback loop. There is "lots of research," Nisbet says, that shows that people who have strong feelings about an issue tend to have difficulty estimating the proportion of people on their side versus the opposition, and they see any news coverage as hostile to their own interests. Journalists play on this, he says, using the culture-war narrative.

As screenwriting guru Robert McKee says, "Conflict lies at the heart of all stories." Journalists are trained to tell stories. They look for conflict to find an angle. The culture-war narrative is conflict on a grand scale. It's journalistic porn—cheap and easy to write, and it stimulates visceral responses that remain the same from issue to issue but carry little intrinsic meaning.

As evidence, Nisbet offers a 2010 *New York Times* article about how antievolution advocates were beginning to gain momentum by linking creationism with opposition to climate science and human cloning, wrapping them all up in a "tarball" of antiscience thinking. The *New York Times*, he says, tends to savor these culture-war stories, which are usually heavily e-mailed among scientists and liberals. A story like that gives no indication of its subject's real strength or threat as a movement, but it sets the reader up for an instant leap in logic: science is under siege and we need to respond!

> This inevitably follows a pattern: Exaggerated perception of the threat. Chatter. Online discussion. People will write opinion pieces. And that gets the other side going. One of the interesting things is that this constant conflict narrative has a quality of a self-fulfilling prophecy. People look to news coverage to make sense of their own opinions. If evolution is constantly portrayed as a conflict of religion versus science, people take that as a cue on what to believe. The more you perceive this conflict in news coverage, and use it as a key to make your own opinion, to that extent your own opinion tends to become more extreme.

This is, of course, exactly what occurs on the right with AM talk radio and cable-news shows. Another example comes from PBS's *Frontline* series. In 2010, the program produced a show on the antivaccine movement and called it "The Vaccine War." There's no war. The press release about the show cast it as doctors and scientists against parents. This suggests there are only two positions: health advocates versus parents. What would many parents tuning in naturally choose? Because they don't identify with the scientists, they would side with the antivaxxers. They would be driven to a more extreme position.

Could it be that we enjoy this conflict? Clearly it sells newspapers, radio shows, television dramas, and political careers, and, at least in my case, books. But is it fair? What is so enticing about this narrative is that it breaks the world down into us versus them. We are right, they are wrong, and knowing that I am not alone in this difficult time makes me feel better. This is of course the baby boomer's life mantra, but it is by no means limited to one generation. The problem is that it is a sideshow. It doesn't actually attempt to deal with the situation by envisioning a new future.

Defusing the Debate: A General's Perspective

Science education advocate Eugenie Scott deals with this kind of polarization on a daily basis. Scott is executive director emeritus of the National Center for Science Education, and has been at the forefront of the legal and policy battles against teaching creationism in public-school science classes for twenty-five years. As Scott and I speak, she pulls out a huge map of the roughly fourteen thousand independent school districts in the United States.

"Now realize," she says, "that each of those fourteen thousand school districts is controlled by a locally elected school board, whose members belong to the local churches and the local chambers of commerce, and who usually set curricula for science classes, usually with little knowledge except what they hear in church or on the radio or in newspapers." They are often committed and caring community leaders who lead busy lives, but typically few are scientists—partially because scientists have, for two generations, been civically disengaged.

Because of this, there are rarely science-educated voices guiding these discussions, and they can get wildly off the mark. Scott has no interest in the "culture wars," even though she is on what could be termed the front lines, since the term tends to cause people to dig in. Instead, she seeks to build bridges by speaking about human origins not as either-or, but rather as falling along a spectrum from evolutionary atheists at one end to young Earth creationists at the other, with most people falling in the middle.

"I think that people should understand science well enough that they don't confuse it with antireligiousness," Scott tells her audiences. "Science is a really good way of knowing about the natural world. It doesn't make any pronouncements about how you should view ultimate reality."

Nisbet says this kind of bridging language helps to break down polarization. He suggests that responsibility for the heightened sense of conflict is not limited to evangelical creationists. Within the community of scientists, Nisbet says, there's a movement of militant atheists, including authors Richard Dawkins and Sam Harris, science blogger P. Z. Myers, and others, who have contributed to the polarization of science.

The Battle's Front Lines

Myers, of course, disagrees. In 2006, the science journal *Nature* listed his blog *Pharyngula* as the top-ranked blog written by a scientist. By 2011, *Pharyngula* had some 2.5 million views per month from about one million unique visitors, rivaling traffic on the websites of some metropolitan newspapers. The blog is named after Myers's favorite stage in embryonic development, the pharyngula stage, in which all vertebrates, including humans, have close similarities, such as a tail and gill arches. A brilliant and popular biology professor at the University of Minnesota, Morris, Myers is known for his sarcastic, incisive, and entertaining style. And it is precisely his outspokenness against antiscience evangelicals that made him so popular.

"[Science journalist and ScienceDebate cofounder] Chris Mooney calls what I do 'the conflict frame,' and he thinks it's bad for science," says Myers. "But conflict is something to talk about. A novel that did not have a narrative, a central conflict, would be boring. Just because I draw a line in the sand, it doesn't mean I expect 100 percent of the people should abandon their faith. That line just gives us a focus."

There is some truth to this. As noted earlier, drama is one of the four horsemen of entertainment, and the narrative or rhetorical argument is the strategy most people use to navigate the world. History shows that social progress is often made through a culture war of conflicting narratives, whether the movement is emancipation, civil rights, women's rights, reproductive rights, or LGBTQ rights. This is confirmed by biologist Simon Levin's findings on opinion dynamics. As mentioned earlier, Levin studies grazing patterns in herd animals, who vote with their feet—or their fins. "It's not the fish that are moving in the right direction that are able to win; in some cases it's those who are the most stubborn, and not responsive to others, that end up controlling the opinion dynamic," Levin says. "Look at George Bush." The opinion tends to move in the direction of those who are the *least* flexible.

This all reminds me of an area we've already discussed—the rise of Nazi authoritarianism. A major debate in America and the United Kingdom at the time was appeasement versus engagement. Levin's research suggests that Myers's principled stand may be the most effective approach if it is broadly communicated and articulated with conviction and leadership. Authoritarians never go down without a fight. Authoritarian and intolerant tendencies like racism, for instance, are held in check in mainstream society largely through force of sanction and stigma, often with legal reinforcement, as well as through empathy and human connection—in other words, the peer review of the crowd. But what makes some views successful and others not? Without science I would suggest it comes down to the moral authority of the argument and the forcefulness, shamelessness, and articulateness of its delivery. The confidence of passionate, visionary leadership.

Nisbet argues, however, that Myers's approach only further polarizes debate. As the most politically vocal opponents of creationism, militant atheists have become stand-ins for the broader science community and contribute to the conflict narrative in news coverage. "The same type of identity formation and opinion extremity has happened on climate change as on evolution," Nisbet says. "It's always stated in binary terms: you're either a supporter or a climate denier. No in-between. It's become wrapped up in Democrat versus Republican in terms of political identity, as a result of activists using these issues to promote their own goals and media using the issues to tell the conflict narrative and sell newspapers and radio ads."

This also seems true, and is likely a function of the atheists confusing two battles: the battle against radical fundamentalism as represented by creationism, and the battle for atheism. While Levin's work supports Myers's approach when it comes to changing opinion, Myers is working against religion, an overwhelming, organized tradition. He is also extending science to do what Scott says it doesn't do—make pronouncements about ultimate reality—by saying there is no God. This is beyond the realm of what it is possible for inductive reasoning to determine, and so it is outside the scope of science and the scope of the argument against radical fundamentalism. This reduces the argument to competing claims—once again, all the way back to Locke.

To Fight or Not to Fight

Myers argues that, since the conflict already exists, he is simply providing a much-needed counter-message to that of the radical authoritarian fundamentalists who have taken over much of the national conversation. "Saying that there's not a conflict, in the face of that, is putting your head in the sand," he says, and much of the recent history of science bears him out.

Right now, the authority of science lies largely in it being sort of arcane and difficult to understand and I don't think that's healthy for us. Science should be something that everybody has access to and everybody can understand. Most scientists are perfectly willing and interested in engaging the public. Lots of them are actually pretty good at it, but there is no incentive in the university system to do that—tenure relies on grants, and then on teaching. Outreach is way down at the bottom and does not influence tenure or promotion decisions. People who do outreach are regarded as second-class citizens in the science world.

Writing in a 2011 editorial, AAAS CEO Alan Leshner echoed this concern:

Unfortunately, traditional reward systems only emphasize publication and grant-getting, at the expense of efforts to promote increased participation in science and engineering. Nationally, university presidents, provosts, deans, and department chairs must find ways to reward faculty members for reaching out to a wider community of potential students and their families. . . . Support like that is essential for innovation because increasing the diversity of the scientific human-resource pool will inevitably enhance the diversity of scientific ideas. By definition, innovation requires the ability to think in new and transformative ways.

Diversity of ideas is precisely the opposite of what occurs with strict adherence to an ideological doctrine, which creates conformity and "dittoheads." With that Nisbet agrees, but he argues that it's important to cast outreach in terms of trust and relationships. Remove the perceived threat or conflict. "People want to feel like you are listening to them," he says. "If you tell, versus listen, you promote mistrust."

In order to bring doubters around to evolution, Nisbet says we should identify topics of discussion that create interest, draw attention, and are relevant to people's lives. One sensible example is the immense contributions evolution has made to agriculture or medical science. So that's the frame. Who, then, is the best communicator? Probably a medical scientist who has local community ties, plus a local teacher or religious leader. Go to places people are familiar with, like churches or schools or malls, and sponsor opportunities for discussion. It's not about promoting information. It's about having a conversation. Communication is context-dependent and political—a valuable insight from postmodernism that scientists have failed to understand.

This is all true, Myers says. But there is also a reasoned argument to be made for what he does. "If I am loud enough and clear enough and convincing enough, some people may—and I repeat, may—begin to reevaluate their thinking," he says. "Look at what Rush Limbaugh or Pat Robertson does. You can't tell me they haven't had influence. If you agree that they have, don't criticize me."

In the end, the recent battle for the legalization of same-sex marriage may provide a useful model. It started as a culture war. That brought people to the table. But then it pivoted to the kind of narrative that Nisbet, Scott, and Watts are talking about, one in which gay people and their families, friends, and allies had personal phone conversations with people, asked them why they got married, listened, and then said, "That's why I want to get married, too."

Lutheran pastor Scott Westphal has seen this work firsthand on science issues. He ministers to a congregation in Scandia, Minnesota, in what was Michele Bachmann's congressional district. "I can't tell you how frustrating it is when parishioners listen to this divisive nonsense on AM talk radio instead of the teachings of their own pastor," Westphal says. "It doesn't get us anywhere as a people. All I see it producing is anger and fear. Jesus questioned. He wanted people to open their hearts and their minds and let go of fear."

Westphal has had scientists and science advocates lead discussions about evolution and the big bang in his church. "Protestants started out by questioning," he says. "These things don't have to be in conflict. By meeting scientists and talking openly about these things, people can let go of their misconceptions and see why scientists think the way they do—and that they are our friends and neighbors, not threats or opponents."

Why the Fight Matters

Eugenie Scott says she is "nontheistic," but that others in her office, such as her colleague Peter Hess, are "theistic," yet still strongly believe that creationism must not be taught in science class. It can be taught at home or in a church setting, but government-funded schools need to teach science so that we have an informed citizenry. "If we're teaching creationism, we're not teaching science," Scott says. "The assumption of creationism is that natural phenomena require supernatural explanations. I'm not saying science is atheistic about ultimate reality. It isn't. To say that you can explain something using natural causes is not the same thing as saying there are no supernatural causes. Science is atheistic in the sense that plumbing is atheistic. It limits itself to the study of natural causes."

This is critically important. The world has advanced because of science and technology. Because of our understanding that even if we haven't figured something out, we can just keep plugging away, looking for those natural causes, and sooner or later we'll find them. The book and movie *The Martian* pivoted on this theme.

But teaching creationism in science classes means teaching a habit of mind that is toxic to the human problem solving that has led these advances. It teaches children to throw up their hands and declare that the problem is unsolvable, particularly if that problem is tough or might have consequences for a particular religious belief. To look instead to God and acquiesce to authority. It teaches them to value not diversity of ideas, but conformity. Not survival, but submittal.

If we do that, we're basically giving up on science and on the probability of finding those answers. That is not going to take the world where we need to go.

Chapter 10

THE INDUSTRIAL WAR ON SCIENCE

The conscious and intelligent manipulation of the organized habits and opinions of the masses is an important element in democratic society. Those who manipulate this unseen mechanism of society constitute an invisible government which is the true ruling power of our country.

We are governed, our minds are molded, our tastes formed, our ideas suggested, largely by men we have never heard of. This is a logical result of the way in which our democratic society is organized. . . . In almost every act of our daily lives, whether in the sphere of politics or business, in our social conduct or our ethical thinking, we are dominated by the relatively small number of persons . . . who understand the mental processes and social patterns of the masses. It is they who pull the wires which control the public mind.

—Edward Bernays, 1928

The Great Fallacy of Scientists

For two generations, scientists labored under the notion that science is not political. They removed themselves from public dialogue, and citizens became less interested in science as a result. And over those two generations, as the public and even academia grew more skeptical of science, the public-policy implications of scientists' insights into nature grew while their ability to communicate those insights with society atrophied.

Meanwhile, using the fruits of science, the world progressed. Population grew exponentially. Development and technology exploded, fragmenting habitats and propelling the world into its sixth mass extinction—and the first caused by humans, with species disappearing about one hundred times faster than the normal rate. Oil use skyrocketed, largely in response to the low-density land-development model the United States instituted as a defense against nuclear

257

attack, and by the creation of new consumer markets in developing countries. A new economic model evolved based on sustained production, low-density development, high consumer turnover, and perpetual market expansion. Science continued to leverage new efficiencies and engineering to produce new conveniences. People responded to the incentives of this bounty. Consumption skyrocketed. Wealth exploded. But much of it, subsequent science has begun to show, was not earned wealth. It was wealth borrowed from the environment through the unsustainable depletion of resources.

Science-driven market economics began to change the planet. Today, the majority of the policy challenges facing the leading governments revolve around science as either the dominant cause or the best hope for solution or both. But the solutions science suggests are often at odds with massive industrial investments and distribution systems. In order to protect their business models from regulatory disruption, several industries have begun to develop business strategies to coopt or create uncertainties about science that does not support their business models. For ideas on how to do this, they turned to the arguments developed by the postmodernist and religious wars on science, and merged them with new insights from the field of public relations.

Propaganda

To understand the modern era of industry-funded antiscience, we have to first look at where it began: with the contribution of Sigmund Freud's nephew, Edward Bernays. Bernays was an Austrian-American journalist-turned-psychologist. He is widely credited as "the father of public relations," and is author of the seminal book in the field, *Propaganda.*

As a young man, Bernays served on the Committee on Public Information in the administration of Woodrow Wilson during World War I, also called the Creel Committee after its chairman, another journalist, the Denver newspaperman and Democratic National Committee publicist George Creel. The committee was charged with influencing US public opinion to support the war effort, "not propaganda as the Germans defined it, but propaganda in the true sense of the word, meaning the 'propagation of faith,'" Creel wrote in his 1947 autobiography. He was referring to the Sacra Congregatio de Propaganda Fide, or Sacred Congregation for the Propagation of the Faith, established by the pope in 1622 to govern the church's missionary work, and from which the modern use of the word "propaganda" originates.

After Wilson was elected, Creel proposed an agency that would fight to convince a divided public of the rightness of a political goal—what Wilson called "the verdict of mankind": the necessity of the United States entering World War I. Creel was authorized to create the committee, and it was charged with combating Wilson's political problem of public dissention about the idea of going to war in Europe. Being a newspaperman, Creel believed that suppression of information was antithetical to the principles of a free press and thus to democracy. Instead, he did an end-run around suppression that still accomplished his main goal: the manipulation of the press, and thus public sentiment. He argued for the expression of counter-information that was based on science and objective facts, but presented in the most positive and favorable light for the war effort. Instead of a frontal assault on the king, Creel's committee would be the court advisor that whispered advice in the king's ear.

Creel hired Bernays and several other bright young men, and in short order the Creel Committee's arguments were found in newspapers across the country. Posters supporting the war effort were on display in public places, and prowar nationalistic shorts played before movies. The committee's own daily "newspaper," the *Official Bulletin*, was distributed to every newspaper, post office, government office, and military base in the United States, providing guidance and suggestions for stories. Hundred of speakers, called Four Minute Men, were recruited to give speeches in more than 5,200 communities across the country, delivering propaganda to an estimated 314 million audience members. Working within businesses, the committee put up posters emphasizing the contributions of American labor to the war effort. Working within minority constituencies, they enlisted prominent community members as spokespersons—an early example of something public-relations professionals would later call the third-party technique.

After the war was won, the committee—whose methods and exaggerations had been criticized by the media and in broad public opinion as manipulative and dishonest—was disbanded. But Edward Bernays had become amazed by the power of what such an immersive approach to message delivery, combined with a purported basis in science and facts, had accomplished. His 1928 book, *Propaganda*, laid out specific methods for controlling public opinion, arguing, "If we understand the mechanism and motives of the group mind, is it not possible to control and regiment the masses according to our will without their knowing about it? The recent practice of propaganda has proved that it is possible, at least up to a certain point and within certain limits."

Bernays was referring to both his service on the Creel Committee and his more recent experience putting his ideas to work at his own public-relations firm, the world's first. In 1919, he wrote the first book on the practice, called *Crystallizing Public Opinion*, and by 1923 he taught the first class in public relations, at New York University. His client list exploded to include massive US corporations like General Electric, Proctor and Gamble, and the American Tobacco Company, which immediately saw this as a way to create new markets and sell more products.

Using press releases, Bernays staged events and treated them like news, often successfully getting them into the news stream. He came to regard the word "propaganda" as too loaded with sinister connotations, particularly after the way the Germans had used propaganda in World War I, so he began calling it "public relations." and is widely credited as the father of the field. Mindful of how masterfully the Americans had deployed propaganda in WWI, the Nazis turned to the ideas of Bernays to develop propaganda to build and justify the Third Reich. Bernays, a Jew, first learned of this at a 1933 dinner at his house:

> Karl von Wiegand, foreign correspondent of the Hearst newspapers, an old hand at interpreting Europe and just returned from Germany, was telling us about Goebbels and his propaganda plans to consolidate Nazi power. Goebbels had shown Wiegand his propaganda library, the best Wiegand had ever seen. Goebbels, said Wiegand, was using my book *Crystallizing Public Opinion* as a basis for his destructive campaign against the Jews of Germany. This shocked me. . . . Obviously the attack on the Jews of Germany was no emotional outburst of the Nazis, but a deliberate, planned campaign.

Just as the combination of the Luther Bible translation and the printing press touched off a cultural revolution in Europe, the combination of propaganda with new mass-media technology proved astoundingly powerful, in Germany, the United States, and elsewhere. Public-relations firms quickly took off and made millions working for US corporate giants, as well as for the US government itself. Among the scores of Bernays's clients was the United Fruit Company (now Chiquita Brands International), which was having problems with the policies of the newly elected president of Guatemala, Jacobo Àrbenz. Bernays worked with United and the US government to build a public-relations campaign to successfully overthrow Àrbenz.

Torches of Freedom

One of Bernays's most historic public-relations campaigns was part of a larger effort he undertook for George Washington Hill, the president of the American Tobacco Company, who was seeking a way to change the public stigma against women smoking. In the 1920s, smoking was associated with prostitutes and other "fallen women," and the act—when performed by a woman—had sexual implications that contributed to views that it was indecent, particularly in public. Several municipalities had passed ordinances against public smoking by women, and heated battles were waged on college campuses between women who saw smoking as a sign of autonomy and administrators who were concerned about preserving social norms.

"Hill called me in," Bernays recalled. "'How can we get women to smoke on the street? They're smoking indoors. But damn it, if they spend half the time outdoors and we can get 'em to smoke outdoors, we'll damn near double our female market. Do something. Act!'"

Remembering what he had learned about invoking patriotism and nationalism to sell ideas during the war, and observing the suffragist movement and the rising social status women had gained during the war period, Bernays turned for ideas to a family friend, A. A. Brill—an Austrian-American psychoanalyst and the first man to translate Bernays's uncle Sigmund Freud's work into English. "Some women regard cigarettes as symbols of freedom," Brill told him.

> Smoking is a sublimation of oral eroticism; holding a cigarette in the mouth excites the oral zone. It is perfectly normal for women to want to smoke cigarettes. Today the emancipation of women has suppressed many of their feminine desires. More women now do the same work as men do. Many women bear no children; those who do bear have fewer children. Feminine traits are masked. Cigarettes, which are equated with men, become torches of freedom.

Based on this psychological feedback, Bernays staged a unit in the 1929 New York City Easter Day Parade and paid "ten young debutantes" to march in it while smoking and waving Lucky Strikes cigarettes. Bernays himself stayed in the background, protecting his and his client's anonymity. This is what Bernays termed the "third-party technique." If the public knows they are being manipulated, they become skeptical. They must take the information in as reality. Bernays employed the third-party technique even with those he hired to march in the parade. As far as they knew, the well-known feminist Ruth Hale had recruited them. "Women!" Hale said in invitations to march. "Light another torch of freedom! Fight another sex taboo!" Bernays then hired a professional photographer to record the event in case newspaper photographers didn't get good shots, and distributed photographs to the press.

The results were staggering. "Our parade of ten young women lighting 'torches of freedom' on Fifth Avenue on Easter Sunday as a protest against women's inequality caused a national stir," he later wrote. "Front-page stories in newspapers reported the freedom march in words and pictures." The coverage touched off a national debate. Feminists rushed to embrace the idea, not realizing they were being manipulated by a corporation, and female smoking shot up as it became associated with the expression of one's individual identity and freedom. "Age-old customs, I learned, could be broken down by a dramatic appeal, disseminated by the network of media," Bernays wrote.

It was an observation that would change history in many ways. Public relations became industry's way of propagating the faith, and it would soon help create the schism over science that came to define today's political parties.

The Birth and Childhood of the Antiscience Industry

Following Bernays's smashing success for the American Tobacco Company, tobacco companies began hiring public-relations firms to come up with more strategies—first to expand market share, and later to fight public perceptions about the emerging science that linked smoking to lung cancer. The Germans discovered the connection in the 1930s, and the Nazi government ran major antismoking campaigns, banned smoking in public transportation, restricted tobacco advertising, and limited its use in restaurants, coffee houses, and other public places. But Nazi science was discredited after the war, and much of that information was lost.

The links between smoking and cancer began to reemerge when US scientists began similar experiments in the late 1940s and early 1950s. The first major postwar study was published in 1953, and within days, wrote science historians Naomi Oreskes and Erik Conway, a new public-relations strategy was hatched to fight science with science—to find tidbits of information that could cast doubt on the public's perception of the link between smoking and cancer.

On that December morning, the presidents of four of America's largest tobacco companies—American Tobacco, Benson and Hedges, Philip Morris, and U.S. Tobacco—met at the venerable Plaza Hotel in New York City. The French Renaissance chateau-style building—in which unaccompanied ladies were not permitted in its famous Oak Room bar—was a fitting place for the task at hand: the protection of one of America's oldest and most powerful industries. The man they had come to meet was equally powerful: John Hill, founder and CEO of one of America's largest and most effective public relations firms, Hill and Knowlton. . . . They would work together to convince the public that there was "no sound scientific basis for the charges," and that the recent reports were simply "sensational accusations" made by publicity-seeking scientists hoping to attract more funds for their research.

One of the first things these men developed was the concept of "balance." They argued that the science was "not settled" and that the details were still open to debate. Since, in science, a detail could overturn an entire theory, as we had seen with Einstein, we needed to hear from "both sides" and have a "healthy debate." They argued that the public should trust the tobacco companies to do the science to see if there were health risks, and, if there were, to figure out how to fix them.

Instead, the tobacco companies cherry-picked data, played games with statistics, and focused on anomalies in the data. They recruited a few scientists outside the mainstream, contrarians who enjoyed the attention and focused on creating a sense of debate. These scientists issued press releases to newspapers and wire services that generated stories creating a sense of doubt about the link between smoking and cancer in the minds of the public, while the companies built up a lobbying presence in Washington to battle regulations.

By 1960, through an aggressive effort, tobacco companies had also developed relationships with doctors, medical schools, and public-health authorities. Oreskes and Conway:

In 1962, when U.S. Surgeon General Luther L. Terry established an Advisory Committee on Smoking and Health, the tobacco industry made nominations, submitted information, and ensured that Dr. Little "established lines of communication" with the committee.

To ensure that the panel was "democratically" constituted, the surgeon general invited nominations from the tobacco industry, as well as from the Federal Trade Commission (who would become involved if restrictions were placed on tobacco advertising). To ensure that the panel was unbiased, he excluded anyone who had publicly expressed a prior opinion. One hundred and fifty names were put forward, and the tobacco industry was permitted to veto anyone they considered unsuitable.

The same year saw the publication of Rachel Carson's *Silent Spring*, which warned of systemic pollution from pesticides, and their health dangers to animals and humans. The book exposed the risks, previously hidden or unknown to the public, of the wide use of pesticides like dichlorodiphenyltrichloroethane, or DDT. Developed in 1939, DDT was first widely used during World War II, clearing South Pacific islands of malaria-carrying insects for U.S. troops, and was used in Europe as an ingredient in delousing powder. The Swiss scientist who identified its insecticide effect, Paul Hermann Müller, was awarded the 1948 Nobel Prize in Physiology and Medicine.

After the war was over, DDT and other chemicals were made available to the civilian market, and began to be used widely in agriculture. By 1958, Rachel Carson began to hear of large bird kills after DDT sprayings, and she started to research the wide-scale application of insecticides in the environment. The result was *Silent Spring*.

The petrochemical and agribusiness industries saw the book as a direct attack on their financial foundations, and they turned to public relations to fight back. The most notable contribution came from another journalist, this one turned press secretary for a politician and then public-relations guy, E. Bruce Harrison. Harrison had recently been hired by the American Chemistry Council, then known as the Manufacturing Chemists Association, to manage the industry's fledgling community-relations program. But when *Silent Spring* became a bestseller in 1962, he was asked to manage the strategic crisis response team, and was given the title "director, environmental information."

The young PR officer threw himself into the task, working with Dow, Monsanto, DuPont, Shell Chemical, and W. R. Grace. "Along the way, they pioneered environmental PR 'crisis management' techniques that have now become standard industry tactics," wrote activist and writer John Stauber in a critical 1994 review of Harrison's book on environmental PR. "They used emotional appeals, scientific misinformation, front groups, extensive mailings to the media and opinion leaders, and the recruitment of doctors and scientists as 'objective' third party defenders of agrochemicals."

The coordinated PR assault against Carson and her book was thundering. Much of it was harsh and vehement, attacking her qualifications, motives, and person, and fighting science with science. *Chemical and Engineering News* ran a review by well-known nutritionist William J. Darby entitled "Silence, Miss Carson," which lambasted her "ignorance or biases." The National Agricultural Chemical Association distributed thousands of negative reviews of the book and doubled its PR budget. Monsanto published a parody of *Silent Spring* called *The Desolate Year*, in which a plague of uncontrolled insects destroyed America. Some five thousand copies were sent to newspapers, gardening editors, science writers, farm journalists, and book reviewers. Like the Creel Committee's approach, it was an all-out immersive campaign.

As the prime architect of the new methods, Harrison became the go-to PR guy for environmental-crisis management and public relations. He developed the now-common practice referred to as "greenwashing," in which a corporate polluter makes a small environmental concession in order to win positive public opinion, allowing it to avoid bigger issues that might affect its bottom line. In 1973, Harrison and his wife Patricia founded their own PR firm to work with corporate clients on environmental issues. Patricia eventually became cochair of the Republican National Committee and, later, CEO of the Corporation for Public Broadcasting.

"Rachel Carson's legacy looms huge today," wrote Stanford conservation biologist Paul Ehrlich to commemorate the fiftieth anniversary of the publication of *Silent Spring*. "Many people have the impression that climate disruption is the worst environmental problem humanity faces, and indeed, its consequences may be catastrophic. But the spread of toxic chemicals from pole to pole may be the dark horse in the race." Harrison and his firm have been intimately involved in crafting public opinion on behalf of corporations on these issues ever since *Silent Spring*.

A year after the book was published, Carson testified on pesticides before a congressional panel. She was dying of breast cancer, as she had been since she started writing the book. She had only months left, but she told almost no one for fear it would undermine her message. She had had a radical mastectomy, her pelvis was riddled with fractures, forcing her to walk with a cane, and she wore a dark wig to hide the fact that she was bald from radiation treatments. None of the legislators noticed, because she spoke so brilliantly and passionately. She might have been dying, but her warning to mankind, she knew, would survive. "Every once in a while in the history of mankind, a book has appeared which has substantially altered the course of history," said Senator Ernest Gruening, a Democrat from Alaska. *Silent Spring* gave focus to a new field of environmental science, a new area of environmental policy, and a new social movement to protect the environment. Within a decade, the United States passed landmark environmental legislation in the Clean Air Act, the Clean Water Act, and the establishment of the Environmental Protection Agency, the formation of which was based on *Silent Spring*, as the agency notes on its own website. "In fact," one article on the site notes, "the EPA today may be said without exaggeration to be the extended shadow of Rachel Carson."

But the bitter battle over Carson's book and its implications for industry, together with the antiscience public-relations efforts, caused a split between, on one side, the old petrochemical and agribusiness industries, whose business models were shaken to their foundations, and, on the other, the government's new environmental science and the social movements that embraced it. It was a split that would become one of the main underpinnings of the modern political schism over science and the role of government today.

The Split That Defines Modern Politics

At the same time *Silent Spring* was giving birth to environmental science, human control over the reproductive cycle was advancing. In 1957, "the pill" was approved by the FDA for women with severe menstrual disorders. That same year, the number of women reporting severe menstrual disorders went through the roof. By 1960, the FDA approved the drug for contraceptive use in all women, and by 1962, when *Silent Spring* was published, 1.2 million women were taking it. Similar developments would continue to play out over the coming decade and a half. 1973's *Roe v. Wade* Supreme Court ruling legalized abortion in the United States and the 1978 conception of the first "test-tube baby," as in vitro fertilization was then described, prompted religious debates over whether babies so conceived would have souls, and a feature story on the cover of *Time* magazine.

Like the theory of evolution half a century before, these advances in the biological sciences resulted in attacks by religious conservatives who were offended by growing human control over reproduction and what it implied about God's plan for when, how, and if a woman might conceive a child. They feared birth control would remove the main drawback to having sex outside marriage and make women promiscuous, leading to the breakdown of society. In addition to the instability and terror wrought by the bomb, it was yet another affront from science.

Suddenly, religious conservatives found themselves in common cause with the oil, chemical, and agricultural industries, who were equally offended by the modern advances of environmental science, and who were advocating for more freedom—corporate freedom, from government regulation. The fundamentalists needed access and legitimacy and the business interests needed passionate foot soldiers. A marriage of convenience was consummated. Biologists and environmental researchers, on the other hand, found a sympathetic ear in the organic farmers, foodies, and health nuts coming out of the hippie and environmental movements, as well as prochoice women and other progressives who appreciated their work. Another marriage of convenience was formed, also around freedom—freedom of choice and freedom from chemical exposure.

On one side of the political schism over the biological and environmental sciences, we see the elements that became the modern US Republican Party, and that influence conservative parties in many Western countries: an anti-regulation, anti-reproductive-control, and pro-corporate marriage of old industry and old religion. On the other side we see the elements that became the modern US Democratic Party, and that influence the attitudes of progressives across the Western world: a pro-environment, prochoice, and anticorporate marriage of government scientists, environmentalists, and activists.

On the right, the antiscience tends to focus on denying known science on issues that often have to do with regulating personal or corporate behavior, such as climate disruption, smoking, evolution, HPV vaccines, abstinence-only sex education, gun control, acid rain, the hole in the ozone layer, and other issues that stem from industrial and religious vested interests. The probusiness conservatives and libertarians don't want regulation, while the social conservatives oppose evolution and want to regulate sex, creating an unusual compromise with many contradictions.

On the left, the antiscience tends to extend worries about health and the environment into areas that are not supported by the evidence, claiming nevertheless that, as in *Silent Spring*, there are hidden dangers to our environment, our health, or our spirits. Examples include the ideas that cell phones cause brain cancer; that Wi-Fi and other electromagnetic fields cause cancer, birth

The Narrative Line of Science Politics

THEME: *Liberal scientists with a socialist agenda*
want to control your life and limit your freedom

MOTIVATION: Anti-Regulation Anti-Reproductive Control Pro-Corporate

Antiscience: "CREEPING SOCIALISM": Climate Change Evolution Reproduction New:Vaccines

Old Industry & Old Religion

Environmental
Biological

Pride Doubt Fear Race Silence Worry Discomfort

Science & Environmentalists

Antiscience: "HIDDEN DANGERS": Mainstream Medicine Cell Phones Vaccines Waste to Energy
EMF Pollution Fluoride in Water GMO Crops

MOTIVATION: Pro-Environment Pro-Choice Anti-Corporate

THEME: *Impersonal doctors, greedy corporations and mechanistic*
scientists hide the real dangers to health, the environment & spirit

defects, or allergies; that vaccines cause autism; that genetically modified crops are unsafe to eat; and that fluoride in water is unsafe to drink. The progressives want to regulate business but don't want to regulate sex. On both sides, generally speaking, conservatives want to protect consolidated power and progressives want to oppose it.

One of the interesting places where these two views overlap is on the issue of nuclear power. By conversion to electric vehicles and electrically powered heating plants such as geothermal systems, most greenhouse-gas emissions could be eliminated, but electrical use will at least triple. Nuclear power is a human-controllable source of electricity that generates almost no greenhouse gases, but it is very much a consolidated power structure in both senses of the term. In a 2015 Pew Research Center poll, 78 percent of US Democrats said the earth is getting warmer because of human activity, yet only 36 percent of them favored building more nuclear plants to generate electricity. The reverse is true among Republicans. Only 10 percent said that the earth is getting warmer because of human activity, yet 73 percent favored building more nuclear plants to generate electricity. This bipartisan non sequitur says a lot about how the opposing ideologies divide over science issues, and perhaps charts an area where common ground may be found if both sides can accept a compromise.

The theme of the right's antiscience can be generally described as: *Liberal scientists with a socialist agenda who are out for more government money want to control your life and limit your freedom.* On the left, the theme of antiscience can be generally described as: *Impersonal doctors, greedy corporations, and mechanistic scientists hide the real dangers to our health, the environment, and the human spirit.*

This split has evolved mostly without the voices of scientists—who, remember, are both conservative and progressive in outlook at the same time.

The Climate Battle

The industrial clash with environmental science came into adulthood over the issue of human-caused climate disruption. It is an issue that has profound existential stakes for the world's most powerful industry (fossil fuels), as well as for the environment, environmental science, the world economy, and democratic governments. The showdown is propelling an industrial war on science the likes of which the world has never seen.

Our understanding that increasing carbon dioxide levels in the atmosphere could change the climate is not new. The greenhouse effect was first identified by famed French mathematician and physicist Joseph Fourier in 1824. Fourier calculated that the planet should be colder than it was if it was only being warmed by the incoming solar radiation, and suggested that the atmosphere might be acting as an insulator of some kind, trapping heat. In 1861, Irish physicist John Tyndall discovered that water vapor, carbon dioxide, and methane all trap heat while oxygen and nitrogen do not. This relationship between carbon dioxide, water vapor, and climate was explored in greater detail in 1896 by the Swedish physicist Svante Arrhenius, who estimated that a doubling of carbon dioxide levels would cause global warming of 4.9 to 6.1 degrees Celsius. In his landmark paper, Arrhenius reported that "a simple calculation shows that the temperature in the arctic regions would rise about 8° to 9°C, if the carbonic acid increased to 2.5 or 3 times its present value," and he suggested that, over time, the burning of fossil fuels—then largely five hundred million tons annually of coal—could have an influence.

This influence began to be noticed in the early part of the twentieth century as Earth began to warm. In 1938, English engineer and inventor Guy Stewart Callendar concluded that atmospheric CO_2 had increased 10 percent in the previous one hundred years, potentially explaining the warming. Scientists began to take notice and when they really had time to focus on the problem after the conclusion of WWII, they became increasingly concerned. At first they argued that Callendar was wrong, and then that the oceans would absorb the extra CO_2 humans were generating. But in 1956, Canadian physicist Gilbert N. Plass showed that the ocean saturation argument was wrong, and that CO_2 increases do in fact increase Earth's energy balance; i.e., they were in fact trapping additional heat. Roger Revelle, the director of the Scripps

Institution of Oceanography in San Diego, followed this up in 1957 by showing that the oceans were absorbing atmospheric CO_2 much more slowly than had been previously thought.

The exact level of atmospheric CO_2 was first documented experimentally by Charles Keeling in the late 1950s. At the time, the US economy was coasting off its postwar boom. Keeling was a young postdoctoral researcher in geochemistry at the California Institute of Technology in 1955 when he figured out how to make a device that would reliably measure the level of carbon dioxide in the atmosphere. He spent three weeks that summer camping out at Big Sur with his wife and newborn son, taking samples and recording measurements.

Revelle thought Keeling's device was important. Public transit was dying and auto use was exploding, and, along with it, suburban home energy use, which was accelerating under planned low-density development. Keeling's precise measurements, if taken in pristine air away from major sources of fossil-fuel emissions, might be able to shed light on the scientific debate over the exact levels and effects of atmospheric CO_2. He persuaded Keeling to join Scripps, and Scripps and the US Weather Bureau (now the National Weather Service) funded a measuring station at Mauna Loa Observatory, high in the mountains of Hawaii, where Keeling could take daily readings.

Keeling's measurements showed that carbon dioxide levels fall as plants take it in during the Northern Hemisphere's growing season, then rise during the fall and winter months as vegetation dies and decomposes, releasing it back into the atmosphere. The planet is essentially breathing. But over time, he also found something troubling: the annual levels, year after year, were steadily rising.

Keeling's first measurements in 1958 showed an average atmospheric carbon dioxide concentration of 310 parts per million (ppm). Every year thereafter, the levels rose by a little more than they had the previous fall and winter, but they fell by only the usual seasonal amount caused by growing vegetation in the spring and summer, creating an overall upward trend. Today, the number is more than 400 parts per million. Keeling's chart of the annual rise and fall in CO_2 levels and their inexorable climb has become one of the most famous images of science, one so foundational that it is carved into a wall at the National Academy of Sciences in Washington, D.C. It is called the Keeling curve.

When scientists saw the plot of CO_2's inexorable rise, they realized it would have major public-policy implications for how we generate and use energy, but they didn't have enough data. They knew CO_2 helps trap heat in the atmosphere, keeping the planet warm enough for life, but they didn't know for certain if the extra CO_2 Keeling's measurements showed was truly enough to affect global temperatures, or if there might be some other mechanism that would possibly counter the effects. Policymakers like certainty, and science is an uncertain endeavor until many independent lines of data can be brought to bear. They began expanding the field of geophysics to get a handle on just how big the problem might be, and month after month, year after year, new landmark papers and reports were presented. In 1965, the US President's Science Advisory Committee, which included Revelle, briefed President Lyndon Johnson, warning that the continued burning of fossil fuels could produce "marked changes in climate, not controllable through local or even national efforts."

Over the following decades, thousands of scientists went into the field and collected massive amounts of data about greenhouse gases, ocean acidification, ice and albedo (Earth's reflectivity), weather patterns, and hundreds of other aspects of Earth's complex climate system. What they found, using many different methods, was that the increasing CO_2 levels were indeed causing global warming, just as Arrhenius had predicted, and we were approaching levels that could change climate patterns in ways that destabilize our economy, our national security, our social structures, and our environment. We had to stop burning fossil fuels.

As had happened to Galileo, making those simple observations once again bumped science up against an established political and economic power elite that viewed it as a threat to their authority. This time it wasn't the church—it was the energy industry. And, just as Galileo had, scientists badly misjudged the situation.

Who Knew What When

The need to limit carbon emissions was something the world's nations, as well as the energy and extraction industries, were aware of at least since 1988, when climate scientist James Hansen first testified before Congress. But the science had been known to oil companies well before that.

A senior Exxon company scientist named James F. Black had informed Exxon's top executives of the same conclusion in July 1977. Delivering a slide presentation, he told them how science showed that carbon dioxide from the burning of fossil fuels was warming the planet and could eventually endanger humanity. "In the first place, there is general scientific agreement that the most likely manner in which mankind is influencing the global climate is through carbon dioxide release from the burning of fossil fuels," Black told Exxon's Management Committee, according to a written version he recorded later.

A year later, Black warned Exxon scientists and managers that independent researchers estimated a doubling of the carbon dioxide concentration in the atmosphere would increase average global temperatures by 2 to 3 degrees Celsius (4 to 5 degrees Fahrenheit), and as much as 10 degrees Celsius (18 degrees Fahrenheit) at the poles. Rainfall might get heavier in some regions, and other places might turn to desert. "Some countries would benefit but others would have their agricultural output reduced or destroyed," Black said, according to his 1977 written summary. "Present thinking holds that man has a time window of five to ten years before the need for hard decisions regarding changes in energy strategies might become critical." As a team of investigative reporters led by Neela Banerjee of the Pulitzer prize–winning *InsideClimate News* found,

> Exxon responded swiftly. Within months the company launched its own extraordinary research into carbon dioxide from fossil fuels and its impact on the earth. Exxon's ambitious program included both empirical CO_2 sampling and rigorous climate modeling. It assembled a brain trust that would spend more than a decade deepening the company's understanding of an environmental problem that posed an existential threat to the oil business.

At the same time, Exxon continued to fund scientific research on global warming and to allow its scientists to publish. Documents show that corporate leadership thought it was the only ethical thing to do. The company's forthcoming approach was recognized by the US Department of Energy. "We are very pleased with Exxon's research intentions related to the CO_2 question. This represents very responsible action, which we hope will serve as a model for research contributions from the corporate sector," said David Slade, manager of the federal government's carbon dioxide research program at the Energy Department, in a May 1979 letter. "This is truly a national and international service."

The research continued. In remarks to a global-warming conference in October 1982, Exxon Research and Engineering Company president Edward E. David identified the problems he foresaw:

> Faith in technologies, markets, and correcting feedback mechanisms is less than satisfying for a situation such as the one you are studying. . . . The critical problem is that the environmental impacts of the CO_2 buildup may be so long delayed. A look at the theory of feedback systems shows that where there is such a long delay the

system breaks down unless there is anticipation built into the loop. The question then becomes how to anticipate the future sufficiently far in advance to prepare for it.

Exxon, he said, was looking at a wide variety of renewable-energy technologies, from producing cheaper solar panels to using advanced solar collectors and nuclear reactors to create cheap hydrogen fuels out of water. "Few people doubt that the world has entered an energy transition away from dependence upon fossil fuels and toward some mix of renewable resources that will not pose problems of CO_2 accumulation," he said. "The question is how do we get from here to there while preserving the health of our political, economic and environmental support systems."

It was a worthy goal, and a way of repositioning Exxon as not only a fossil-fuel company but as the world's leading *energy* company. But then business and politics intervened, as they so often do in the global energy market. Saudi Arabia had seen its oil revenues fall from $119 billion in 1981 to $26 billion in 1985. The revenue decline drove the country's government into a large budget deficit, and the reduced share of global oil sales also meant Saudi Arabia was losing its global clout. At the same time, the Saudis saw expensive oil production from American and British offshore rigs taking more of the global market share. They concluded that by supporting high oil prices they were propping up this more expensive production, enabling their own competition, while Margaret Thatcher went on about free markets and how she didn't care about the price of oil.

In December 1985, Saudi Arabia declared its intention to regain market share. It broke with the other Organization of the Petroleum Exporting Countries (OPEC) nations and began producing more oil, causing a glut. The other OPEC nations balked at first, but after a hastily called meeting they soon joined in, declaring that market share, not price, was what they wanted. Oil prices began to decline, falling from $31.71 per barrel in November 1985 to a low of $10.42 a barrel in March 1986. The more expensive US and British production collapsed, hurting Exxon's bottom line, and the Saudis regained their leading role in the world's oil market.

In June 1988, the US energy industry was still reeling. The domestic drilling that had resumed in the early 1980s was all but shuttered. Exxon had made deep cuts during the glut, including to much of its climate-science staff. Then, on June 23, James Hansen, head of NASA's Goddard Institute for Space Studies in Manhattan, went before the US Senate's Energy and Natural Resources

Committee and told them that global warming was not some peril in the distant future—it was already happening. "The first five months of 1988 are so warm globally that we conclude that 1988 will be the warmest year on record unless there is a remarkable, improbable cooling in the remainder of the year," Hansen told the Senate committee. "Global warming has reached a level such that we can ascribe with a high degree of confidence a cause and effect relationship between the greenhouse effect and observed warming. It is already happening now."

Senator Tim Wirth, who presided over the hearing, responded to the scientific evidence Hansen and other climate scientists were producing, and said Congress needed to take policy action to slow the problem. "As I read it, the scientific evidence is compelling: the global climate is changing as the earth's atmosphere gets warmer," he said. "Now, the Congress must begin to consider how we are going to slow or halt that warming trend and how we are going to cope with the changes that may already be inevitable."

The publicity surrounding the hearing must have alarmed Exxon executives. An internal draft memo from August 1988, titled "The Greenhouse Effect," shows a change in strategy. A company public-affairs manager described the new "Exxon Position." Toward the end of the document, after an analysis that noted scientific consensus on the role fossil fuels play in global warming, the PR expert wrote that the company should "emphasize the uncertainty in scientific conclusions regarding the potential enhanced greenhouse effect," and "urge a balanced scientific approach."

The threat of regulatory action based on new knowledge from climate science was not limited to the United States. Other nations saw the need for action as well, and, since the problem was global, solving it would require global ingenuity and cooperation. On December 6, the United Nations announced plans to form an intergovernmental panel to study climate change science, as well as its public-policy implications and possible solutions.

Exxon knew full well what the science said from its own decade-plus of research into the problem, and knew where this kind of exploration would inevitably lead. "Arguments that we can't tolerate delay and must act now can lead to irreversible and costly Draconian steps," Duane LeVine, Exxon's manager of science and strategy development, warned the board of directors in 1989. In other words, climate science was threatening Exxon's vested interests at a time when they could least afford it. A culture change had occurred at Exxon following the oil glut and the "Greenhouse Effect" memo—one that would affect the debate over climate science and broader environmental concerns for the next three decades. Exxon abandoned its good-citizen approach of funding research, allowing its publication, and discussing ways to transition to a new energy economy. It began to fund climate denial aimed at obfuscating the science and slowing its regulatory response.

One of the first things Exxon did was help found and fund a front-group coalition of energy-industry companies called the Global Climate Coalition (GCC). Mobil's climate scientist, Leonard S. Bernstein, a chemical engineer, wrote a report on behalf of the group that was circulated widely among industry executives. Bernstein noted that the potential for human-caused global warming was "based on well-established scientific fact, and should not be denied." The first draft of the report even examined the common contrarian or "climate-denier" arguments against man-made global warming, including temperature anomalies, solar variability, and the role of water vapor, and debunked each of them in turn, saying "they do not offer convincing arguments." But at the request of the committee, this section was removed before the report was widely circulated, and those very same denier arguments would later be promoted by Exxon's CEO, Lee Raymond.

The public was aware of the need for action as well, and supportive by wide margins. In a 1998 Ohio State University poll of American opinion, 79 percent of people believed global warming was happening, 80 percent thought reducing air pollution would reduce global warming, and 91 percent thought the US government should limit air pollution by businesses—numbers that energy and extraction industry executives likely found alarming.

The Global Climate Science Communications Plan

The poll had been conducted following the December 1997 talks in Kyoto, Japan, to forge an international treaty to limit the world's carbon emissions. Called the Kyoto Protocol, the treaty was tied to an earlier United Nations effort in 1992, the United Nations Framework Convention on Climate Change.

The American Petroleum Institute (API), whose members included BP, ConocoPhillips, Chevron, Exxon, Mobil, Shell, and other energy and extraction companies, wanted to prevent the United States from ratifying the Kyoto Protocol, and their members rallied support from other oil producers around the world. "Let's agree there's a lot we really don't know about how climate will change in the twenty-first century and beyond," Exxon CEO Lee Raymond told the World Petroleum Congress in Beijing in October 1997.

Today, concern about the environment focuses on the issue of global climate change. In December, representatives from some 160 nations will meet in the beautiful city of Kyoto, Japan, to decide on legally binding agreements that would have the effect of gutting the use of oil and other fossil fuels. Clearly, all of us here today have a big stake in the decisions that will be made. Proponents of the agreements say they are necessary because burning fossil fuels causes global warming. Many people—politicians and the public alike—believe that global warming is a rock-solid certainty. But it's not.

Raymond then departed from what past Exxon and other industry scientists and executives had known, and laid out the major denial talking points that would be used to combat climate science for the next twenty years.

We know that natural fluctuations in the Earth's temperature have occurred throughout history—with wide temperature swings. The ice ages are a good example. In fact, one period of cooling occurred from 1940 to 1975. In the 1970s, some of today's prophets of doom from global warming were predicting the coming of a new ice age. Some measurements suggest that the Earth's average temperature has risen about half a degree centigrade since the late nineteenth century. Yet sensitive satellite measurements have shown no warming trend since the late 1970s. In fact, the earth is cooler today than it was 20 years ago. We also have to keep in mind that most of the greenhouse effect comes from natural sources, especially water vapor. Less than a quarter is from carbon dioxide, and, of this, only 4 percent of the carbon dioxide entering the atmosphere is due to human activities—96 percent comes from nature. Leaping to radically cut this tiny sliver of the greenhouse pie on the premise that it will affect climate defies common sense and lacks foundation in our current understanding of the climate system. Forecasts of future warming come from computer models that try to replicate Earth's past climate and predict the future. They are notoriously inaccurate. None can do it without significant overriding adjustments. Even then, 1990's models were predicting temperature increases of two to five degrees Celsius by the year 2100. Last year's models say one to three degrees. Where to next year? As one climate modeling researcher said in the May issue of the prestigious magazine, *Science*, "The more you learn, the more you understand that you don't understand very much." So the case for so-called global warming is far from airtight. . . . To achieve the kind of reduction in carbon dioxide emissions most advocates are talking about, governments would have to resort to energy rationing administered by a vast international bureaucracy responsible to no one. This could include the imposition of punishing, high energy taxes. This heavy burden of taxes and regulation would take its toll in many ways—in slower economic growth, lost jobs, and a profound and unpleasant impact on the way we live. Companies in industrialized nations that compete in world markets would be seriously handicapped. Currently, most proposals exclude developing nations, including China, Indonesia, and many other countries here in the Far East.

In the following months, the API commissioned a "Global Climate Science Communications Plan" (GCSCP)—a major public-relations plan designed to engage climate scientists and their conclusions in an all-out battle. The plan was developed by a group akin to an oil-industry Last Supper, made up of twelve energy-industry apostles, many of whom would make lucrative and high-profile careers out of preaching the gospel of climate-change denial over the next several decades:

- John Adams, another journalist-turned-public-relations-expert, and his firm John Adams Associates (now KellenAdams Public Affairs)
- Candace Crandall, the wife of climate-science denier Fred Singer, and his partner in the Science and Environmental Policy Project
- David Rothbard, founder of denialist group the Committee for a Constructive Tomorrow (CFACT)
- Lee Garrigan of the Environmental Issues Council, an industry trade group working to battle "ill-conceived environmental regulation"
- Jeffrey Salmon, the executive director of the denialist George C. Marshall Institute and a speechwriter for Vice-President Dick Cheney
- Myron Ebell of Frontiers of Freedom, a denialist who went on to become director of the Center for Energy and Environment at the

Competitive Enterprise Institute and chair the Cooler Heads Coalition, and was a player behind the electoral defeat of Republicans who supported climate-change legislation, including Bob Ingliss

- Peter Cleary, yet another journalist-turned-public-relations-expert, who was then the communications manager and coalitions director for Grover Norquist's Americans for Tax Reform
- Randy Randall, a lobbyist and senior environmental advisor for ExxonMobil
- Robert Gehri, a research specialist for the Southern Company, one of the nation's largest utility holding companies, with major investments in coal-fired power plants. In 1991, the Southern Company had been part of forming the Information Council on the Environment, a coal industry group that developed a public-relations strategy to "reposition global warming as theory (not fact)" and purchased some of the first climate-denial ads, which ran on the Rush Limbaugh show in select areas. Gheri has also been Willie Soon's liaison for receiving "research" funding from the Southern Company.
- Sharon Kneiss, the federal relations manager for Chevron
- Steve Milloy, the denialist executive director of the tobacco-industry front group the Advancement of Sound Science Coalition, and now the director of external policy and strategy at Murray Energy Corporation, the largest privately-held coal company in the United States
- Joseph Walker, a public-relations consultant for the API

Under the guidance of their public-relations experts, the twelve anti-climate apostles identified five measures of success in an industry public-relations campaign to manipulate public perception and policy outcomes on climate disruption, built around the core principle Exxon's public affairs department had articulated: "uncertainties." The plan's goal was that a "majority of the American public, including industry leadership, recognizes that significant uncertainties exist in climate science, and therefore raises questions among those (e.g., Congress) who chart the future of the US course on global climate change." The group's plan stated this effort in terms of a war on science, saying that "Victory Will Be Achieved When":

- Average citizens 'understand' (recognize) uncertainties in climate science; recognition of uncertainties becomes part of the "conventional wisdom"

- Media "understands" (recognizes) uncertainties in climate science
- Media coverage reflects balance on climate science and recognition of the validity of viewpoints that challenge the current "conventional wisdom"
- Industry senior leadership understands uncertainties in climate science, making them stronger ambassadors to those who shape climate policy
- Those promoting the Kyoto treaty on the basis of extant science appear to be out of touch with reality.

The apostles noted the specific purpose of this effort: "Unless 'climate change' becomes a non-issue, meaning that . . . there are no further initiatives to thwart the threat of climate change, there may be no moment when we can declare victory for our efforts. It will be necessary to establish measurements for the science effort to track progress toward achieving the goal and strategic success." They then laid out several specific proselytizing "strategies and tactics" to achieve that victory:

National Media Relations Program: Develop and implement a national media relations program to inform the media about uncertainties in climate science; to generate national, regional and local media coverage on the scientific uncertainties, and thereby educate and inform the public, stimulating them to raise questions with policymakers.

The tactics for this strategy included:

- Identify, recruit and train a team of five independent scientists to participate in media outreach. These will be individuals who do not have a long history of visibility and/or participation in the climate change debate. Rather, this team will consist of new faces who will add their voices to those recognized scientists who are already vocal.
- Develop a global climate science information kit for media including peer-reviewed papers that undercut the "conventional wisdom" on climate science. This kit also will include understandable communications, including simple fact sheets that present scientific uncertainties in a language that the media and public can understand.
- Conduct briefings by media-trained scientists for science writers in the

top 20 media markets, using the information kits. Distribute the information kits to daily newspapers nationwide with offer of scientists to brief reporters at each paper. Develop, disseminate radio news releases featuring scientists nationwide, and offer scientists to appear on radio talk shows across the country.

- Produce, distribute a steady stream of climate science information via facsimile and e-mail to science writers across the country.
- Produce, distribute via syndicate and directly to newspapers nationwide a steady stream of op-ed columns and letters to the editor authored by scientists.
- Convince one of the major news national TV journalists (e.g., John Stossel) to produce a report examining the scientific underpinnings of the Kyoto treaty.
- Organize, promote and conduct through grassroots organizations a series of campus/community workshops/debates on climate science in 10 most important states during the period mid-August through October, 1998.
- Consider advertising the scientific uncertainties in select markets to support national, regional and local (e.g., workshops/debates), as appropriate.

The second strategy focused on ways to further undercut the scientific consensus:

Global Climate Science Information Source: Develop and implement a program to inject credible science and scientific accountability into the global climate debate, thereby raising questions about and undercutting the "prevailing scientific wisdom." The strategy will have the added benefit of providing a platform for credible, constructive criticism of the opposition's position on the science.

The tactics for this strategy included:

Establish a Global Climate Science Data Center . . . in Washington as a non-profit educational foundation with an advisory board of respected climate scientists. It will be staffed initially with professionals on loan from various companies and associations with a major interest in the climate issue.

The plan noted that it would be important to include "funding for research projects that may be deemed appropriate to fill gaps in climate science (e.g., a complete scientific critique of the IPCC research and its conclusions)."

The plan's idea was to provide sympathetic scientists "with the logistical and moral support they have been lacking. In short, it will be a sound scientific alternative to the IPCC." Its functions were to include:

- Providing an easily accessible database (including a website) of all mainstream climate science information.
- Identifying and establishing cooperative relationships with all major scientists whose research in this field supports our position.
- Establishing cooperative relationships with the other mainstream scientific organizations (e.g., meteorologists, geophysicists) to bring their perspectives to bear on the debate, as appropriate.
- Developing opportunities to maximize the impact of scientific views consistent with ours with Congress, the media and other key audiences.
- Monitoring and serving as an early warning system for scientific developments with the potential to impact on the climate science debate, pro and con.
- Responding to claims from the scientific alarmists and the media.
- Providing grants for advocacy on climate science, as deemed appropriate.

The plan had other, more nefarious aspects, including "a direct outreach program to inform and educate members of Congress, state officials, industry leadership, and school teachers/students about uncertainties in climate science." Denier organization the Heartland Institute engages in this activity, regularly sending its legitimate-appearing propaganda reports to thousands of policymakers and developing climate-denying curricula to distribute to science educators.

The plan called for the indoctrination of students by working with the National Science Teachers Association to "develop school materials that present a credible, balanced picture of climate science for use in classrooms nationwide," and to "distribute educational materials directly to the schools and through grassroots organizations of climate science partners."

Following the plan, energy companies successfully pressured the US government not to ratify the Kyoto treaty, and built an unprecedented, ongoing campaign to sow public "uncertainties" about the potential for human-caused climate disruption—a process that their own internal documents show they already knew was an "established scientific fact, and should not be denied."

An Explosion in Third-Party Dark-Money Funding

This plan was implemented by coopting the developing network of conservative think tanks and allied nonprofits, which used ideas about science developed by postmodernist academics and religious fundamentalists to turn the political tides in their favor. Their targets were regulations derived from environmental science and the perceived successes of the environmental movement. They sought to build a full-scale conservative counterculture to oppose these measures in both science and regulation and urge members of Congress to "have the courage to do nothing."

"Hard-core environmentalist activists like the Natural Resources Defense Council have been highly effective for years in utilizing the court system to enact policy, effect change, and generate significant exposure for their cause," said one conservative think tank. "The same opportunities exist for those who advocate a free-market approach, and we have an impressive track record in the courts despite being significantly overmatched by those promoting more regulation, and government-based solutions."

One of these groups' earliest and most enduring tactics was borrowed from the vast fundamentalist publishing enterprise. Libertarian groups began publishing and promoting "environmentally skeptical" books downplaying the case for environmental regulations that could hurt corporate bottom lines. The authors of one study of these books noted:

> Environmental scepticism denies the seriousness of environmental problems, and self-professed "sceptics" claim to be unbiased analysts combating "junk science." This study quantitatively analyses 141 English-language environmentally sceptical books published between 1972 and 2005. We find that over 92 per cent of these books, most published in the US since 1992, are linked to conservative think tanks (CTTs). Further, we analyse CTTs involved with environmental issues and find that 90 per cent of them espouse environmental scepticism. We conclude that scepticism is a tactic of an elite-driven counter-movement designed to combat environmentalism, and that the successful use of this tactic has contributed to the weakening of US commitment to environmental protection.

Energy companies and their allies began pumping large amounts of funding into these think tanks, steering them toward more and more engagement on the climate issue. For example, longtime Exxon public affairs adviser and lobbyist Walter F. Buchholtz joined the board of directors of the Heartland Institute, as did Thomas Walton, General Motors' director of economic policy, and James L. Johnston, a former senior economist for Amoco.

As a part of this campaign, energy and extraction companies, with the help of their PR firms, also created and funded scores of front groups engaged in activism, media outreach, lobbying, and producing pseudoscience, implementing an all-out propaganda operation. But the growing scale of industry involvement, and the heightened visibility of these think tanks in the press and the political dialogue, meant the third-party technique needed to be more closely observed. Until 2007, Koch Industries–affiliated foundations and the ExxonMobil Foundation were heavily involved in funding climate-change-denial organizations. In 2003, for ninety-one of the 118 organizations engaging in climate denial for which IRS data was available, ExxonMobil Foundation giving ran to about 5 percent of the $640 million in total foundation funding. But in 2007 and 2008, the Union of Concerned Scientists and Greenpeace began criticizing both ExxonMobil and Koch Industries as funders of climate denial, highlighting their attempts to manufacture uncertainties. Following this negative publicity, Exxon experienced a 2008 shareholder revolt over its climate stance. Big energy foundations switched strategies and stopped making publicly traceable contributions to most of the climate-denial groups. ExxonMobil Foundation's traceable giving fell to zero by 2008. Koch-affiliated foundation funding, which provided 9 percent of the funding of climate-denial organizations in 2006, fell to just 2 percent by 2010.

Coinciding with this decline in traceable funding was a rise in dark money given to denial organizations by the Donors Trust and the related Donors Capital Fund. Donors Trust is a donor-directed foundation whose funders cannot be traced, shielding them from accountability for the funding of controversial enterprises. The organizations' funding of climate-denial outfits increased dramatically, from just 3 percent of the total in 2003 to 24 percent in 2010, when total annual revenues for these organizations climbed to nearly $1.2 billion. Donors Trust now provides about 25 percent of total foundation funding used by organizations engaged in denial of climate disruption and arguing against climate regulation, according to Drexel University sociologist Robert Brulle.

Brulle conducted a broad study of IRS filings and found that, aside from Donors Trust and Donors Capital, the largest and most consistent visible funders of organizations involved in climate-change denial are now a number of well-known conservative foundations, such as the Searle Freedom Trust (Donald Rumsfeld served as president and CEO of the life-sciences company G.D. Searle, which later merged with Monsanto), the John William Pope Foundation, the Lynde and Harry Bradley Foundation, the Howard Charitable Foundation, and the Scaife Foundations (which are controlled by the family that owned Gulf Oil—which became Chevron—and several other major companies). These foundations promote ultra-free-market ideas and fund the groups Americans for Prosperity (cofounded by Charles and David Koch, AFP is currently three times the size of the Republican National Committee), the American Enterprise Institute, the Heritage Foundation, the Cato Institute (originally the Charles Koch Foundation), the American Legislative Exchange Council (ALEC), the Heartland Institute, the George C. Marshall Institute, and other organizations that promote climate-change denial, and whose combined annual funding soon swelled to hundreds of millions of dollars.

It was a marshaling of financial resources aimed at a specific policy objective, and at discrediting unsupportive science, at a level never before seen. Not surprisingly, pro-environmental voting among conservatives in the US Congress declined beginning in 1990.

The Politics of the Climate

California congressman George Miller had learned about the GCSCP plan to create "uncertainties" in the public's mind about climate science from a report he'd read after it was leaked to the media. Miller was well aware of the evidence scientists had. He knew oil and coal companies were concerned, but what he read alarmed him, and, in April 1998, he spoke against their plans on the House floor.

> We have seen this all before, my colleagues. We saw it when the tobacco companies got together to try to convince the American public that there was no link between tobacco and cancer. . . . They spent millions of dollars to undermine the scientists who were saying there is a link, to undermine the evidence.
>
> Now we see an effort where some industries do not like the scientists, independent scientists. They do not like what they have

come up with on global warming. So what they want to do is, they want to establish what they would consider an independent global climate science data center, and from this center would flow information to members of Congress, to the public, to state legislatures, to the mayors, city council people. But this independent center reportedly would be initially staffed . . . with professionals on loan from the various oil companies and associations of the major interests in climate change.

So here we are going to have a bunch of people who work for oil companies as scientists who are now going to tell us what the independent science is on global warming, as opposed to the independent scientists who have been out there now for a number of years working for universities and foundations and others to try to find out what is happening. They want to create the impression that they have scientists who radically disagree with the prevailing science about the harms of greenhouse gases and the consequential global warming.

Mr. Speaker, we have to understand that there is something going on in business in America. . . . We see the tobacco companies, they set up their spin organizations; the health care corporations, they set up their spin organizations; and now the oil companies are going to set up their spin organizations to tell us that all of this we have heard about climate change, greenhouse gases, global warming is nothing for us to be concerned about.

Scientists were largely unaware of the public-relations hurricane that Miller saw brewing, and their two generations of civic and social disengagement was about to leave them almost defenseless before the approaching gale.

Early in the Bush administration, the writing was already on the wall, as scientific conclusions were being held back or altered by ideologically motivated appointees. Still, the climate scientists held out hope. Former president George H. W. Bush had publicly acknowledged climate change and the need to do something about it in 1990. So, during the administration of his son George W. Bush, they appealed to Congress and urged the president to finally sign the Kyoto Protocol. They had no idea what they were in for.

Bush's family members were oil investors and his vice-president was a former executive at Halliburton, an oil services company. Probably aware of the political hurricane brewing in the energy industry, Bush was circumspect. Instead of signing on right away, he asked the National Academies to look into the matter, likely knowing that they tend to take a long time to assemble a panel, investigate evidence, and write their blue-ribbon reports.

When they finally did, in 2001, they concluded that (I am summarizing):

- Climate change is real.
- Satellites have detected no increase in solar energy reaching Earth since they were placed in orbit in the 1970s, yet the global climate is warming, so the warming cannot be attributed to the sun.
- By analyzing the isotopes of the hydrogen and oxygen atoms that make up the mountain and polar ice caps, we can infer past temperatures in the region.
- By analyzing tree-ring density and calibrating it with past temperature records and with ice-cap isotopes, we can estimate annual temperatures going back more than a thousand years.
- We can observe that carbon dioxide is a greenhouse gas that traps heat.
- We can measure the amount of carbon dioxide created by burning fossil fuels because it is a different isotope from environmental carbon dioxide.
- Atmospheric and oceanic carbon dioxide levels are increasing from burning fossil fuels, and the rate of this increase correlates with the warming temperatures. By the year 2000, when the report was being researched, the levels had increased from the 315 ppm measured in 1958 to 370 ppm.
- By analyzing air bubbles trapped in ancient ice, we can observe that carbon dioxide levels are at their highest level in at least 400,000 years and are continuing to rise.
- A warmer atmosphere holds more moisture and is stormier. Water vapor is also a very powerful greenhouse gas, leading to a positive feedback loop that can accelerate warming.
- We would logically expect this to result in a drier climate as more moisture remains trapped in the atmosphere and in more violent storms because of the extra moisture and heat. Heavy rainfalls tend to run off before they can be absorbed, so we would logically expect less moisture in the soil and more flooding.
- The most recent decade (then the 1990s), when global temperatures are averaged, is the warmest in at least a thousand years.

- As logically expected, there is more rapid warming over land and the north polar region than over the oceans, which take longer to warm. We also expect some areas will have local cooling as weather patterns change. Weather should not be confused with climate.
- Vast increases in methane gas production by large-scale ranching and farming, coal mining, landfills, and natural gas handling are also adding to the warming, as methane is a very potent greenhouse gas. Increases in emissions of nitrous oxide, a by-product of the use of nitrogen fertilizers, are also contributing.

Bush got the report and responded on June 11 with a major speech in which he said he would not sign onto Kyoto because China and India, as developing countries, didn't have to reduce their emissions, and because we still didn't know enough—effectively adopting the Global Climate Science Communications Plan and its focus on uncertainties.

Like much of science during the bottom-wing Bush administration, the exhaustive National Academies report was ignored in favor of the political wishes of powerful vested interests. National Academies reports and testimony delivered to President Bush and to Congress in every subsequent year of his presidency further developed these data with new research, showing, for example, that the climate was warmer than at any time in the last 650,000 years, and that increased warming may lead to the release of large amounts of methane currently frozen under the ocean and in northern Russia, possibly inducing a positive feedback loop of global warming that we may not be able to stop. These reports met with a similar response.

Of Polar Bears and Profits:
A Case Study in Antiscience PR

What wasn't ignored was the polar bear. One of the many data sets scientists had been tracking during this time was wildlife counts, to see if and how wildlife were reacting to climate disruption, particularly in the polar regions, where its effects are most pronounced. By 2005, scientists had ten years of data showing declining numbers of polar bears. Because polar ice was retreating earlier in the season, polar bears had a narrower time frame in which to hunt during the spring, when seal pups are born. The seals are a primary spring food source for both adult bears and their cubs. The retreating ice also made it more difficult for the bears to hunt for other prey. The polar bears observed were thinner and producing less milk, causing bear cubs to die and reproduction rates to decline. As a result, overall polar-bear populations in the Hudson Bay area were down by about 20 percent. Based on this data, a young environmental lawyer named Kassie Siegel at the Tucson, Arizona–based Center for Biological Diversity petitioned the US Fish and Wildlife Service (FWS) to list the polar bear as threatened under the Endangered Species Act due to global warming.

"We wanted to force the Bush administration to acknowledge the reality of climate change, specifically via its effects on the polar bear," says Siegel. "The Endangered Species Act requires the government to judge cases only on the basis of the best available science, not economic impact, not political pressures, not industry compromises. We could use that to say, 'Okay, Bush administration, either go ahead and protect the polar bear, or deny the petition and we can go to court and litigate the best available standard on climate change, and you don't want that either.'"

The petition generated considerable publicity, most of it sympathetic to the bears, which the public thinks of as cute, fuzzy, or majestic. They were even Christmas mascots for Coca-Cola and Marshall Field's. FWS director H. Dale Hall said the agency received some six hundred thousand public comments. The FWS asked the US Geological Survey to prepare nine detailed studies of the polar-bear population to better understand what was happening.

But then a strange thing happened. Scientists began to speak up and challenge the idea that polar bears were threatened. It began with David Legates, then the state climatologist for Delaware. Legates published a May 2006 study that critiqued a report by scientists from eight arctic nations and six indigenous peoples documenting how climate disruption was causing melting of the polar-ice cap, concluding that the study's claims were "not supported by the evidence." The only problem is that Legates's "study," which was widely quoted in conservative publications, was an unscientific policy paper prepared and published by the National Center for Policy Analysis, a think tank in Dallas whose "mission is to seek innovative private sector solutions to public policy problems." Legates was listed as a senior fellow and was also working for numerous other climate-denial outfits, many funded by ExxonMobil and the Koch foundations. His paper offered a rhetorical argument couched as science.

The same day, a polar-bear biologist named Mitch Taylor from the Department of the Environment of the government of Nunavut, Canada, published an opinion piece in the *Toronto Star* arguing that the polar-bear population was increasing, not decreasing. The population, Taylor argued, had grown by 25 percent during the previous decade. Energy-industry groups and conservative media outlets reported that overall polar-bear populations had exploded—from five thousand in 1972 to twenty-five thousand in 2007. The actual estimate was a low of five thousand to ten thousand and a 2007 population of twenty thousand to twenty-five thousand. As well, the report neglected to mention that polar bears were recovering from a period of overhunting in which sport hunters took 85 to 90 percent of the polar-bear kill, which stopped only when Canada established quotas in 1968 and Alaska banned sport hunting in 1972. The government of Nunavut had just increased its polar-bear sport-hunting quota by 20 percent in January, and Taylor, their polar-bear expert, was "concerned that listing the bears as threatened could lead to a ban on sports hunting." The sport hunting of polar bears brings about $2 million to Nunavut communities each year.

Next came a paper in an obscure science journal called *Ecological Complexity*, coauthored by Legates, Harvard-Smithsonian Center for Astrophysics professor Wei-Hock "Willie" Soon, Markus Dyck of Nunavut Arctic College, and others. Wait a minute, it argued. Polar bears aren't declining because of loss of Arctic ice due to climate disruption—Arctic temperatures aren't changing that much. And if polar bears are declining at all, it's probably because of their interactions with humans, or because they are competing with other polar bears. Any claim that the decline has to do with climate change is irresponsible science done by people with a political agenda.

Again, conservative bloggers pounced on the paper and mainstream journalists became interested. But there was a problem that few of the journalists covering the story noticed: this "scientific" paper actually wasn't. Like Legates's policy paper, it did no original research at all. Instead, the paper's authors cherry-picked bits and pieces of research from other papers to support an a priori conclusion. Thus the paper was posing a rhetorical argument as a scientific one, a classic "vulgar Induction," as Bacon called it, producing "a case, or several cases, wherein their proposition holds." But *it was not science*.

Not being science, there was nothing to peer-review, so it was accepted for publication the day after it was submitted and run in the journal as a "Viewpoint" article—scientists' term for something that's interesting and perhaps informative, but is "but faith, or opinion."

Soon There Will Be Lies

The paper cited 101 sources and so appeared to be scientific. Willie Soon had authored other discredited, industry-funded climate-denial papers in the past, including with Legates. He is not a climate scientist or a polar-bear expert. He is an aerospace engineer, and was one of the "new faces" the GCSCP public-relations plan had identified would be strategic in a disinformation campaign that created "uncertainties" about climate science. In fact, Soon would come to be paid very handsomely by several of the original creators of the plan.

Soon is educated enough to understand that what he is doing is not science, but rather the rhetoric of "vulgar Induction" or pseudoscience, and that he is using the credibility of his position at the Harvard-Smithsonian Center for Astrophysics to fool people.

Soon is also a frequent speaker at climate-change-denial conferences, such as those sponsored by the Heartland Institute. The Heartland Institute is a libertarian tobacco- and climate-denialist outfit, located in Chicago and headed by Joseph Bast. In addition to speaking at the institute's annual conferences, Bast speaks to lawmakers at meetings of ALEC, the American Legislative Exchange Council, a conservative advocacy organization of state legislators and industry, and publishes the propaganda report of "The Nongovernmental International Panel on Climate Change" (NIPCC) which Heartland prepares along with Fred Singer. The NIPCC is a ripoff of the official United Nations scientific organization, the Intergovernmental Panel on Climate Change (IPCC). This trick is common in climate-denier PR, and is designed to create confusion. In addition to speaking regularly at Heartland conferences, Soon is a contributor to the NIPCC report, which is sent to thousands of legislators across the country.

Ecological Complexity has ethical rules about disclosure of conflicts of interest. Because of this, in the back matter of the polar-bear paper, Soon acknowledged receiving some portion of funding from ExxonMobil, as well as from the Charles G. Koch Charitable Foundation and the American Petroleum Institute. But perhaps journalists didn't read that far. And, as it turns out, he didn't disclose all his funding sources.

Contracts, proposals, reports, letters, and other documents obtained in 2015 via a Freedom of Information Act request showed that, while Soon ostensibly works at the Harvard-Smithsonian Center for Astrophysics, he has relied exclusively on grants from the extraction and energy industries for his entire salary and research budget. He received more than $1.2 million from such companies, including ExxonMobil, the American Petroleum Institute, the Charles G.

Koch Charitable Foundation, the Donors Trust, and Southern Company, a large electric utility in Atlanta that generates most of its power from coal. Soon has published at least eleven papers since 2008 that have omitted disclosure of this funding. In at least eight of those cases, he appears to have violated the ethical guidelines of the journals that published his work, which require such disclosure in order to prevent just the sort of agenda-driven pseudoscience in which Soon engages.

"ExxonMobil is pleased to provide the enclosed contribution to the Smithsonian Astrophysical Observatory in the amount of $76,106.00 for General Support," said a letter from an Exxon-Mobil senior director accompanying a check in March 2009, one of many examples. The letter made no reference to Soon.

The check was in response to an itemized "request for payment" listing Soon's salary of $49,305 for seventy-five days plus nearly $23,000 in other benefits, plus $3,355 in travel, publication, and indirect costs—a total of $76,106.00. This was just one of many such exchanges between the Harvard-Smithsonian Center for Astrophysics and various extraction and energy companies, including proposals for "deliverables"—papers Soon would write that created "uncertainties" about mainstream climate science—and subsequent requests for payment. These work-for-hire payments were kept quiet according to the classic public-relations dictum of protecting the anonymity of the client using the third-party technique, as laid out by Edward Bernays.

Trading on the respected names of Harvard and the Smithsonian made Soon's work particularly valuable. His many obfuscations over the years have been promoted by groups and bloggers aligned with ExxonMobil, the Koch foundations, and the conservative think tanks to which they donated. Many news outlets then reprinted versions of these stories without critical examination, their reporters using their control-C keys in a way that BBC business journalist Waseem Zakir has referred to as "churnalism."

AM talk-radio hosts picked up the new "findings" in a second wave of denialism. On the basis of these articles and radio pronouncements, conservative and industry groups said there were simply no data to support the decision to list the polar bear under the Endangered Species Act. These groups included the American Enterprise Institute, the Heritage Foundation, the National Association of Manufacturers, and the George C. Marshall Institute, which was also known for its antiscience public-relations efforts on behalf of tobacco companies, and which was a creator of the GCSCP.

How to Short-Circuit Democracy

With the GCSCP now put into practice, a more virulent and aggressive climate-denial method developed. Not only was it possible to create "uncertainties" in the public's mind, but complete, defiant rejection by a large portion of the public, which could be mobilized to stop policy in its tracks. This more aggressive approach has now become a well-established seven-stage method of cloaking rhetorical arguments in the language of scientific legitimacy in order to influence public perception and effect a desired policy objective—the original goal of public-relations campaigns as first developed by the Creel Committee. Often, since these industry-funded campaigns have been about forestalling regulation that could affect corporate profits, they employ the same logical fallacies used by fundamentalists and the antiscience militia: *Lacking certainty, we should do nothing* and *Since the conclusion is not certain, we should get a balanced perspective from both sides.*

Step one of the seven-step method starts with phony science, in the form of either a paper or another event that creates "uncertainties" about the accepted views of mainstream science. This is usually done by cherry-picking bits of real knowledge that appear to contradict established science. These bits are then taken out of context and used by trumpeting them ("high-pointing," as public relations defines the focusing of attention) or by limiting the statistical scope of the analysis. Sometimes, they are also employed in outright lies, as we'll see more of in a minute, similar to the arguments used to battle both tobacco and pesticide regulation.

Step two follows with slanted press materials spoon-fed to journalists by industry-affiliated nonprofits and bloggers. Remember, for greatest effect, PR firms work hard to maintain their anonymity in the process via the third-party technique. These press releases and canned stories promote a narrative about the supposed "controversy." If there is a controversy, there must be legitimate disagreement, which suggests the science is uncertain.

A concurrent aspect of the PR battle is the third step: building and financing industry-aligned front groups (fake public-interest organizations) and astroturf groups (fake grassroots organizations). Using phony science and controversy, these outfits "whip up grassroots anger and frustration," to quote Tim Phillips, the president of the Koch-aligned Americans for Prosperity, by holding earned media stunts and rallies, and by purchasing paid media. This may also include recruiting, supporting, or promoting candidates to challenge offending policymakers in endorsement battles or primary races, where the outcome can be affected by relatively few voters and a relatively small investment. Another ploy has been to forge letters. During the battle over climate change in the US House of Representatives, members received letters purporting to be from the National Association for the Advancement of Colored People (NAACP). The letters urged members of Congress to "protect minorities and other consumers in your district from higher electricity bills" by voting against the proposed cap-and-trade bill. In fact, these letters were from Bonner and Associates, a Washington, D.C., public-relations firm that specializes in astroturfing. Bonner was a subcontractor for the Hawthorn Group, another PR firm, whose client was the American Coalition for Clean Coal Electricity (ACCCE), an alliance of coal companies, utilities, and railroads that ship coal and that in 2012 became headed by Mike Duncan, the former chairman of the Republican National Committee. This PR campaign was likely part of the $11.8 million the ACCCE spent on lobbying between April 1 and June 30, 2009, when the bill was being debated, according to the Office of the Clerk of the US House of Representatives.

Step four is to proselytize the intelligentsia who fund political campaigns and shape opinions in the press. Outlier scientists are recruited to publish in phony journals and speak at conferences of physicians, lawyers, and other professionals, emphasizing the controversy and sowing "uncertainties" and denial, thus using peer-pressure to create true believers among the influential opinion leaders.

Once laundered in the legitimate press and supported by front groups and opinion leaders (all third-party techniques), the story is then picked up, in step five, by industry-aligned, or otherwise sympathetic talk-radio and cable-news purveyors, who reference these mainstream sources, react with outrage, and call for policy action.

The first five stages of the propaganda campaign lay down political cover in a sort of suppressive fire for the sixth stage of the attack: legislative or other policy action by partisan allies in government, who are now politically supported to act along the desired lines. In the polar-bear case, the sixth stage began with a major speech by the US Senate's leading climate denier, Oklahoma senator James Inhofe, who attacked the petition. This was followed by Alaska governor Sarah Palin's effort to prevent the polar bear's "threatened" listing.

All of these conspire to support the seventh and final stage. Now seemingly supported by science, the press, the government, the public, the business and professional opinion leaders, and sometimes by actors purporting to represent religion—in other words, by all other major houses of power in society—industry representatives can step safely out from behind the curtain for the main act of the culture-war drama and plead their case to policymakers or in advertisements to their constituents, or both. They don't want to appear negative; they are on our side, protecting us from perhaps well-intentioned but poorly thought-out policies that are simply (they say) not fully supported by science, as one can see from the controversy and uncertainties. In the polar-bear example, five energy-industry organizations filed suit in August 2008 against then–interior secretary Dirk Kempthorne and FWS Director Dale Hall, joining Palin's effort to reverse the listing.

The strategy is designed to neutralize the primacy of objective knowledge and slowly move public opinion toward accepting the industry's position as the only truly reasonable one, subverting the democratic process just as Bernays, and Creel before him, intended. In certain circumstances it is fully implemented; in others, like the polar-bear example, there may not be the time or the need for an astroturf grassroots campaign or candidate recruitment and support (which wasn't applicable in the polar-bear case), so those elements may be skipped.

The coordination between energy-industry companies and their allies in the seven-point PR attack on the polar-bear listing was well-orchestrated. The documentation accompanying Palin's April 9, 2007, letter to Kempthorne referenced Willie Soon's polar-bear paper—seven days *before* it was actually published.

The Part the Public Didn't Hear

Kassie Siegel describes copies of e-mails between Hall and Kempthorne over the decision of whether or not to list the polar bears. "Hall says in one e-mail, 'I got this call from Lyle Laverty,' who was Kempthorne's assistant, 'and he told me the secretary has decided not to list the polar bear and I told him of course that's his decision to make but I don't want to be at the press conference because I can't support it and I won't be helpful to you.' Then they have all these notes back and forth, reasons to not list: we want to be in the driver's seat on policy, why not let the courts decide. There was really a lot of deliberation, they didn't want to do it, but they realized they would lose because the legal standard was the science. Then on May 15, 2008, Kempthorne gets up there and says he's listing the polar bear."

Kempthorne made it clear he wasn't pleased. "While the legal standards under the ESA compel me to list the polar bear as threatened," he said, "I want to make clear that this listing will not stop global climate change or prevent any sea ice from melting. Any real solution requires action by all major economies for it to be effective. That is why I am taking administrative and regulatory action to make certain the ESA isn't abused to make global-warming policies."

Hall—who had been appointed by President George W. Bush—testified before Congress that numerous studies showed there was no significant scientific uncertainty over the fact that polar bears were endangered by global warming. But the public didn't hear *that* story.

Reverse Highpointing: The Hockey-Stick Attack

After the polar-bear attack, energy companies stepped up their investments in the infrastructure supporting this attack strategy, from funding anti-climate-change nonprofits and activist groups to paying scientists like Willie Soon, Fred Singer, and others to create "uncertainties" over the real science, all the way to massive lobbying, advertising, and political-contribution campaigns. Public data from the Securities and Exchange Commission and charitable organizations' reports to the IRS show that, between 2005 and 2008, when a new US president would be elected, ExxonMobil gave about $9 million to groups linked to climate-change denial, while foundations associated with Koch Industries gave nearly $25 million. The third major funder was the American Petroleum Institute. But this was a drop in the bucket compared to what these companies and organizations would spend overall.

Between 1999 and 2010, the energy industry spent more than $2 billion fighting climate-change legislation, more than a quarter of it—$500 million— from January 2009 to June 2010, when President Obama's cap-and-trade bill was in play in Congress. That amounts to almost $1,900 per day in lobby expenditures for every single US senator and representative in Washington—and those numbers don't include nonreportable expenses like public relations, publicity, earned media, paid media, nonprofit donations, rallies, and polling. They spent an estimated $73 million more on anti-clean-energy ads from January through October 2010, and Koch family foundations gave $48 million overall to groups engaged in climate-change denial between 1997 and 2008.

It was an assault that climate scientists were wholly unprepared for, even though it had been building for some time. In 2003, Willie Soon had coauthored another phony-science paper, this one published by the George C. Marshall Institute. That paper attacked the work of prominent climate scientist Michael Mann, developer of the famous "hockey stick" graph. Mann's graph charted average temperatures for the last thousand years, and the sharp increase it shows in the last century makes it resemble a hockey stick. It was used prominently in the 2001 report issued by the United Nations' Intergovernmental Panel on Climate Change. It was also used, more famously, by Al Gore in the documentary film *An Inconvenient Truth.*

Tree scientists have correlated tree-ring density measurements with annual average temperature data. Mann used this correlation to analyze tree rings going back a thousand years and infer unknown temperatures. He then plotted them on a chart using blue and, being a scientist, he added gray bars to represent the statistical probability of error in his estimates, which increased with time. Finally, he added current known temperature measurements in red.

The findings by Mann and two other climate scientists were originally reported in a paper published in *Nature* in 1998, and further refined the following year. They represented classic observational empiricism. By proceeding logically from observations, they were able to create knowledge that could be tested and verified by others, which it was. The results of all of the various methods and observations overlapped in relative agreement, creating confidence that the temperature estimates were reliable. Mann's work had the simple, iconic power of Hubble's four-page paper placing Andromeda outside the Milky Way. And that power would soon make it the target of a much, much larger attack.

Science's Rightful Place

After the election of President Barack Obama in November 2008, many scientists hoped that science was entering a new era in which it would be restored, as the president promised in his inaugural address, "to its rightful place," which was the mission statement of ScienceDebate.

Obama's science advisor, John Holdren, was outspoken on population and climate disruption. Energy secretary Steven Chu was a Nobel Prize–winning nuclear physicist and a prominent climate scientist who supported transitioning from fossil fuels to a low-carbon economy. NOAA administrator and under secretary of commerce for oceans and atmosphere Jane Lubchenco was a renowned marine biologist and a past president of the AAAS who was vocal about dealing with climate disruption, as was former Clinton chief of staff John Podesta, chief of Obama's transition team, who backed a transition to a low-carbon economy. National Cancer Institute director Harold Varmus had pushed Obama to focus on science during the campaign. US Geological Survey director Marcia McNutt had run the Monterey Bay Aquarium Research

Institute and was similarly outspoken. The oceans are currently absorbing more than half of the extra carbon dioxide being produced by humans. They are now 30 percent more acidic than they were at the beginning of the industrial revolution, inhibiting the formation of corals and shells, and threatening the foundation of the entire aquatic food chain. All of these presidential appointees had stated that the next president needed to prioritize major science issues, and all had publicly backed the call for a science debate during the campaign.

Anatomy of an Antiscience Takedown: Cap-and-Trade

Once in office, Obama took a more cautious approach. The economy had collapsed in the final days of the campaign and, like President Kennedy in the space race, Obama found himself caught between his broader vision and financial limitations, this time caused by the Great Recession. He pulled back, prompting some to call him "timid." He had to choose which initiatives he would put his full force behind. Strategically, he chose the one he thought would have the most immediate positive impact for voters and the economy: health-care reform. This afforded opponents of climate legislation the space to mount a massive PR campaign to change the dialogue in Congress.

The way it went down is illustrative of the effectiveness of a full-out execution of the seven-stage antiscience PR strategy—phony science, controversy, grassroots astroturfing, proselytizing, outrage, policy response, and pleading in support—in circumventing the democratic process.

"I ask this Congress to send me a market-based cap on carbon pollution and drive the production of more renewable energy in America," President Obama challenged Congress. It was February 24, 2009, just a month into his presidency, and his first address to Congress. But he didn't actively lobby for it in the same way he did health care, and at the same time, reacting to political pressures, he began expanding offshore drilling to a greater extent than any previous president, sending a very mixed message.

On June 26, the House passed the American Clean Energy and Security Act, a cap-and-trade bill, by a vote of 219–212, with eight Republican votes. The US Senate was controlled by a Democratic majority, and the bill looked as if it would get through.

But Americans for Prosperity, the advocacy organization started by David Koch and Richard Fink, a member of Koch Industries' board, got busy whipping up public sentiment. Koch Industries is America's largest private oil refiner, and in 2014 *Forbes* named it the largest privately held company in the United States, with annual revenues of nearly $135 billion. AFP was one of the lead organizations behind the Tax Day Tea Party protests of 2009.

The group also led the energy industry's all-out PR assault against the cap-and-trade bill. AFP president Tim Phillips described the effort: "We certainly did radio ads, TV ads, social media. We did rallies, events. We launched something we called Hot Air." Hot Air consisted of a hot-air balloon that Americans for Prosperity drove around the country, doing PR stunts and rallies led by Phillips, which were filmed to make public service–style television ads and internet spots. "I'm Tim Phillips, president of Americans for Prosperity," one ad went, "and we're here in Billings, Montana, . . . and we're sending a message to Senator Baucus and Senator Tester to vote no on this job-killing, tax-increasing cap-and-trade." Behind Phillips was the hot-air balloon, which bore a banner saying, "Global warming alarmism: lost jobs, higher taxes, less freedom."

"We went all over the country stirring up grassroots anger, and frustration, concern," said Phillips. Another ad showed a lighter labeled "Cap and Trade" lighting a bomb labeled "Unemployment" as an announcer declared, "Cap-and-trade will slow the economy and cost American jobs." The ad claimed the bill could cost $200 billion. Other ads claimed the science behind global warming was "a hoax" and called it "the greatest scam in history," quoting climate-change denier John Coleman, founder of the Weather Channel.

There was a reason Phillips was targeting certain races. "People in this country were really not concerned one way or the other on this issue until the economy started going south in 2008," said Wisconsin Republican congressman James Sensenbrenner Jr., the ranking Republican on the House Select Committee on Energy Independence and Global Warming and a longtime global-warming critic. He said the economic downturn was a game changer, much as it had been in 1988 when Exxon made its fateful turn against climate science after the oil glut. Economic downturns are good for climate denial.

Soon the antigovernment, anti-tax, antiregulation rhetoric at Tea Party gatherings, which were running at a fever pitch in 2009 and 2010 after the election of Obama, the first black US president, started including language ripping on cap-and-trade as "taxing America to death."

The negative economic mood was also affecting the White House's views on environmental issues. Obama was still grappling with staving off a depression. Top officials forbade administration members from using the words "climate change" in their messaging. They called leading environmental activists to the White House to tell them the words "climate change" did not represent a winning message and could open Obama up to attacks from industry and conservative groups opposed to regulation. Instead they focused on the "clean-energy economy," causing a dearth of discussion about climate change until Hurricane Sandy blew it back onto the agenda in the closing days of Obama's 2012 reelection campaign. Still, during that campaign, there was no mention of climate change in the six hours of televised debate. Debate moderators failed to bring up the question, and Obama and Romney were happy not to talk about it. It was ScienceDebate that finally got the candidates to answer questions about it.

While Tim Phillips was out in the field whipping up anger and frustration and the Obama administration was backing away, climate-denial spin doctors were testifying before Congress in hearings on the cap-and-trade bill. They included Christopher Monckton from the innocuous-sounding Science and Public Policy Institute, E. Calvin Beisner from the Cornwall Alliance, and Patrick J. Michaels of the Cato Institute. All of these men were advisory board members of CFACT—the Committee for a Constructive Tomorrow—a libertarian think tank and one of the original creators of the GCSCP—the public-relations plan to create "uncertainties" in the mind of the public about climate science.

At Tea Party rallies, cap-and-trade opponents were gearing up for "a new civil war, and the greatest part of that battlefield is the global-warming battle," according to Monckton, a British journalist with no particular expertise in climate science who traveled America calling global warming "bullshit."

Phillips and other public-relations experts and climate deniers soon had members of Congress hearing from them both at home and in committee hearings. Now it was time to influence opinion leaders—members of the professions that strengthen communities, like business owners, doctors, and lawyers—and who also tend to give to political campaigns.

Americans for Prosperity, the Competitive Enterprise Institute, CFACT, and others sent out their hired scientists—people like physicist and contrarian Fred Singer—who spoke to opinion leaders at conferences. "In the last ten years, there hasn't been a warming," Singer said. "We don't know why that is. But one doesn't see any warming in the observations. There simply is no trend." Singer has a calm, reasoned, grandfatherly demeanor, and is the sort of person listeners might imagine a scientist to be. But he uses statistics to fool audiences when he says there hasn't been any warming in the last ten years, and this is one of the most common tricks in science-denial PR.

Statistics, as discussed, are important to science because that's how scientists quantify the relative probability that a statement is true. We need statistics to do this because of the inductive nature of scientific reasoning. We cannot say all swans are white because we cannot observe the whole universe at once. So we have to say that, based on observations, all swans are probably white, holding out a chance that the statement could be wrong. We use statistics to express relative confidence that the statement is correct. That is also one of the main reasons we try to quantify things in science. The more data, the better the reliability.

In the case of climate disruption, the global temperature anomaly, or warming, rises with the sort of jagged up-down-up progression that one might see in a chart of a rising stock price. The price may fluctuate up and down, but the overall trend is up. When Fred Singer says, "In the last ten years, there hasn't been a warming," he's choosing a limited statistical selection of the overall record. Because of the jagged, up-and-down increase in the temperature record, one can pick *any* ten-year period and show no warming, as long as one starts on a spike and ends in a trough. But overall, when we consider all the data without cherry-picking, the trend is up. Playing with statistics in this way is an excellent strategy to fool smart people with numbers, which is why it worked on Singer's audience of opinion leaders.

With phony science, controversy, grassroots, and wealthy donors and opinion leaders covered, with bloggers and partisan media figures like Rush Limbaugh whipping up outrage and demanding action, AFP, the Competitive Enterprise Institute, and other front groups began seeking to influence congressional races, their adherents supporting Tea Party primary opponents running against Republicans who had either voted for or spoken in favor of the cap-and-trade bill. Such efforts helped to knock out at least two well-respected longtime Republicans: Bob Inglis of South Carolina and Mike Castle of Delaware, both of whom had supported climate legislation. "That was a very key factor in ten to fifteen congressional districts in the 2010 election, where strong supporters of climate-change legislation ended up being defeated," said Sensenbrenner.

Why did the PR campaign focus on the House when the next vote was in the Senate? Because the House candidates (and some Senate candidates) were up for election in 2010, and the bill wasn't scheduled to be taken up by the Senate until after the spring primary election season—a strategic blunder. As then-senator John Kerry put it, "There's nothing like a loss in an election to promote fear in the survivors. And that's exactly what happened in the United States Congress."

Senators were getting nervous. Both Obamacare and cap-and-trade were large-scale economic reforms, and the economy was still recovering from the 2008 meltdown. "I often say, if you ever wonder, as an activist on our side, are you making a difference, just look at cap-and-trade," Phillips told John Hockenberry in a 2012 *Frontline* interview. "January of '09, we had a president with sixty votes in the Senate and then Speaker Pelosi with a fifty-plus-seat majority. And cap-and-trade was at the top of their agenda. And in the end, they were beaten."

In July 2010, Senate Majority Leader Harry Reid announced that he no longer had the votes to pass the legislation before the August break. The momentum towards action on global warming had all but vanished.

Climategate

But science was another matter. While politicians can be intimidated, scientists focus on data. The energy companies had known for decades that the data was not good and was only going to get worse. The United Nations was planning to host a major climate summit in Copenhagen, which would exert increasing pressure for global action. Concurrently with the PR campaign in US legislative districts and in Congress, another PR campaign was needed in the scientific community—one that would provide evidence for the outrage stage of the plan. This is where the antiscience PR efforts got really ugly.

On November 17, 2009, not quite five months after the passage of cap-and-trade in the US House and days before the start of the Copenhagen summit, an unidentified hacker posted a sixty-one-megabyte file on a Russian FTP server. It contained e-mails stolen from servers at England's University of East Anglia Climatic Research Unit (CRU). The hacker then posted a link to the file on the climate-skeptic blogs the *Air Vent* and *Watts Up with That?* as well as the blog *RealClimate*, which is run by several leading climate scientists, including Michael Mann.

The CRU is one of the world's leading centers for climate research and a hub of global climate-science communication. The uploaded file contained thousands of private e-mails exchanged by top climate scientists over more than thirteen years. The *Air Vent* blogger Jeff Id quickly began highlighting ones that made it appear as though scientists were cooking their data to reach an a priori conclusion—which, ironically, was exactly what energy-industry-funded scientists like Willie Soon were doing. Anthony Watts cobroke the story on *Watts Up with That?* and even though he was traveling he too managed to identify key e-mails and quickly post them on his blog.

The leak was then broadly publicized, with reporters' attention being drawn to the most damning and easily reframed e-mails. One of these, reproduced by Watts in his first post on the hack, was between CRU director Phil Jones and the three original authors of the hockey-stick graph—Michael Mann, Raymond Bradley, and Malcolm Hughes.

From: Phil Jones <p.jones@xxxxxxxxx.xxx>
To: ray bradley <rbradley@xxxxxxxxx.xxx>,
mann@xxxxxxxxx.xxx, mhughes@xxxxxxxxx.xxx
Subject: Diagram for WMO Statement
Date: Tue, 16 Nov 1999 13:31:15 +0000
Cc: k.briffa@xxxxxxxxx.xxx,t.osborn@xxxxxxxxx.xxx

Dear Ray, Mike and Malcolm,
Once Tim's got a diagram here we'll send that either later today or first thing tomorrow.
I've just completed Mike's Nature trick of adding in the real temps to each series for the last 20 years (i.e. from 1981 onwards) amd [*sic*] from 1961 for Keith's to hide the decline. Mike's series got the annual land and marine values while the other two got April-Sept for NH land N of 20N. The latter two are real for 1999, while the estimate for 1999 for NH combined is +0.44C wrt 61-90. The Global estimate for 1999 with data through Oct is +0.35C cf. 0.57 for 1998.
Thanks for the comments, Ray.
Cheers
Phil

Prof. Phil Jones
Climatic Research Unit Telephone +44 (0) 1603 592090
School of Environmental Sciences Fax
+44 (0) 1603 507784
Email p.jones@xxxxxxxxx.xxx
University of East Anglia
Norwich
NR4 7TJ
UK

As in the polar-bear case, a seven-stage propaganda attack was employed: phony science; grassroots organizing; controversy via canned stories by bloggers for the benefit of the mainstream press; proselytizing opinion leaders; talk-radio and cable-news reaction and outrage; government intervention; and hand-wringing by industry actors, our much-maligned and patriotic heroes. Suddenly, scientists found themselves not just fending off junk science, but directly in the crosshairs.

"Phil Jones has gone on record saying that he was using the term 'trick' in the sense of 'a clever way of solving a problem,' like when you divide both sides of a math problem to isolate a variable. It's a way to make things apparent, like a magic trick," Mann says. "In referring to our 1998 *Nature* article, he was simply pointing out that the proxy record of tree-ring data we used to estimate historical annual temperatures ended in 1980, so it didn't include the warming of the past few decades. So in our *Nature* article we also showed the thermometer data that was available after 1980 and through 1995, which we clearly labeled in a different color so that the reconstruction could be viewed in the context of recent instrumental temperatures." The validity of Mann's temperature reconstruction was subsequently independently verified by the National Academy of Sciences.

Mann says Jones's reference to adding data to "hide the decline" referred to work by their colleague Keith Briffa, who was being copied on the e-mail. "It had nothing to do with hiding a supposed decline in global temperatures," he says. That was a reframing of it by antiscience opponents. The "decline" referred to a well-known decline in the response of high-latitude tree-ring density to temperatures after about 1960.

In an article published in *Nature* in 1998 that correlated tree-ring density to temperature, Briffa and his colleagues explained that the close correlation began to decline after about 1950 and was unreliable after 1960. "Decline" was being used in that specific scientific sense.

"Hide" probably hadn't been the best word choice, since the existence of the decline was a major point in the original Briffa *Nature* paper. Jones was simply saying that he was substituting instrumental data. "The irony," says Mann, "is that Phil was trying to be precise and careful, and was criticized as if he were trying to be just the opposite."

Exploiting the different meanings of a word—scientific versus colloquial— to foment public mistrust was becoming a familiar strategy, one taken from the playbook of fundamentalist evolution deniers and their talk about "theory."

Manipulating Churnalists with a Sexy Nut Graf

Journalists who don't understand science and believe that there is no such thing as objectivity can easily be manipulated by propaganda campaigns (often run by former journalists), because they think taking a position on evidence is advocacy. That this could happen may at first seem confusing. If these journalists don't believe in objectivity, why would they have a problem with advocacy? But the refusal to believe in objectivity means that matters of science—or more precisely, matters of objective knowledge—present such believers with an intellectual paradox where they are unqualified to endorse the validity of a scientist's claim, and therefore too often disregard it in favor of presenting "both sides" of scientific versus antiscience controversies from the frame of a rhetorical dispute rather than an evidentiary one. This is not objective reporting, when an effort is made to establish the objective truth or falsehood of a statement based on evidence; neither is it subjective in every sense, where the reporter is a gonzo who acknowledges his role and biases. It is, rather, a third thing, a sort of Pontius Pilate, laissez-faire washing of the hands brand of journalism, applied when knowledge issues are in play. No effort is made to determine which "side" is objectively correct because there is no such thing as objectivity and therefore attempting to pursue it would be an inappropriate interjection of subjective bias into the story. Thus the best they can do is report both sides and wash their hands of it.

To understand the difference between journalistic and scientific communication, the way statistics can be manipulated to fool journalists, and the care journalists must take to avoid this, consider a German study conducted by John Bohannon. Bohannon has a PhD in molecular biology but he works as a science reporter. He wanted to expose the junk science of the diet and nutritional-supplement industry, so he worked with Gunter Frank, a German doctor, to conduct a study that showed that eating bitter dark chocolate daily helps one lose weight. Frank said he chose dark chocolate because it was a favorite of the whole-food fanatics. "Bitter chocolate tastes bad, therefore it must be good for you," he says. "It's like a religion." The results fooled millions. Here's how.

Frank recruited fifteen test subjects on Facebook who volunteered for a twenty-one-day study on weight loss. Each person was paid 150 euros to participate. Frank ran blood tests to rule out diabetes and other health issues, then divided the volunteers into three groups. One group followed a low-carbohydrate diet. Another followed the same low-carb diet plus a daily 1.5-ounce bar of dark chocolate. And the rest, a control group, were instructed to make no changes to their diet. The subjects weighed themselves each morning for twenty-one days, and the study finished with a final round of questionnaires and blood tests.

The results? The subjects who didn't change their diet bounced around zero weight loss. The group that dieted without chocolate lost weight. But the group that dieted and ate a 1.5-ounce bar of dark chocolate daily lost 10 percent more weight, a statistically significant finding. Bohannon had his trick. He wrote up the study and submitted it to twenty journals-for-hire—a new PR-driven set of journals in which scientists pay to be published and that, as Bohannon had shown in an earlier sting for *Science*, do little to no peer review. Within twenty-four hours, the paper was accepted by the *International Archives of Medicine*. The publisher's CEO, Carlos Vasquez, e-mailed Bohannon to let him know that he had produced an "outstanding manuscript," and that for just 600 euros it "could be accepted directly in our premier journal." Bohannon paid the fee, then called a friend who worked in science public relations, who coached him on how to manipulate journalists:

> The key is to exploit journalists' incredible laziness. If you lay out the information just right, you can shape the story that emerges in the media almost like you were writing those stories yourself. In fact, that's literally what you're doing, since many reporters just copied and pasted our text.

The story was picked up by newspapers, magazines, and television shows around the world. For a few days, it was the science and health story everyone was talking about. And why not? It was true that the people who ate chocolate as they dieted lost weight 10 percent faster.

It was all a hoax, à la Alan Sokal. However, this was a sting not against postmodernist posers but for-profit journals and lazy journalists, one designed to expose the manipulability of the media using statistics. Bohannon had Frank ask the volunteers for eighteen different measurements: weight, cholesterol, sodium, blood protein levels, sleep quality, well-being, and so on. When we take that many different measurements of such a small number of people, the chances that we can find something that changes in a statistically significant way go way up, simply due to natural variability. It's like buying lottery tickets; if one buys a lot of them, one's chances of getting a winner improve. The statistically significant finding could have been that the subjects dropped cholesterol faster, or that they had better sleep quality, but it just happened to be

10 percent weight reduction. This is another way of stacking the deck toward a predetermined conclusion by massaging experimental design to achieve the desired outcome. A magic trick, if you will—the same trick climate deniers use: limit the statistics and pick your endpoints to make natural variability look like a trend when it's not.

The sad thing was that no one even googled Bohannon. If someone had, he or she would have found out that the pseudonym he'd used—Johannes Bohannon—didn't exist, and that there was no such place as the Institute of Diet and Health, where he claimed to work. Nor did any journalist contact any outside scientists to review the study or get an opinion on it. In many cases, they just reprinted the press release, or excerpts from it, as a story—classic churnalism. "Take a look at the press release I cooked up," Bohannon said in a write-up announcing the hoax:

> It has everything. In reporter lingo: a sexy lede, a clear nut graf, some punchy quotes, and a kicker. And there's no need to even read the scientific paper because the key details are already boiled down. I took special care to keep it accurate. Rather than tricking journalists, the goal was to lure them with a completely typical press release about a research paper. (Of course, what's missing is the number of subjects and the minuscule weight differences between the groups.)

Health and diet news changes like the weather—salt is bad or good; chocolate is bad or good; red wine is bad or good—and a lot of that noise is caused by bogus studies paid for by people hawking a product or a fad. Legitimate science—in which it is often difficult to make any uniform conclusions—gets lost in all that noise. In the meantime, all of that conflicting advice about the science issues people are most interested in—their diet and health—gives the public less reason to trust science. Climate deniers and science contrarians like Fred Singer, in turn, take advantage of this dynamic.

To see how easy it is to fool journalists with a statistics magic trick about climate science, consider the year 2008. News headlines about global warming in 2008 could have accurately said:

A. "2008 Warmer Than Any Year in the 20th Century Except 1998"
B. "Global Temperatures Cooled over the Last Decade" (because 1998 was warmer)

C. "2008 the Coldest Year of the Century" (which started in 2001, with 2001 through 2007 all warmer than 2008)

D. "8 of 9 Warmest Years Ever Recorded Since 2000" (1998 was the ninth)

Or, as Rush Limbaugh put it on his March 27, 2009, show: "More people are starting to consider the notion that we actually may be in a cooling phase, 'cause there hasn't been any significant warming in years."

Each of these headlines and statements is true, but each gives a very different impression about global warming. As the saying goes, "Figures don't lie, but liars can figure." And therein lies the problem. Language, the tool of the press, is imprecise, and journalists, by and large, don't understand statistics.

Media outlets ran the peddled, predigested, and packaged angle that the stolen e-mails seemed to show: that a small cadre of politically motivated climate scientists was not using "the real temps" and was trying to "trick" the public and "hide the decline" in global temperatures.

Citing this reframing, Sarah Palin stepped up her criticism of climate science, calling it "snake-oil science" and "junk science." Instead of investigating the truth or falsehood of the claims, which could have been easily established, mainstream media outlets lapped up the "controversy," much as they lapped up the chocolate diet hoax, and called it Climategate. "Too many lazy journalists simply uncritically parroted what they had read from dubious sources, such as climate-change-denial outfits and blogs," said Mann.

These stories highlight the problem that arises when most of the journalists in a democratic society whose major policy challenges revolve around science have no training in it. It's simply not good enough to write he-said she-said stories or to churn out regurgitated press releases after adding a few quotes or changing a few sentences. Life in a democracy depends on the press informing citizens about what's really going on based on knowledge, not on warring opinions that can have little or nothing to do with the underlying science driving the issues that policymakers ultimately have to grapple with. But that takes work, money, training, and a belief that there is objective knowledge in the first place. If one side presents knowledge and the other opinion, simply reporting both sides is not journalism. It constitutes malfeasance.

Outrage

After the phony-science stage of the PR campaign was in place and grassroots anger was whipped up around the supposed Climategate controversy, talk radio stepped into the attack. Rush Limbaugh's November 24, 2009, show wrapped all the talking points together, using the same kinds of messaging that had been developed by the coal PR campaign ad he had done on his show eighteen years earlier:

> The people who have been preaching to us about global warming have been doing so, as the left usually does, from the "crisis mode" standpoint. "We've got twenty years." "We got ten years." Remember Ted Danson in 1988? "Ten years to save the oceans." Ten years to this; twenty years for that. "We're killing ourselves. We're killing the polar bears!" Except it hasn't warmed in ten years. And now we've got the hoax fully exposed. Wouldn't you think that people genuinely believing in manmade global warming and its destructive results would be happy that it isn't happening? They're not. They are distressed, and they're trying to cover up the hoax, and they're going to try to weather the storm—'cause it isn't about global warming like health care is not about health care, like cap-and-tax is not about cap-and-tax, like Obama is not who he is. They're all frauds. They are all liars. They are skunks, and they ought to be held up for public ridicule. Obama said he wants to "restore science" to its rightful whatever? Then he ought to be leading the way to find out who these people are, what they've done, who they've infected, who went along with them—calling them out by name—making sure that every scientist at every university in this country that's been involved in this is named and fired, drawn and quartered or whatever it is. Because this is a worldwide hoax, and it's primary target was you. The people of the United States of America.

Limbaugh's rhetoric was reminiscent of Nazi rhetoric: the claims of a vast secret hoax, the castigation of his targets as liars and frauds, the idea of publicly targeting individual scientists for firing and perhaps even torture.

The Climategate deployment had all the characteristics of a successful psy-ops campaign and became "one of the best-funded, most highly orchestrated attacks against science we have ever witnessed," says Mann. "The evidence for the reality of human-caused climate change gets stronger with each additional year. Greenhouse gases don't care whether you're a Democrat or a Republican. Nor do the ice sheets. Unfortunately, some have found it convenient to politicize

the science, as others have with the science of tobacco, acid rain, ozone depletion, stem-cell research, human health, environmental contaminants, etc."

A Failure of Authority

On December 4, 2009, twenty-nine of the country's most prominent climate
scientists, led by the conservative, careful, and highly respected AAAS board
chairman James McCarthy, sent a joint letter to Congress.

> In the last few weeks, opponents of taking action on climate change
> have misrepresented both the content and the significance of sto-
> len emails to obscure public understanding of climate science
> and the scientific process. . . . Observations throughout the world
> make it clear that climate change is occurring, and rigorous sci-
> entific research demonstrates that the greenhouse gases emitted
> by human activities are the primary driver. These conclusions are
> based on multiple independent lines of evidence, and contrary
> assertions are inconsistent with an objective assessment of the
> vast body of peer-reviewed science. . . . If we are to avoid the most
> severe impacts of climate change, emissions of greenhouse gases
> must be dramatically reduced.

This was a noble statement, and prominent scientists frequently use this
overwhelming-authority approach of issuing open letters. It is very impressive,
and to people who trust science it is influential in the same way that instruction
from a group of bishops can be influential for devout Catholics. But such an
approach has two problems that doomed it to failure in this instance.

First, it relies on people having respect for the unquestioned authority of
science leaders. But, as we've seen, that authority had been called into question. If one's not Catholic, it doesn't matter how many bishops agree on something. Thus, Palin was able to flippantly but effectively preempt the appeal with
a Facebook post calling concerns over global warming "doomsday scare tactics
pushed by an environmental priesthood."

Second, unquestioned authority isn't the message of science. The best science has never rested on "trust me" or "take it on faith." Authority was the message of kings and popes, the thing Locke was trying to escape. Science's strong
suit has always been its focus on observation and process: *See, look for yourself.
Don't take my word for it.* The authority approach inadvertently reinforces the

perception that science is just another expert opinion. In this way, the letter played into the energy industry's PR campaign, which was casting science as a political opinion instead of knowledge. Not surprisingly, it fell on deaf ears among those who didn't believe in global warming.

Scientists were losing the debate because they didn't understand the terms of the battle. "It's not about science literacy," says social scientist Matthew Nisbet. "It's about competing mental models." Once those models become polarized, science gets tossed around like any other opinion. How to avoid that? Emphasize the process.

The Foxgate E-Mail Scandal

That this was a full-blown battle between climate scientists and PR operatives became abundantly clear later, when the curtain was lifted for a brief instant to expose the workings of partisan cable news. On December 8, 2009, Fox News's then-White House correspondent, Wendell Goler, delivered a live report from Copenhagen on the daytime news show *Happening Now*. Goler was asked by host Jon Scott about "UN scientists issuing a new report today saying this decade is on track to be the warmest on record."

Goler said yes, 2000 to 2009 was "expected to turn out to be the warmest decade on record." He said it was a "trend that has scientists concerned because 2000 to 2009 [was] warmer than the 1990s, which were warmer than the 1980s." He then said, "Ironically, 2009 was a cooler-than-average year in the United States and Canada," which was "politically troubling because Americans are among the most skeptical about global warming."

When Scott brought up the Climategate e-mails, Goler said that although people had raised questions about the CRU data, "the data also comes from the National Oceanic and Atmospheric Administration and from NASA. And scientists say the data across, across all three sources is pretty consistent." Goler then quoted a statement from UN Secretary-General Ban Ki-moon, saying, "Nothing in the data, quote, casts doubt on the basic scientific message that climate change is happening much faster than we realized and that human beings are the primary cause."

This unbiased reporting must have prompted *someone* to chew out Fox News's Washington managing editor Bill Sammon, because, within fifteen minutes, Sammon had fired off a strident e-mail to the staffs of *Special Report*, the Fox White House Unit, *Fox News Sunday*, FoxNews.com, and several other reporters, producers, and executives:

From: Sammon, Bill
To: 169 -SPECIAL REPORT; 036 -FOX.WHU; 054 -FNSunday; 030
-Root (FoxNews.Com); 050 -Senior Producers; 051 -Producers; 069
-Politics; 005 -Washington
Cc: Clemente, Michael; Stack, John; Wallace, Jay; Smith, Sean
Sent: Tue Dec 08 12:49:51 2009
Subject: Given the controversy over the veracity of climate change
data . . .

. . . we should refrain from asserting that the planet has warmed (or
cooled) in any given period without IMMEDIATELY pointing out
that such theories are based upon data that critics have called into
question. It is not our place as journalists to assert such notions as
facts, especially as this debate intensifies.

The directive ignored the fact that the Climategate e-mails had no bearing on the data Goler mentioned, which had been accumulated over fifty years of research, a fact that had already been established by a wide variety of the nation's leading scientists, academies, and journal editorial boards. Around this time, to their credit, the *New York Times*, Politifact.com, FactCheck.org, the Associated Press, and McClatchy Newspapers all released statements saying that the e-mail exchanges did not change the credibility of the fundamental science, which was based on a wide variety of sources. Sammon's statement, on the other hand, was an ideological directive that placed knowledge on an equal footing with opinion and was not in keeping with the standards of good journalism. It would set the tone of junk-science skepticism for all Fox News reportage and GOP messaging on climate change that was to follow.

Congress Harasses the IPCC

Enter stage six of the propaganda strategy: the government allies who present a proposed policy response. On December 7, 2009, Congressman Jim Sensenbrenner sent a letter to Rajendra Pachauri, the chair of the IPCC, demanding that Mann and the other top climate scientists included in the e-mails be barred from participating in the next IPCC assessment of climate change for having "caused grave damage to the public trust in climate science in general, and to the IPCC, in particular." Concern appropriation—the act of appearing to adopt the concerns of one's opponents while using their language against them—is another classic tactic that frames the appropriator as the reasoned centrist and his or her opponents as the extreme fringe.

On December 9, Palin also used concern appropriation in a major op-ed published in the *Washington Post* in which she said that the Climategate scandal

> exposes a highly politicized scientific circle—the same circle whose work underlies efforts at the Copenhagen climate change conference. . . . The e-mails reveal that leading climate "experts" deliberately destroyed records, manipulated data to "hide the decline" in global temperatures, and tried to silence their critics by preventing them from publishing in peer-reviewed journals. What's more, the documents show that there was no real consensus even within the CRU crowd. Some scientists had strong doubts about the accuracy of estimates of temperatures from centuries ago, estimates used to back claims that more recent temperatures are rising at an alarming rate.
>
> This scandal obviously calls into question the proposals being pushed in Copenhagen. I've always believed that policy should be based on sound science, not politics.

Except, of course, when it comes to the teaching of creationism in science class.

When You're 'Splainin', You're Not Gainin'

The industry's PR campaign wasn't the only problem facing climate scientists. On December 1, another story had broken that further undermined Copenhagen, climate science, and the IPCC. Graham Cogley, a geologist at Trent University in Peterborough, Ontario, who apparently had no connection to the propaganda campaign, noticed that the IPCC's fourth assessment contained an error. The assessment, which was released in 2007, the year the IPCC and Al Gore had won the Nobel Prize, said, "Glaciers in the Himalayas are receding faster than in any other part of the world," and that all the glaciers in the central and eastern Himalayas could disappear by 2035. The IPCC classed the statement as "very likely," meaning that it had a probability of being true of 90 to 99 percent, implying that it was the result of measurement and observation and was supported by inductive reasoning and statistics. It also said the glaciers' "total area will likely shrink from the present 500,000 to 100,000 km2 by the year 2035." But it was wrong.

The statement was sourced to a World Wildlife Fund report, which in turn had drawn it from a 1999 article published by science journalist Fred Pearce in the *New Scientist*. But the scientist who was quoted in the *New Scientist* article, Syed Hasnain of Jawaharlal Nehru University in New Delhi, was speaking offhandedly, offering an opinion, and had never repeated the prediction in a peer-reviewed journal. He has since called the comment "speculative." When Cogley went back to check previous statements, he found a separate 1996 study by a leading hydrologist, V. M. Kotlyakov, that said that "the extrapolar glaciation of the Earth will be decaying at rapid, catastrophic rates—its total area will shrink from 500,000 to 100,000 square kilometres by the year 2350." It now appeared the IPCC error was the result of a simple transposition—2350 to 2035—that no one had caught, even when the 2035 prediction had garnered worldwide media attention going into Copenhagen.

This happened at the same time as Climategate. The press smelled blood in the water, and coverage of the error went worldwide. The retreating glaciers' iconic beauty had inspired press stories, but so had the fact that an estimated one billion people depend on Himalayan glacial melt for water. That such a dire prediction was not supported seemed to confirm the right's characterizations of climate scientists as doomsayers and shattered the public's confidence in their credibility.

This situation was worsened by a study by India's environment minister, Jairam Ramesh, that similarly suggested Himalayan glaciers might not be melting as much as was widely feared and accused the IPCC of being "alarmist." Instead of exercising caution and scientific open-mindedness, the embattled Pachauri dismissed the Indian study as "voodoo science," recalling Shapley's response to Humason.

From a tactician's perspective, there is only one way out of this trap: set ego aside, admit the mistake as quickly as possible, accept full responsibility, and take immediate steps to make things right. The row with India further damaged the IPCC's and Pachauri's credibility. Since the IPCC's statement was unsupported, *it* was voodoo science, and Pachauri came off as arrogant.

By now, the mainstream press was actually gathering news, combing through the report's three thousand pages, looking for other possible biases and inconsistencies. They had an angle. The idea that an elite and arrogant group of scientists was working a political agenda had begun to stick.

A few days later, this conscientious investigation brought more errors to light. The report also wrongly stated that 55 percent of the Netherlands was below sea level because the IPCC had failed to independently confirm information supplied by a Dutch government agency.

It was a spin doctor's dream, the gift of a scandal that keeps on giving. It couldn't have come at a worse time or been more poorly handled by the IPCC, which should have pulled the report, corrected it diligently, and thanked the critics for being such good citizens and strengthening the report's credibility. Instead, over the rest of the month, several more minor errors were identified in a death-by-a-thousand-cuts nightmare. This is another classic campaign strategy: dribble out bad news to make one's opponent explain him- or herself over and over, until all credibility with the public has been lost. The adage in politics is "when you're 'splainin', you're not gainin.'"

After all the publicity, here's the most shocking thing: the errors were not in the scientific section of the IPCC report, published separately as *Climate Change 2007: The Physical Science Basis*, which is subject to peer review. They were in the section published as *Climate Change 2007: Impacts, Adaptation and Vulnerability*, which speculated about possible impacts of climate change and was not subject to peer review. They did not purport to be scientific. But journalists and the public didn't seem to notice that distinction.

"Neither of these statements were important enough to even make it to the technical summaries or the summary for policymakers, so they really had no effect whatsoever," Mann says. "And let's not forget, we're talking about a couple mistaken sentences in thousands of pages. What's amazing to me is that these are the only errors that were found in a thousands-of-pages-long document. Were that most sources had such a low error rate."

All that is true—but it didn't matter. The IPCC errors had nothing to do with Mann or the science. But they were conflated with the CRU e-mails and Mann's hockey-stick graph, and the narrative of climate science began shifting under his feet.

The New McCarthyism

Recognizing what was happening, Mann took the bull by the horns and embarked on an aggressive publicity campaign, appearing on CNN and other news outlets to counter the attack by doing what scientists should have been doing for decades: breaking it down and talking process.

In response, climate-change deniers renewed their criticism of the hockey-stick graph, which they argued was an inaccurate depiction based on faulty science. The graph had self-evident power of the sort scientists often dream of but rarely achieve. Invalidating it was the holy grail of climate denial, the mortal blow they badly wanted to inflict upon climate science.

With a climate bill finally taking shape in Congress, the attacks escalated. At the same time, the energy industry was spending half a billion dollars lobbying Congress, trying to discredit Mann and other scientists, and running television ads and pushing news stories to influence public opinion and policymakers. In keeping with the idea that politics is narrative, they needed a bad guy to tar and feather. Mann fit the bill.

In February 2010, Senator James Inhofe of Oklahoma, the ranking Republican on the Senate Committee on Environment and Public Works and a long-time climate denier, went on the attack. Following Rush Limbaugh's on-air suggestion, Inhofe targeted Mann personally in a Senate minority report.

> The CRU controversy is about far more than just scientists who lack interpersonal skills, or a little email squabble. It's about unethical and potentially illegal behavior by some of the world's leading climate scientists. The report also shows the world's leading climate scientists acting like political scientists, with an agenda disconnected from the principles of good science. And it shows that there is no consensus—except that there are significant gaps in what scientists know about the climate system. It's time for the Obama Administration to recognize this. Its endangerment finding for greenhouse gases rests on bad science. It should throw out that finding and abandon greenhouse gas regulation under the Clean Air Act—a policy that will mean fewer jobs, higher taxes and economic decline.

Inhofe's next act should be morally repugnant to every member of free society. In an act of McCarthyism, he publicly named the scientists he wanted investigated for possible referral to the US Justice Department for prosecution: Raymond Bradley, Keith Briffa, Timothy Carter, Edward Cook, Malcolm Hughes, Phil Jones, Thomas Karl, Michael Mann, Michael Oppenheimer, Jonathan Overpeck, Benjamin Santer, Gavin Schmidt, Stephen Schneider, Susan Solomon, Peter Stott, Kevin Trenberth, and Thomas Wigley.

The bullying of scientists by a sitting US senator was picked up by the right-wing-media echo chamber. The scientists began getting hundreds of attack e-mails accusing them of falsifying science for political purposes—and then they started getting death threats. Unless one has been through that kind of public shaming and national shunning, it's hard to imagine the fear and the gross sense of injustice. It strikes against every value that Americans hold dear. "People said I should go and kill myself," Phil Jones said. "They said they knew where I lived. I did think about it, yes. About suicide. I thought about it several times, but I think I've got past that stage now."

Jones, who has since been exonerated by the British House of Commons, said the worldwide scandal—over nothing, it would turn out—was something for which he was totally unprepared. "I am just a scientist. I have no training in PR or dealing with crises," he said. That lack of training is a major problem now that scientists find themselves increasingly involved in political battles, often against opponents whose hired scientists do have media training, as outlined in the GCSCP. Climate change is just the beginning.

Climate Science as Fraud Against the Taxpayers

The attacks were taking a decidedly more sinister turn. On April 23, 2010, Virginia attorney general Ken Cuccinelli subpoenaed the papers and e-mails of Michael Mann and the University of Virginia, where Mann once worked, demanding all "data, materials and communications" of Mann's related to any grants he may have sought or received in which any funds came from the Commonwealth of Virginia. He cited a 2002 law, the Fraud Against Taxpayers Act, that gives the attorney general the right to demand documents and testimony in cases in which tax dollars have allegedly been obtained falsely by state employees.

An engineer and an attorney, Cuccinelli has said he doesn't believe in climate change. He argued that he was simply exploring whether there were any "knowing inconsistencies" Mann made when applying for grant money. "The revelations of Climate-gate indicate that some climate data may have been deliberately manipulated," Cuccinelli's office said in a statement. "The legal standards for the misuse of taxpayer dollars apply the same at universities as they do at any other agency of state government. This is about rooting out possible fraud and not about infringing upon academic freedom."

"It's totally unacceptable for Cuccinelli to put forward allegations of fraud against Dr. Mann simply because he doesn't agree with the results of the research," said Francesca Grifo, then-director of the Scientific Integrity Program at the Union of Concerned Scientists. In fact, Grifo argued, it can have a chilling effect on research and academic freedom across America.

That was, of course, the intent, which is what makes Cuccinelli's abuse of power so pernicious. Who will be the next target when an elected official wants to quash the results of research he or she doesn't like? Stem-cell researchers? Geneticists? Nutritionists when an official has ties to the food industry? It's difficult to see how that helps a country succeed in a highly competitive, science-driven global economy.

Bullying and FOIA Abuse

Mann's battle with Cuccinelli highlighted many of the ways climate scientists are being targeted and harassed by groups funded by the energy industry and its allies. But this goes well beyond the sort of claim brought by Cuccinelli. Stephen Schneider, another of Inhofe's targets, was already familiar with the oil companies' tactics. He wrote the book *Science as a Contact Sport*. He shared one of the e-mails he received during this period:

> You communistic dupe of the U.N. who wants to impose world government on us and take away American freedom of religion and economy—you are a traitor to the U.S., belong in jail and should be executed.

The e-mail was by no means unique. Most climate scientists get slews of them. Katharine Hayhoe of Texas Tech University, who also does climate-science outreach to evangelical Christians, receives up to two hundred e-mails and letters after a media appearance, many of them telling her she's a fraud and a liar, or threatening her and her family. "One e-mail I got said something like, 'I hope your child sees your head in a basket after you've been guillotined for all the fraud you climate scientists have been committing,'" Hayhoe told science journalist Katherine Bagley.

Lawrence Livermore National Laboratory climate scientist Ben Santer said he often receives hate mail—and worse. Santer answered the doorbell late one evening to find a dead rat on his doorstep. He looked up and heard a driver shouting obscenities from a departing Hummer. Michael Mann opened a letter at his office one day to find a suspicious white powder wafting into the air that may have been anthrax poison. He closed the door, washed his hands, and then called the police. The FBI determined it was cornstarch, but the intent was clear: intimidation. Prominent German climate scientist John Schellnhuber was accosted by a demonstrator who a held a noose in front of him at a climate conference in Melbourne, Australia. Climate scientists have been relocated after threats to their spouses or children. "Certainly I'm concerned about the threats," said Australian climate scientist Will Steffen. "I've had a few myself. Some of my colleagues have had even more than I've had, and it is fairly widespread in the scientific community. . . . A few of them are threats of violence. A very few of them are very direct threats of violence, to the point where they have to be taken seriously."

Sometimes, the attacks aren't just physical or psychological, but legal as well. One increasingly common tactic used by industry-funded "institutes" (and by other conservatives against political opponents) is the abuse of the Freedom of Information Act (FOIA)—not as a vehicle for public transparency, but as a weapon to harass and cripple scientists' operations, political reputations, professional focus, and relationships with their universities. One example lies with a group called the Energy and Environment Legal Institute (E&E), which specializes in this sort of harassment. The group intervened in the Cuccinelli case with a follow-on lawsuit. E&E is a political advocacy group primarily backed by members of the energy industry. It was first registered as Western Tradition Partnership (WTP) as a Colorado nonprofit in 2008 by Scott Shires, a Republican operative who pleaded guilty that same year to fraudulently obtaining federal grants to develop alternative fuels.

In 2010, WTP changed its name to American Tradition Partnership, and announced that it had launched the American Tradition Institute, a think tank that would be "battling radical environmentalist junk science head on." The "junk science" ATP seemed most concerned with was what the US National Academy of Sciences says should now be regarded as "settled facts"—that the earth is warming and humans are the primary cause. After a move to Washington, D.C., the group was renamed again, this time as the Energy & Environment Legal Institute. "E&E Legal's Board of Directors voted to to [sic] refine its focus primarily to the area of strategic litigation, and to change its name in order to reflect more accurately its work in the legal arena," the group said.

In 2011, E&E fought for a Colorado referendum allowing voters to opt out of the state's renewable-energy standard. The referendum's backers missed the filing deadline, but E&E sued Colorado over the standard, and targeted similar standards in Delaware, Minnesota, Montana, New Mexico, and Ohio.

In 2011 and 2012, E&E fellows developed a public-relations plan to subvert the wind-energy industry "so that it effectively becomes so bad no one wants to admit in public they are for it." The lengthy document lays out specific tactics and "goals of the PR campaign."

During the 2010 elections, the Montana Commission of Political Practices found that E&E broke state campaign laws by failing to register as a political committee or report its donors and spending. The state suggested the group was involved in corruption and money laundering, saying that it solicited unlimited contributions to support candidates and then passed them through a "sham organization," a political action committee called the Coalition for Energy and the Environment that ran attack ads against Democrats. E&E told corporations that it aimed to combat "radical environmentalists" and "beat them at their own game" and that, as with the Donors Trust's third-party technique, their contributions would remain secret.

E&E's case against the University of Virginia used FOIA requests to demand much the same level of exhaustive and personal information that Cuccinelli had, including Mann's personal notes and papers. Mann filed a response in order to protect his own papers under the principles of privacy and academic freedom, arguing that such items were nonpublic. Without this filing, Mann would have had no say in what papers the UVA might ultimately release to E&E.

The AAAS said lawsuits like E&E's "have created a hostile environment that inhibits the free exchange of scientific findings and ideas." Grifo, then of the Union of Concerned Scientists, said, "Scientists should be able to challenge other scientists' ideas and discuss their preliminary thinking before their analyses are complete and published." She added that "scientists shouldn't have to worry if political opponents of science will be sifting through emails" for bits and phrases to spin, which "has a chilling effect" on research. The American Association of University Professors sent a letter to the UVA president arguing that the Virginia public-documents statute exempts scholarly data of a proprietary nature that has not yet been publicly released, published, copyrighted, or patented—in other words, Mann's e-mails. And US copyright law suggests that the documents may be covered under common copyright.

In the end, Mann won this case, too, but not without undergoing considerable personal hardship and legal expense—things that threatened to hamper his ability to focus on his work. Scott Mandia and John Abraham, cofounders of the Climate Science Rapid Response Team and of an equally important group of climate scientists and bloggers, also founded the Climate Science Legal Defense

Fund in order to provide financial and legal support to climate scientists who find themselves in the crosshairs of industry-financed legal attacks. What was at stake in Mann's case—and in similar FOIA abuses—is something much larger and more precious than papers, e-mails, or partisan politics. What is at stake is the freedom to investigate, debate, and express ideas that run counter to the interests of corporations and their political allies. Attacks on this basic freedom hide behind the guise of transparency but, in reality, are a step toward tyranny.

The New Denialism

In November 2010, the energy-industry financing and organizing of the Tea Party movement through FreedomWorks, Americans for Prosperity, American Majority, and other front groups paid off, and Inhofe and other congressional outliers gained new allies who placed ideology ahead of knowledge. The political tide swept in a class of eighty-seven freshmen GOP US representatives and thirteen senators. Ninety-four of these one hundred new lawmakers either explicitly denied climate change, signed the FreedomWorks "Contract from America" to reject cap-and-trade, or signed the "No Climate Tax" pledge promoted by Americans for Prosperity. Not a single freshman Republican publicly accepted the scientific consensus that anthropogenic greenhouse gases are an immediate threat.

"I am vindicated," said Inhofe, who was ridiculed by environmentalists in 2003 when he declared that man-made global warming was the "greatest hoax ever perpetuated on the American people." Forty-six of the freshmen had explicitly denied anthropogenic climate change is even happening in public statements they made on the campaign trail.

"Today's global warming doomsayers simply lack the scientific evidence to support their claims," said newly elected congressman Bill Huizenga (R-MI) in a typical comment. "A host of leaders in the scientific community have recognized that the argument for drastic anthropogenic global warming is no longer based on science, but is being driven by irrational fanaticism."

But the newbies were not alone. Forty-four incumbents professed what were sometimes even more strident antiscience views, always with the common themes of elevating the critics, creating "uncertainties" and branding global warming a hoax. Incoming house speaker John Boehner (R-OH) led the charge on national TV, speaking to George Stephanopoulos on ABC's *This Week* in April 2009. "George, the idea that carbon dioxide is a carcinogen that is harmful to our environment is almost comical. Every time we exhale, we exhale carbon dioxide. Every cow in the world, you know, when they do what they do, you've got more carbon dioxide."

The willful rejection of science as a GOP plank is lamented by former Republican congressmen like Vernon Ehlers, John Porter, and Sherwood Boehlert. Boehlert, who was chair of the House Committee on Science, wrote about this in a *Washington Post* piece shortly after the 2010 election.

> I call on my fellow Republicans to open their minds to rethinking what has largely become our party's line: denying that climate change and global warming are occurring and that they are largely due to human activities. . . .
>
> There is a natural aversion to more government regulation. But that should be included in the debate about how to respond to climate change, not as an excuse to deny the problem's existence. The current practice of disparaging the science and the scientists only clouds our understanding and delays a solution. . . .
>
> We shouldn't stand by while the reputations of scientists are dragged through the mud in order to win a political argument. And no member of any party should look the other way when the basic operating parameters of scientific inquiry—the need to question, express doubt, replicate research and encourage curiosity—are exploited for the sake of political expediency. My fellow Republicans should understand that wholesale, ideologically based or special-interest-driven rejection of science is bad policy. And that in the long run, it's also bad politics.

Experimenting with the Economy

By 2010, the climate bill was long and full of compromises, directives, and exceptions that irked conservatives—at more than 1,200 pages, it was roughly double the length of the Clean Air Act Amendments of 1990 on which it was modeled. As it moved forward, some pundits didn't deny the science, but rather argued that "cap and tax," as cap-and-trade came to be called, was risky "experimenting with the economy." They argued that the government should not make major changes in people's lives on the basis of a few years of research. That kind of quick action doesn't allow time for the debate to get organized, they said, especially when long-term trends need to be studied. And, they cautioned, new research methods can give unexpected results. For example, remember "Vegetables will prevent cancer, wait, sorry, they won't." Science is uncertain.

This argument forgets that we have been accumulating data for not just a few years but for sixty years, and that we experiment with the economy constantly. Any time we cut taxes for the wealthy on the grounds that it will trickle down, that's an experiment. Deregulating banks was an experiment that didn't have the best results. Bailing out those same banks was another experiment. We experimented with the economy—plus millions of lives—when the United States unilaterally declared war on Iraq. Raising and lowering interest rates is an experiment. Democracy is an ongoing experiment; that is the nature of the democratic process. In fact, we know a lot about bringing externalities into the market with fees and taxes. We can quantify those things better than the schemes some economists have proposed for Social Security reform.

But these arguments failed to move the new Tea Party activists, who, for all their rhetoric about freedom, appeared to be interested only in lockstep adherence to the team message. By contrast, conservative economists generally supported cap-and-trade—and some argued a revenue-neutral carbon tax was the simplest and best solution.

Douglas Holtz-Eakin is the former head of the Council of Economic Advisers to President George W. Bush. Holtz-Eakin went on to become the director of the Congressional Budget Office, making his Republican economic credentials nearly impeccable. In 2008, he was director of domestic and economic policy for John McCain's presidential bid. After that, Senate Minority Leader Mitch McConnell (R-KY) appointed him to the Financial Crisis Inquiry Commission, and, as president of the conservative think tank American Action Network, he actively opposed the 2010 health-care bill.

Holtz-Eakin says the entire cap-and-trade argument is misplaced. "There needs to be some reeducation," he says. "Conservatives *invented* cap-and-trade, to battle acid rain. They were leaders in overthrowing liberals on it under Reagan. Before that it was all command-and-control approaches, and *we* brought market forces in to bear through cap-and-trade and it saved a ton of money." Indeed, what was expected to cost between $4 billion and $6 billion annually wound up costing a quarter of that, and has saved an estimated $70 billion annually in quantifiable health-care expenses—a return on investment of more than forty to one. Holtz-Eakin says conservatives have forgotten that, and lost their roots:

"They've taken positions that are divorced from any reality on the policy and from their own history."

Triumph of the Epithet

By calling the approach "cap and tax" and turning it into an epithet, the energy-industry PR campaign poisoned the water. This is a common propaganda tactic because it destroys critical thinking. In a democracy, authority comes from the ballot box, and the fastest way to shape a vote's outcome is to use epithets to demonize the opposing side. Thus we see the rise of negative campaigning and the use of epithets and talking points: "Liberal." "Cap and tax." "Socialism." After that level of emotion has been reached, a reeducation of the kind Holtz-Eakin has in mind is not going to be easy. "You have to explain how it really works," he says. "These are ideas that matter to conservatives, and previous conservatives embraced them, so you use some strength by affiliation."

Many of the Tea Partyers were propelled into office by a genuine fear of big government overreach. "That's not unjustified," says Holtz-Eakin.

> There was a lot of overreach. The health-care bill was enormous overreach. But it doesn't mean there shouldn't be some reach, and there's a conservative principle that says government can do good by intervening with appropriate regulations, like cap-and-trade. It's gotten to be such a loaded word in Washington you can't even say it, but it's a conservative, free-market idea. You either pick the price or set the quantity, and let the price fall out. Let the top-line policy drive the outcome and you let private business innovate and get out of the line completely.

Kassie Siegel of the Center for Biological Diversity says that, if Republicans have rejected cap-and-trade, that's their loss. "We already have all these laws set up and ready to go," she says. "The Clean Water Act, the Clean Air Act, the Endangered Species Act. Maybe they're not perfect, but they're the world's best, most successful laws. A theoretical cap-and-trade system is not incompatible with them." She says the only way to counter special-interest money is for the administration to use the bully pulpit: "We need the government involved, and the Obama administration has really dropped the ball on climate. They should have been out there with a full-court educational press, a World

War II–type mobilization to counter the misinformation and educate the public, and they haven't done it."

Demonizing cap-and-trade, Holtz-Eakin says, threatens conservative principles. If it becomes completely discredited, conservatives won't have it in their toolbox for managing environmental regulations in the future. "We don't want to go back to the old, big-government approach of one size fits all of the 1970s," he argues. "Conservatives need to figure out what they stand for. Is it just the highest bidder, or are there principles in there that mean something? I've always felt it's in their political interest to not deny the science, that's where the votes of the future are. But it awaits a teachable moment."

Truth Was Twisted

Republicans' mad dash into the intellectual wilderness picked up speed in March 2011, when a Republican-sponsored bill intended to prevent the Environmental Protection Agency (EPA) from regulating greenhouse-gas emissions passed out of committee. In 2009, the agency had issued an endangerment finding declaring the gases a threat to public welfare, which served as the EPA's legal basis for regulation. Repealing the finding would eliminate its authority over greenhouse gases.

The finding was scientifically sound, but at the hearing the Republicans on the House Energy and Commerce Committee's Subcommittee on Energy and Power directed anger and distrust at scientists and respected scientific societies.

The British journal *Nature*, one of the two leading periodicals of science, grieved the day as a new low point for science in America: "Misinformation was presented as fact, truth was twisted and nobody showed any inclination to listen to scientists, let alone learn from them. It has been an embarrassing display, not just for the Republican Party but also for Congress and the US citizens it represents."

Scientists were described at the hearing as "elitist" and "arrogant" and said to be hiding behind "discredited" institutions. One lawmaker cited melting ice caps on Mars as evidence contradicting anthropogenic warming on Earth. Antarctica was falsely reported to be gaining ice. Also mentioned was the myth that, in the 1970s, the scientific community had warned of an imminent ice age—a classic example of cherry-picking, using the work of a few scientists who published papers suggesting a cooling hypothesis, which ran contrary to the vast number of evidence-based publications that were beginning to indicate warming even then. Not surprisingly, the subcommittee chairman, Ed Whitfield (R-KY), a former oil distributor who touted his home state's $3.5 billion coal industry on his official website, said he "will continue to be a leader in the development and responsible use of new coal technologies and innovations."

At the request of Democrats on the subcommittee, several scientists were on hand to answer questions, but the Republican lawmakers weren't interested in facts. They rammed the bill through despite all the evidence to the contrary, severing the fundamental relationship of democracy to knowledge. If knowledge does not have primacy in public decision making, then no truth can be said to be self-evident, and we are left with the tyranny of ideology, with shots called by the wealthy and enforced by might. As has happened in the United States in the past, these lawmakers were behaving in as un-American a manner as the tyrant King George.

The Public-Relations Spillover

Ralph Cicerone, president of the US National Academy of Sciences, talked to several top scientists about what to do. He worried that the collapse of the public's confidence in climate science could spill over into other sciences. It was a legitimate concern, with legislatures in three US states—Louisiana, South Dakota, and Texas—telling their schools to teach "alternatives" to the scientific consensus on climate change.

South Dakota's resolution "urges" schools to take a "balanced approach" to teaching climate change, because the science is "unresolved" and has been "complicated and prejudiced" by "political and philosophical viewpoints." Whose?

> The South Dakota Legislature urges that instruction in the public schools relating to global warming include the following:
> (1) That global warming is a scientific theory rather than a proven fact;
> (2) That there are a variety of climatological, meteorological, astrological [sic], thermological, cosmological, and ecological dynamics that can effect [sic] world weather phenomena and that the significance and interrelativity of these factors is largely speculative.
> (3) That the debate on global warming has subsumed political and philosophical viewpoints which have complicated and prejudiced the scientific investigation of global warming phenomena.

The embarrassment of a formal South Dakota state resolution that pronounces that "astrology," as in horoscopes, can "effect" weather was overshadowed by the tragedy of the collapse of public trust in science. But this story was by no means unique. Driven by massive PR investments in climate denial,

2011–2012 was beginning to look like a tipping point in the war on science. It was happening across the United States, as a new wave of climate-denying Tea Party state legislators found themselves in power.

In Minnesota, home of the Boundary Waters Canoe Area Wilderness, Lake Superior, the headwaters of the Mississippi, and the land of more than ten thousand lakes—where, in 2008, voters had passed a constitutional amendment to protect water quality—legislators were paving the regulatory way for a proposed copper-nickel mining industry whose acid discharge could kill the state's famed wild-rice harvest. But science was in the way. "No large stands of rice occur in waters having a SO_4 [sulfate] content greater than 10 p.p.m.," state fisheries scientist John B. Moyle had written in the *Journal of Wildlife Management* in 1944, "and rice generally is absent from water with more than 5 p.p.m." But legislators passed a bill that arbitrarily increased the environmental standard for sulfates in water from 10 ppm to 50 ppm.

North Carolina lawmakers were thumbing their noses at science in a different way. There, the legislature passed a law banning state and local authorities from basing coastal policies—including zoning, building, and road elevation—on the latest scientific predictions of how much the sea level will rise. In a move reminiscent of Maoist China in the days leading up to the great famine, the only permitted numbers were politically approved projections based on past measurements instead of those based on current science.

In 2015, Florida Department of Environmental Protection employees said that they were ordered to refrain from using the terms "climate change" and "global warming" in official communications, e-mails, or reports, shortly after Governor Rick Scott, a global-warming "skeptic," took office.

Similarly, Wisconsin's Board of Commissioners of Public Lands, a three-member panel overseeing an agency that benefits schools and communities in the state, enacted a staff ban on discussing climate change or even responding to e-mails about the topic. The agency receives income from the timber industries and from other sources that use public lands, and it manages the funds for the benefit of public schools. The effects of climate disruption on those assets—particularly on forests, which are under stress in the Midwest—makes the topic worthy of consideration from a planning point of view. "It's not a part of our sole mission, which is to make money for our beneficiaries," said newly elected state treasurer Matt Adamczyk, a Republican who sits on the board. "That's what I want our employees working on. That's it. Managing our trust funds."

The Pope

In June of 2015, Pope Francis released an encyclical that has the potential to break the climate-change logjam by dismantling the decades-old Western antiscience alignment between anti-reproductive-science religious conservatives and anti-environmental-science business interests. "Doomsday predictions," the pope said, "can no longer be met with irony or disdain." Citing the scientific consensus that global warming is happening and human-caused, he called out the culprits: energy and extraction companies, short-sighted politicians, denialist scientists, free-market economists, indifferent citizens, fundamentalist Christians, and a press lost in subjectivity. "The Earth, our home, is beginning to look more and more like an immense pile of filth," he said. He referenced the works of his namesake, Saint Francis, whose observations of nature were similar to those of all natural philosophers—that is, scientists—and led to the birth of modern science. Francis himself has the equivalent of a two-year degree in chemistry.

"What is more," the pope said, "Saint Francis, faithful to Scripture, invites us to see nature as a magnificent book in which God speaks to us and grants us a glimpse of his infinite beauty and goodness. 'Through the greatness and the beauty of creatures one comes to know by analogy their maker' (Wis 13:5); indeed, 'his eternal power and divinity have been made known through his works since the creation of the world' (Rom 1:20)."

The encyclical, which covered issues related to both climate disruption and environmental degradation, is to be taken seriously by the world's Catholics. If they do, conservative business interests may find themselves in a significant minority, forcing the US Republican Party and other conservative political parties to fundamentally rethink their platforms around climate, the environment, and science in general.

Evidence that this was a frightening possibility for climate deniers could be seen in their immediate, outraged responses to a leaked draft of the pope's encyclical while at the Heartland Institute's tenth annual climate-change-denial conference, held in Washington, D.C., on June 11–12, 2015. "Oh, everyone's going to ride the pope now," said denialist US senator James Inhofe (R-OK). "The pope ought to stay with his job and let us stay with ours." What his job may be as a US senator in attending a science-denial conference was not clear. "He's very ill-informed, and he's ill-judged in doing this," complained the Competitive Enterprise Institute's Myron Ebell, one of the original strategists in the early climate-denial PR strategy.

Prescott Crocker, a retired bond trader, thought that "the pope's irrelevant on this issue, and my feeling is the Catholic Church has been losing its support, losing its congregation, and he's just one more opinion."

Dennis Bussel, a retired engineer, could barely contain his anger. "The pope is seriously, terribly misinformed," Bussel said. "He's doing a terrible disservice to the entire Catholic theology. He is way, way out of line."

Steve Milloy, of climate-denier website JunkScience.com, said "I think it's going to be a big mistake for him. He's going to come up with a document that is going to be ripped to shreds."

Marc Morano, of climate-denial website ClimateDepot.com, a project of CFACT, and the former communications director for Inhofe, felt similarly, arguing that the encyclical was "going to confuse Catholics, ultimately. The pope is basically being horribly ill-served."

James Taylor, communications director and vice-president of the Heartland Institute, said that the pope was focusing on a myth. "I think if he were to see the science as it is, then he would understand that it's more important to raise people out of poverty and give them that opportunity, rather than fighting a mythical global-warming crisis."

Beyond the disruption it could pose to the conservative political marriage of old energy and old religion, the pope's encyclical signals a possible new phase of political leadership and reflection for the Catholic Church—and perhaps for other world religions—when it comes to pressing international issues driven by science. Neither individual countries nor the United Nations seem to be particularly good at tackling these sorts of problems in the time frame needed. Both, because of their democratic structures, are vulnerable to public-relations campaigns that forestall or circumvent policy action. This provides both an opportunity and a need for leadership by pan-national moral organizations like the world's major religions.

This could be both good and bad. It could signal a time of increasing religious conflict, as power struggles occur to sort out a new world order. Or it could signal the beginning of a fruitful partnership between science—particularly environmental science and the environmental movement—and religion in rethinking our approach to economics, morality, and our shared heritage.

The PR Oil Spill Spreads

The 2016 US presidential election continued to be influenced by the war on science in several ways. As previously discussed, several candidates took an antivaccine position that wasn't as radical as the one Michele Bachmann paid so dearly for in 2011, but was coded in the language of parent choice.

At the same time, the massive investments in antiscience PR around environmental science, and particularly climate disruption, continued to drive much of the Republican conversation. A Center for American Progress analysis of contributions and lobbying data from the Center for Responsive Politics and advertising spending data from Kantar Media Intelligence/CMAG, published by the Atlas Project, showed that the fossil-fuel industry had continued its massive spending on antiregulation efforts in the 2014 US campaign cycle.

The industry had directly invested $721 million, and had certainly paid hundreds of millions of dollars more through contributions to 118 anti-climate front groups, astroturf organizations, and aligned think tanks. Of these investments, the fossil-fuel industry directly contributed more than $64 million to candidates and political parties, spent more than $163 million on television ads across the country (many of them on climate change), and paid almost $500 million to Washington lobbyists.

As a result, climate denial continued to be a party litmus test in the GOP field. Beyond climate denial, there was an even broader rejection of environmental regulations and the science they were based on. The industry-funded, anti-environmental conservative counter-movement was paying off. In 1990, Democrats in the US Senate scored an average of 65 percent from the League of Conservation Voters, while their GOP counterparts averaged 32 percent. By 2014, Senate Democrats scored 73 percent, but Senate Republican votes had fallen to an average score of only 9 percent. The House was worse. In 1990, House Democrats scored 68 percent and Republicans scored a respectable 40 percent, higher than their Senate peers. But by 2014, Democrats scored 87 percent while Republican scores had tanked to just over 4 percent.

The effects of the fossil-fuel industry's PR also, of course, spilled over into the 2016 presidential race. Here, too, the pope's encyclical on climate disruption provided fertile opportunity for candidates to please major GOP donors like the Koch brothers and their aligned front groups by doubling down on antiscience positions.

"The church has gotten it wrong a few times on science, and I think we're probably better off leaving science to the scientists and focus on what we're really good on, which is theology and morality," said Rick Santorum, an evolution-denying Catholic. "When we get involved with political and controversial scientific theories, then I think the church is probably not as forceful and credible." Santorum has called climate change "a hoax" and "junk science."

Jeb Bush, a former governor of Florida, the US state most vulnerable to climate disruption, struck a similar note the day after announcing his candidacy for president. Like Mitt Romney four years earlier, he attended a town-hall meeting in New Hampshire shortly after his announcement, where, like Romney, one of the first questions was about climate change—in this case, the upcoming release of the pope's encyclical. He replied that religion "ought to be about making us better as people and less about things that end up getting into the political realm." The irony that his brother's reelection campaign had depended on whipping up the evangelical voters that still formed much of the Republican Party base seemed lost on him.

"The last fifteen years, there has been no recorded warming," Ted Cruz told CNN's Dana Bush, using a nonsense denier talking point. "Contrary to all the theories that—that they are expounding, there should have been warming over the last fifteen years. It hasn't happened." Cruz would also repeal Environmental Protection Agency regulations on climate change, and forbid the EPA from instituting any regulation that "has a negative impact on employment in the United States unless the regulation, rule, or policy is approved by Congress and signed by the president."

Florida senator Marco Rubio said he didn't agree "with the notion that some are putting out there, including scientists, that somehow, there are actions we can take today that would actually have an impact on what's happening in our climate. Our climate is always changing. And what they have chosen to do is take a handful of decades of research and—and say that this is now evidence of a longer-term trend that's directly and almost solely attributable to manmade activity. . . . I have no problem with taking mitigation activity. What I have a problem with is these changes to our law that somehow politicians say are going to change our weather. That's absurd."

Rubio also signed Americans for Prosperity's "no climate tax" pledge, promising not to support legislation that would increase government revenue to combat climate change—as did candidates Rand Paul, Ted Cruz, Rick Perry, Scott Walker, and Mike Pence, as well as then-house speaker John Boehner. Rubio has also been a keynote speaker at the Heartland Institute's conferences.

Retired neuroscientist Ben Carson thinks "we may be warming. We may be cooling." He says that "we need to focus on the right thing here. Our

Environmental Protection Agency should be told to work in conjunction with business, industry, and universities to find the most eco-friendly ways of developing our energy resources."

In an interview with David Axelrod, Rand Paul said the earth goes through periods of time when the climate changes, but he's "not sure anybody exactly knows why."

Donald Trump denies warming is happening. "I'm not a believer in global warming," he said when radio host Hugh Hewitt asked about his views on climate change.

> And I'm not a believer in man-made global warming. It could be warming, and it's going to start to cool at some point. And you know, in the early, in the 1920s, people talked about global cooling. I don't know if you know that or not, they thought the earth was cooling. Now it's global warming. And actually we've had times where the weather wasn't working out and so they changed it to extreme weather and they have all different names, you know, so it fits the bill. But the problem we have, you look at, you know, our energy costs and all the things we're doing to solve the problem that I don't think in any major fashion exists.

That's false. There was no talk of global cooling in the 1920s. Most deniers point to a 1975 *Newsweek* article by Peter Gwynne arguing that the world could be cooling. Gwynne has since acknowledged the scientific consensus, saying, "In retrospect, I was over-enthusiastic in parts of my *Newsweek* article. Thus, I suggested a connection between the purported global cooling and increases in tornado activity that was unjustified by climate science. I also predicted a forthcoming impact of global cooling on the world's food production that had scant research to back it."

Of the Republican candidates, Senator Lindsey Graham (R-SC) and former Ohio governor John Kasich were the only ones, as of this writing, who agree that humans are causing climate change—but Kasich didn't think we should do anything about it. Graham worked with then-senator John Kerry to pass the cap-and-trade bill in 2010. "I think there will be a political problem for the Republican Party going into 2016 if we don't define what we are for on the environment," Graham said. "I don't know what the environmental policy of the Republican Party is."

Some Republicans are trying a new approach, since the idea of cap-and-trade remains toxic. Carbon-fee-and-dividend advocates argue that returning all revenue from a carbon fee to households would accomplish two things: keep the federal government from getting bigger and add jobs by putting money into the pockets of people who will spend it. It sounds like an interesting solution, another version of the carbon tax, but it fails to consider the billions the energy industry is continuing to spend to fund climate denial and block mitigation.

Some billionaires are trying to change this. North Carolina billionaire Republican Jay Faison announced a plan to spend $165 million on a campaign to get Republicans to face the facts on climate change in the 2016 elections. California billionaire Democrat Tom Steyer has pledged a similar amount and is funding the NextGen climate activist group. But the Koch brothers' network of climate-denial groups announced plans to spend $889 million on the 2016 race, and the rest of the industry-aligned groups engaging in climate denial have a combined budget that will likely exceed $1.5 billion annually.

Racketeers

In September 2015, twenty prominent scientists sent a joint letter to the US president, attorney general, and White House science advisor, encouraging them to follow up on an idea first broached by US senator Sheldon Whitehouse (D-RI): that the Department of Justice initiate a RICO (Racketeer Influenced and Corrupt Organizations Act) investigation of corporations and other organizations that have knowingly deceived the American people about climate disruption. RICO was originally passed and used to prosecute organized-crime operations, but has since been used against tobacco companies to force them to stop disinformation campaigns.

That same month, *InsideClimate News* reporters Neela Banerjee, Lisa Song, and David Hasemyer broke a story uncovering internal documents that showed Exxon's own climate scientists informed company executives in the 1970s that continued burning of fossil fuels would warm the climate and disrupt the global economy. But in subsequent years Exxon instead engaged in a PR campaign of climate denial to cast doubt on the science and protect its vested interests. R. L. Miller, Hunter Cutting, and Brad Johnson of the Super PAC Climate Hawks Vote took up the issue, urging other members of Congress to support a RICO investigation. After a similar report by the *Los Angeles Times* and the Columbia Journalism School corroborated the *InsideClimate News* allegations, Whitehouse reiterated his call for a RICO investigation. In the following days, other senators joined the call, as did the top three demo-cratic candidates for president: Bernie Sanders, Martin O'Malley, and Hillary Clinton. If the Department of Justice or state attorneys general prevailed in any such cases, the judgments could fund mitigation efforts.

Other nations besides the United States have laws that function similarly to RICO. Interpol uses a standardized definition of RICO-like crimes, but the implementation varies across the nations Interpol polices. Some efforts have been made to standardize some aspects of fraud fighting under Europol, which focuses on stopping "terrorism, international drug trafficking and money laun-dering, organised fraud, counterfeiting of the euro currency, and people smug-gling." Whether US RICO law can be used by other nations is also an open question. Since the tobacco cases, several similar attempts at RICO prosecu-tions have failed in federal court, significantly weakening the law. In 2015, a federal court denied rehearing a 2104 decision in which the EU and twenty-six of its member states filed suit against RJR Nabisco and its related entities, alleg-ing that the companies had engaged in mail fraud, wire fraud, money launder-ing, and other RICO violations, damaging the member nations' economies. Several of the judges on the panel filed dissenting opinions, saying the issue warrants further review. Canada and Australia have powers based on some-what similar regulation. Canada's Royal Mounted Police and the Office of the Superintendant of Financial Affairs both handle fraud cases. The Australian Crime Commission has responsibilities under the Public Service Act 1999 and the Public Governance, Performance and Accountability Act 2013 that give it some RICO-like powers.

One US state attorney general I spoke to about the RICO approach was skeptical about its long-term success because of the aforementioned broad use of RICO in a wide variety of complaints that have accrued a bulk of case law that has considerably weakened it. This attorney general suggested that state-level consumer fraud or financial fraud laws may be a more effective approach. Such state-level laws have won multibillion-dollar settlements against the tobacco industry, and could be used for prosecution of energy companies that have engaged in a knowing pattern of defrauding the public.

Other possible avenues include both federal and state financial fraud laws, such as Title VIII of the Sarbanes-Oxley Act, which deals with fraud in a cor-poration's representations to investors. Certainly the evidence of a major risk to the market for continued fossil-fuel burning should be accounted for, as it sun-sets the current business model. And more generally, among other exposures, the US Securities and Exchange Commission requires companies to disclose risks from climate change. On a state level, New York's powerful stock-fraud statute, the Martin Act, which forbids "any fraud, deception, concealment, suppression, false pretense" or "any representation or statement which is false," has already been used by a prior New York attorney general to force other fossil-fuel companies to disclose more about the financial risks they face from climate change. Several weeks after the *InsideClimate News* articles broke, New York attorney general Eric Schneiderman announced that his office had issued a subpoena under the Martin Act asking for forty years of records about what and when Exxon executives and scientists knew about climate change. In January 2016, California attorney general Kamala Harris began a similar investigation into what Exxon Mobil executives knew about climate change and what they told investors. Clearly, based on the communications efforts the entire industry engaged in, including information about climate change, it's not just Exxon. The disclosures of the conflict between what Exxon's scientists were telling its top executives and what those executives said and did in public—including the funding of aggressive climate-denial campaigns—is likely the tip of the iceberg, and any prosecutor's goal would in part be to discover more.

In the end, this is where the rubber meets the road: legal and policy action based on extending existing legal theory to apply sanctions when companies promote antiscience propaganda—that is, propaganda that flies in the face of a substantial body of known evidence from science—for the purposes of defrauding investors or forestalling public regulation that affects their business models. So far, attorneys have viewed private litigation as risky and expensive, just as they did in the early decades of the tobacco lawsuits, which were met with extremely aggressive responses by tobacco companies who refused out of court settlements. But public attorneys general have the duty to defend the government—a duty that extends to safeguarding democracy and the people. There is reason to be hopeful that such an action under one of the above or a different legal theory could eventually create a precedent for stopping the industrial war on science.

The Top Ten Climate-Denial Talking Points

Public-relations firms working for the energy industry and their hired scientists have tried for decades to create what their early strategy memo called "uncertainties in climate science." Here are a few of the most common climate-denial arguments used to create those uncertainties, and their rebuttals:

1. **"The climate has always varied"**

   The climate reacts to whatever forces it to change at any given time, and humans are now the dominant forcing. There is no other known natural forcing that fits all the independent lines of evidence pointing to increased CO_2 from human activity.

2. **"It's because of changes in the sun"**

   Scientists have used satellites since the 1970s to measure the amount of solar radiation reaching Earth, and it has slightly decreased during that time, while the world has gotten warmer.

3. **"There's been no warming since 1998"**

   The fifteen years since 2000 have been the warmest period ever recorded. 1998 was an especially hot spike due to an El Niño, and subsequent years were slightly cooler, but there have now been years that were hotter than 1998 in 2005, 2010, 2013, and 2014, which was the hottest year on record until, with a strong El Niño, 2015 shattered all records as the hottest year since records started being widely kept in the nineteenth century.

4. **"Greenland used to be green and grow crops"**

   Greenland's ice sheet is a mile thick and between three hundred thousand and five hundred thousand years old. There were no crops there in recorded history except a few spots on the periphery near the ocean.

5. **"Antarctica is gaining ice"**

   Satellites in orbit over Earth measure ice cover and show Antarctica is losing ice cover at an accelerating rate, not gaining it.

6. **"Scientists predicted an ice age in the 1970s"**

   Science works by exploring all options and seeing where the evidence goes. Even in the 1970s, the overwhelming majority of scientific papers predicted warming, but a few scientists were exploring the possibilities of cooling, and journalists like to cover the outliers because they are more dramatic stories.

7. **"E-mails show scientists concocted their data to hide the decline"**

   Scientists' e-mails were illegally hacked and portions of them were posted out of context to make them look suspicious. "The decline" was actually the divergence between proxy records of tree ring growth and temperatures in modern times, which have higher carbon, and so the rings weren't matching as well any more. This decline in agreement between proxy measurements and actual temperature data was "hidden" by substituting actual temperature records—an *improvement* in accuracy. Several independent investigations cleared all scientists involved of any wrongdoing. There was no conspiracy. It was a PR reframing by climate deniers.

8. **"Humans contribute less than 2 percent of CO_2 emissions"**
   Plants dying off in the winter contribute the majority of CO_2 emissions, but they absorb them back in the spring as new plants grow. Human emissions are on top of that and are not absorbed. Each year the extra emissions from humans accumulate in the atmosphere and oceans, driving up CO_2. Just like interest, year over year, this starts to add up in a big way.

9. **"Water vapor is a far more powerful greenhouse gas"**
   Yes, it is. And CO_2 causes the atmosphere to hold more water vapor, increasing warming. That's how global warming works, actually.

10. **"It's not worth wrecking the economy"**
    The environment is the foundation of the economy, and by threatening major environmental disruptions, including the inundation of three-quarters of the world's major cities, climate disruption is already starting to wreck the economy. Plus, major droughts are causing political instability. Considerable evidence shows the four-year drought over Syria and Northern Iraq contributed to political instability in the area, contributing in turn to the rise of ISIL. Switching to a carbon-free energy system is an enormous economic opportunity in which gigantic fortunes will be made and new activity will be stimulated by people and countries who invest early enough.

You can find more of the common climate-denial myths, as well as both short and long answers debunking them, along with the associated science, at the excellent website SkepticalScience.com.

The Stakes

Knowledge is power, so it follows that suppression of knowledge to protect vested interests ultimately weakens government. It weakened the Roman Catholic Church's credibility when it indicted Galileo, and Italy lost its place in the intellectual and economic vanguard.

It similarly weakened Russia in the Lysenko affair, when a political ideo-logue and former peasant named Trofim Lysenko ingratiated himself to Communist leaders and was placed in charge of national agriculture because of his ideological conformity. He denounced and suppressed scientists who questioned his odd schemes as "fly lovers and people haters," and his unedu-cated methods decimated Soviet agriculture. Soviet scientists who opposed him were persecuted, jailed, and shot. Similar to modern Republican charac-terizations of climate science as "junk science" by an "environmental priest-hood," Soviet geneticists, physicists, and chemists were characterized as "caste priests of ivory-tower bourgeois pseudoscience." Soviet agriculture, biology, and genetics were held back for forty years, weakening the Soviet Union and helping lead to its eventual downfall.

Suppression of knowledge also weakened China during Mao's Great Leap Forward. Mao prided himself on his peasant roots and considered intellec-tuals arrogant, dangerous counter-revolutionaries. Mao was concerned that scientists could take over as a "technical elite," so he demanded that ideology take precedence over science, effectively silencing scientists. In 1957, he set forth a plan to transform China into a modern industrialized society. It would overtake Britain in fifteen years while simultaneously feeding its own people and exporting grain to other nations. Mao had no knowledge of metallurgy, but based on a single demonstration he mobilized millions of peasants to smelt steel in "backyard furnaces." They burned trees, doors, and furniture as fuel and melted scrap metal like their pots and pans. At the same time, peasants were given outrageously optimistic grain production quotas based on Lysenko's assumptions from the USSR. Because ideology and the appearance of success mattered more than the facts of their meager harvest, the peasants gave more grain to the state than they could spare. Meanwhile, millions of other peasants were diverted from the farms into large-scale public-works projects needed to industrialize the country, and grain crops were left to rot in the fields. Scientists and others who suggested that Mao's plans were unrealistic were "struggled against," sentenced to hard labor, and often executed. The furnaces failed, the steel was unusable, and forty million Chinese people died in the greatest fam-ine in human history.

Science is inherently political. If authoritarians with vested interests who disagree with its findings are allowed to intimidate scientists or quash those results, democracy loses.

PART IV

Winning the War

Chapter 11

FREEDOM AND THE COMMONS

The only way to have real success in science, the field I'm familiar with, is to describe the evidence very carefully without regard to the way you feel it should be. If you have a theory, you must try to explain what's good and what's bad about it equally. In science, you learn a kind of standard integrity and honesty.

In other fields, such as business, it's different. For example, almost every advertisement you see is obviously designed, in some way or another, to fool the customer: The print that they don't want you to read is small; the statements are written in an obscure way. It is obvious to anybody that the product is not being presented in a scientific and balanced way. Therefore, in the selling business, there's a lack of integrity.

—Richard Feynman, 1988

Science's Challenge to Democracy

We've now exposed what is happening in the three major fronts of the war on science:

1. The postmodernist front being fought by academics, activists, and journalists on the secular left, who elevate subjectivity at the expense of objectivity, which they deny exists. Their actions provide philosophical support for:

2. The ideological front being fought by religious fundamentalists, who object to the emerging scientific mastery of reproduction and the life cycle, and seek to redefine scientific terms according to their own values and to debate science as if it were an opinion. They are often the foot soldiers for:

3. The industry and public-relations front, financed by corporations and conducted by PR experts, shills, and front groups, who take advantage of journalists' naivety about objectivity and truth in order to manipulate the media, thereby shaping public opinion using "uncertainties," deception, personal attacks, and outrage to move public policy toward an antiscience position that supports the funders' business objectives.

Add to this the aloofness of many scientists toward civic engagement; the general complexity and inaccessibility of modern science and technology; the many problems with industry-skewed science; agribusiness food additives and farming practices that wound up not being so healthy; environmental pollution from chemicals; the not-unjustified suspicion of government scientists left over from a history of unethical experiments on the public; the cooption of scientific societies by government operatives, like the American Psychological Association's collusion with the Bush-era CIA's torture policies; and government engineers working for agencies like the National Security Agency, which have leveraged technology to collect more data on private citizens than in all of recorded history; and it becomes easy to see why we are having problems with democracy in the age of science.

Clearly we have reached a crisis point. Science already affects every aspect of life, yet voters are increasingly unable to make decisions grounded in it. Our quality of life, our economy, the existence of the middle class, and the ongoing viability of the planet hang in the balance. The breakdown in our policy-making apparatus presents perhaps the greatest moral and political dilemma in human history. How are these factors likely to play out in the near term? Understanding this is the first step in winning the war.

Evidence suggests that the problem is only going to accelerate as the century progresses. Most of the points of conflict in the war on science are going to pivot on the growing battle between the rights of the individual and the responsibilities of the collective on a finite planet.

Like business, science has gone global. It owes allegiance to no nation, and knowledge is always a double-edged sword. Advanced science education of a world-class quality is increasingly accessible in many places outside the developed Western nations where it originated, and research labs rarely have particular geopolitical requirements beyond liberty. Scientists are connected with one another via the Internet and are sharing information at the speed of light in huge volumes, across fields, and with unprecedented interconnectivity.

Technology is flattening power structures and weakening nation-states as knowledge spreads through broadly distributed communications and social-media networks or is leaked by technology-empowered whistleblowers like Edward Snowden and the contributors to WikiLeaks. Even as technologically advanced governments like the United States use their prowess to initiate new spying programs, they are ever more vulnerable to hackers and the overall forces of democratization and of instability that technology offers by putting increased power in the hands of the individual. Snowden singlehandedly exposed the massive level of spying done by the National Security Agency, an act that ultimately forced changes in US policy. At the same time, capitalism is slowly restructuring the Internet into a shopping mall. Nations and corporations are using this same massive computing power to spy on their own citizens and customers—and those of other nations and corporations—at a rate that dwarfs anything previous. In its ability to create and publicly reveal knowledge about the individual, this is an authoritarian power. But technology also allows individuals to expose this authoritarianism and fight back against it in its various guises.

New power structures based on non-state leaderless groups are emerging even in nondemocratic societies, fueled in part by science's shifting of knowledge and power into the hands of individuals through increased computing power and the Internet. This restructuring expands opportunity for freedom, but also increases terrorists', anarchists', and ideologues' access to tools. The Arab Spring provides an excellent example. Twitter was used as an organizing platform for largely leaderless groups who sought—sometimes successfully—to topple autocratic regimes. The United States was supportive of the idea of democracy in the Middle East, and President Obama "upended three decades of American relations with its most stalwart ally in the Arab world, putting the weight of the United States squarely on the side of the Arab street" when he called for Egypt's autocratic ruler and longtime American ally Hosni Mubarak to step down, the *New York Times* reported.

But few foreign-policy experts seemed to know how to help the situation, or to understand that democracy is not truly democratic unless it's secular. Simply holding a vote in a religiously conservative society will return religious authoritarianism to power, as Egypt's Arab Spring showed with the election of Mohamed Morsi. The formula for an antiauthoritarian, and thus egalitarian and stable, democratic republic is predicated not merely on the vote, but on the secular vote, ensured by the separation of church and state as defined in a constitution. That was Benjamin Franklin's great insight. As of this writing, the only country in the Arab Spring to successfully transition to a democracy was also the only one that put the focus on secularism.

In January 2014, Tunisia adopted a constitution that protects the rights of women and minorities; guarantees freedom of expression, religion, and an independent judiciary; and declares that all citizens have a right to health care and a living wage—all while affirming Islam as the state religion. This came out of a unique national consensus developed by a quartet of social institutions: the Tunisian General Labour Union; the Tunisian Confederation of Industry, Trade, and Handicrafts; the Order of Lawyers; and the Tunisian League of Human Rights, which worked together following the revolution to create a civil national dialogue.

In October of that year, the country held parliamentary elections in which—for the first time—an Islamic party voluntarily relinquished power to a secular one, the Nidaa Tounes party. A few months later, on December 21, Tunisia held its first democratic presidential election, and, in what *Time* magazine called a "watershed moment," elected Beji Caid Essebsi, the leader of Nidaa Tounes. Tunisia's success, tenuous as it may still be, contrasts sharply with elections in other Arab Spring nations in the Middle East, such as Egypt, that did not emphasize secularism. The Quartet was awarded the 2015 Nobel Peace Prize for its part in laying the groundwork for this remarkable transition.

The US Defense Department's 2014 *Quadrennial Defense Review Report* examines how US armed forces are adjusting to the changing environment driven by technology and the Internet. "Unprecedented levels of global connectedness provide common incentives for international cooperation and shared norms of behavior," the report states. But "powerful global forces are emerging. Shifting centers of gravity are empowering smaller countries and non-state actors on the international stage. Global connections are multiplying and deepening, resulting in greater interaction between states, non-state entities, and private citizens." One key factor is the very real danger of the striation of society in terms of scientific and technological literacy.

In reaction to these unprecedented changes, we are seeing a staunch embrace of anti-intellectualism among those who favor a more authoritarian social order, a position naturally opposed by science's antiauthoritarianism. This is happening worldwide, in each of the major world religions and in many of its leading nations. At the same time, moral direction has also been lost by a class of Western leaders unknowingly steeped in postmodernism and Descartes subjectivism.

As a consequence, the Western democracies increasingly find themselves lacking an effective framework for solving our most pressing problems. We are increasingly paralyzed. We need a new set of political and ethical approaches for the twenty-first century. Among these essential new tools are science debates, which focus policy attention on knowledge-based discussion of the world's most pressing problems. But there are many others.

At the center of many of the arguments over international economic and environmental problems are differing views of how individuals relate to the commons—the common property of humankind. These questions are central to the relationship between science and democracy, and between individual freedom and shared responsibility. How we come to think about them will determine a great deal about our future.

The Tragedy of the Commons

In December 1968, a little-known University of California, Santa Barbara, biologist named Garrett Hardin published a paper in *Science* that would change the way we look at these questions. The core dilemma it identified, which came to be called "the tragedy of the commons" after the paper's title, lies at the heart of the unresolved economic and environmental dilemmas of the twentieth century, among them climate disruption, ocean acidification, overfishing, biodiversity loss, habitat fragmentation, overdevelopment, pollution, exploding population, and unsustainable energy use. The dilemma suggests that politicians are paralyzed by a fundamental conflict between the environment and the economy that arises from the deeply held but mistaken belief that freedom and regulation are incompatible.

Hardin's paper was remarkable because it offered such a sound rebuttal to the ideas of the Scottish economist Adam Smith, whose collaborator and mentor was David Hume. Their ideas heavily influenced the Founding Fathers from Benjamin Franklin to George Washington, and was central to the founding arguments for the United States, where both government and the economy would be ruled by the invisible hand. In 1776, Smith argued in *The Wealth of Nations* that in a shared economy, an individual, who "intends only his own gain," was in effect "led by an invisible hand" to promote the greater public interest, since willing buyers and willing sellers will always arrive at a natural price for things, and the highest value and efficiency will be obtained. "Nor is it always the worse for the society that [altruism] was no part of it," he wrote. "By pursuing his own interest he frequently promotes that of the society more effectually than when he really intends to promote it." The argument of the invisible hand was so well made that it became an axiom of economics: just get out of the way and let the market work.

But, Hardin asked, did the same reasoning still hold true in the economics not of the 1770s, when the world seemed unlimited, but of the 1970s, when it didn't? Imagine a situation where village herdsmen share a common pasture, he offered. Over time, factors like disease, war, poaching, periodic famine—market inefficiencies, if you will—keep both the herds and their herdsmen at well below the carrying capacity of the pasture. In this situation, we could consider the pasture limitless, as the world appeared to be in Adam Smith's time, and his argument would hold true.

But then, one day, improved farming practices permit social stability, and these losses or inefficiencies are minimized. At this very point of highest social good, or perhaps of maximum market efficiency, the logic of the commons creates a tragedy. Each herdsman thinks, "What is the utility *to me* of adding one more animal to my herd?" Since the herdsman receives all the proceeds from the sale of the animal, the positive benefit is +1. However, because the loss of grass, more weeds, increased erosion, etc., caused by the additional animal's grazing is shared equally by all of the herdsmen, the detriment to the herdsman is only a fraction of −1.

In this way, each herdsman is motivated by the only "rational" economic conclusion: add one animal, and another, and another. But what seems rational when the problem is looked at within the individual's frame of market reference becomes grotesquely irrational when the frame used is the collective market of all the herdsmen—and, indeed, of their economy and society at large, which now must cope with an overgrazed pasture that can sustain only a fraction of the number of cattle it did prior to the bubble. By failing to account for the context in the transaction, the environment and the economy both collapse.

Hardin concluded that, in this circumstance, "each man is locked into a system that compels him to increase his herd without limit—in a world that is limited. Ruin is the destination toward which all men rush, each pursuing his own best interest in a society that believes in the freedom of the commons. Freedom in a commons brings ruin to all." Technically, it wasn't freedom in a commons, but freedom in a bounded commons, that brought ruin. Smith's ideas were based on the assumption of geographically limitless freedom and resources. But did a free-market economy still work in the larger framework when that assumption was no longer true? Or were we in need of a new model?

Freedom versus Tyranny

The simple dilemma that drives the tragedy of the commons is writ large in the greatest political argument of our time: the clash between individualism and collectivism at the heart of every environmental issue. In the political realm, this first became a clash between capitalism and communism, and more recently between supercapitalist libertarianism and democratic socialism. Politics is narrative, and every narrative argument has an underlying value at stake. In this case, the value is freedom versus tyranny. But which road leads to freedom, and which to tyranny?

Capitalism is a more moral system if it offers more freedom to individuals. What do we mean by freedom? Liberty means freedom to choose, as David Hume, Smith's mentor, first defined it—the definition the US Founding Fathers contemplated in writing the Declaration of Independence. But there are qualifications. In a capitalist and democratic system, my freedom to do as I wish is moral and just to the degree that it does not reduce your freedom to do the same. Neither of us is the king, and both of us are. We are a society of equal kings.

To mediate a fair compromise, each of us must accept limitations equally to receive the equal benefits of having freedom from the tyranny of *might makes right*. Thus society has created titles to land and other private property, one of the most basic functions of Western governments. My freedom is bounded by the sanction of government-issued property titles, as is yours, based on how much treasure we can each muster for the purchase of land and the payment of the taxes that protect us and provide common services. What we have each gained through this self-imposed limiting and taxation—to pay for roads, a sheriff, a judge—is freedom from the conscienceless, who would use brutality and authoritarianism to take our lands by force. Freedom from tyranny proceeds from laws and regulations. The alternative is the assertion of the mighty. Thus we hear talk of the rule of law. This seems just and reasonable to most people as the price of enjoying the benefits of living together in a democratic society.

But what happens when the argument is extended beyond private property to the use of the commons, such as public parks and lands, lakes and rivers? To carrying concealed weapons in public? To smoking in public, or not using a seat belt in a car, or riding a motorcycle without a helmet? To the atmosphere, the oceans, the rain forests? To the planet? To paying or not paying taxes to maintain these common things? What is the proper equilibrium between my individual rights and the rights of the collective of everyone else—now and intergenerationally into the future? How do we balance freedom and our rights to the commons?

Today's Herdsmen

Because we have a limited planet, today's herdsmen—nation-states, supra-national corporations, leaderless non-state networks, and individuals—all have powerful economic incentives to pursue their rational self-interests until the tragedy of the commons occurs on a global scale. This is the nature of an economic bubble, but the assumption of limitless growth is what our economic model is currently based on.

The purpose of democracy is to find a balance that protects the equal rights of all individuals by using the antiauthoritarian rule of law—or regulation—and the vote to maximize the overall level of freedom and minimize tyranny. Thus, each herdsman's rational self-interest is moderated in a forum with all others, resulting in common regulations to individual behavior. Through regulation, democracy affords the opportunity to govern the engine of individualism in the commons before it races to ruin. Democracy, in that sense, is self-regulation. That is its purpose.

Understanding this relationship, teasing out the logical fallacy that underlies the tragedy of the commons, ties into the fundamental relationship between economics, democracy, and knowledge gained from science. Finding ways to create more freedom and more wealth that allow for progress and economic growth while also sustaining or improving the environment is the great task of the twenty-first century.

Freedom through Regulation

Marine geophysicist Marcia McNutt witnessed this dilemma firsthand. As director of the US Geological Survey, McNutt was in charge of measuring the flow rate of the Deepwater Horizon oil leak and attempting to assess the damage to the marine ecology of the Gulf of Mexico. I spoke to her about her frustrations at the time.

> In the Gulf I see this playing out very concretely every day . . . the
> short-term extraction issues of oil and gas versus the long-term
> restoration issues of the Gulf coastline. And many of the policies

that need to be put into place that would allow a long-term res-
toration plan that would restore wetlands, stop coastal erosion,
allow for more prosperous fishing, more bountiful fishing, more
abundant wetlands, better sediment supply to nourish beaches,
and therefore better tourism, hunting, recreational values, quality
of living, better human health, better water quality, better air quality . . . all of that is on the line because of concerns by the Gulf
Coast citizens that the policy that will allow all that will come at
the expense of the oil-extraction activities. For example, the chan-
nels that are cut through the marshes that allow wellhead access,
and many of the other dredging policies that allow ships access for
maintenance of oil wells and pipelines that allow this to continue,
are done with the thought of "one more pipe won't matter, one
more ship channel won't matter, one more bulwark by the Army
Corps of Engineers won't matter"—yet it all disrupts the ecological
balance that protects the coastlines and coastal cities.

Throughout history, the dilemma has repeated itself over and over again
with the quality of a mirage: as we get close, it seems to evaporate. Is regulatory tyranny an illusion? The answer seems to be yes. The evidence shows that
successful regulations that define a fair trade in the commons do not reduce
freedom. Instead, they increase it by leveling the playing field and preventing
unfair abuses, to the shared benefit of all.

Consider sewage disposal. In the Middle Ages, we soiled the commons
by throwing feces out into the streets and into the water sources from which
we also drew drinking water—until science showed us that this is how cholera
spreads. There was a time, not many years ago, when a factory owner whose
land abutted a river thought it was well within his rights to dump factory by-
products into what seemed an endless flow, just as the world itself seemed
practically endless. The factory owner would have called antipollution regulations infringements of his freedom. A flowing river cleans itself every twenty
miles, the old saying went.

Today, with increased awareness from science, we would view those acts
as equally stupid and unconscionable. So have the stricter regulations and laws
reduced or increased our freedom? To the extent that the common waters
are cleaner and we are less likely to contract cholera, are we each deprived or
enriched? It's pretty clear we are each as individuals healthier, wealthier, and
freer because of fair and equal regulation based on the knowledge from science.

As Hume pointed out, freedom is the liberty to *choose*. Its sources are objective knowledge, science, democracy, and fair and equitable regulation under the rule of law—all of which work to maximize the liberty to choose.

In a similar way, we accept limitations on our individual freedoms to gain greater freedom in the forms of regulations that reduce smog, acid rain, ozone destruction, the use of DDT, backyard burning of garbage, driving while intoxicated, noise pollution, lead in paint and gasoline, certain carcinogens, water pollution—and, more recently, exposure to secondhand smoke, injuries caused by not wearing seat belts, and texting while driving. Generally, most people appreciate these laws and regulations for the freedom they provide from the tyranny of others' stupid decisions, which impose upon us bad health, environmental destruction, devastating injury, higher insurance rates, capricious death, and other economic "takings" that reduce the quality—and quantity—of our lives.

We have not yet seen fit to extend our personal freedom to regulating other pollutants that limit our choices and those of our children, such as the freedom to a stable environment by regulating the release of extra carbon dioxide and methane; or regulating new pesticides like neonicotinoids that are systemically absorbed by plants and expressed in their leaves and pollen, killing monarch butterflies and pollinators.

How can we give up freedom to roam land without fences; freedom to dump sewage where we please; freedom to dump gaseous by-products into the commons and the air; freedom to use more advanced technology to promote agricultural productivity and industrial value; and to reap the "limitless bounty of the sea"? These are the acts of our ignorant past, but once we build knowledge of environmental effects, once we "know better," these things become the acts of tyrants that deprive everyone of freedom and remove choices; they impose a tyranny of trash—of ignorance. Ignorance is not bliss. It's tyranny. Tyrants take private property; they take health, life, and clean water; they take clean lungs and fresh air; they take fish by depleting the oceans, money by raising the cost of insurance. The shifting—or, as economists say, the externalizing—of private costs and risks onto the commons takes from everyone, and in fact reduces wealth throughout the economy.

Tyranny on the Commons

All of this ties back in to the ideas of conservative economics, particularly those of its father, the American economist Milton Friedman. In every economic transaction there is a willing buyer and a willing seller, Friedman theorized, and they agree on a price that benefits both. But there are spillover effects in many economic transactions—costs and/or benefits that are transferred to third parties. Friedman called these spillovers neighborhood effects. Today, most economists call them externalities because they are external to the transaction.

At their most basic, externalities don't have to involve buying and selling. If you smoke in a restaurant instead of stepping outside, it's easier for you, but it's worse for everybody else. That's an externality. If you throw your McDonald's bag out the car window, it makes life easier for you, but it's worse for everybody else. That's an externality, too. If a pretty girl walks by in a low-cut top, she's probably a little colder, but the added happiness of the boys in the neighborhood is a positive externality. (For some reason, economists seem to love that example.)

Now let's add in money. If your power company sells you coal-fired electricity at three cents per kilowatt-hour, that's a good deal. But if it does it by burning cheap, dirty coal and doesn't have a smoke scrubber, the soot that's dumped on the neighboring town is a negative externality, because it's costs aren't reflected in the transaction. The people in that town are subsidizing your cheap power by paying the cost in terms of property damage, extra cleaning, poor health, environmental pollution, and aggravation. And the next generation is also subsidizing your price break if the utility isn't removing the carbon dioxide from its emissions. That's an externality, too. In terms of freedom, you are forcing a tyranny on the commons because the third party (in this case, everyone else) is deprived of a choice about whether or not to partake in the transaction that gives you cheap power. It's not fair or equal freedom. You're being a tyrant.

Now let's say someone in your town starts a company to produce a new kind of biofuel. This biofuel is so efficient that everybody wants it, and the company's business skyrockets. It needs to hire more workers and expand its plant. As a result, the town starts to boom economically and your home's value soars. That's a positive externality: some of the wealth the company is generating has spilled over onto you, even if you don't buy the biofuel or work at the plant. Friedman would say this is not fair either, and that you should pay some of this wealth back to the plant to balance the freedom in the transaction. Again you're being a tyrant.

A science-based example of a positive externality is vaccinations. Vaccinations work based on the number of people in the population who are vaccinated. Once a certain threshold is reached, the disease can't spread effectively and is essentially eliminated. As long as enough people are vaccinated—i.e., conform with the regulation requiring vaccinations—others who choose not to be still get to enjoy that positive externality at no cost. These are what economists call freeloaders.

Economists like the golden rule: do unto others as you would have them do unto you. They don't like externalities because they represent inefficiencies in the marketplace that skew the real value of transactions and so are implicitly unfair. By reducing choice, they reduce freedom. Those who suffer by exposure to external costs do so involuntarily, while those who enjoy external benefits do so at no cost. Friedman thought that overcoming neighborhood effects "widely regarded as sufficiently important to justify government intervention" was one of the key roles of government.

Government can do this in two ways: by imposing taxes, and through laws (and regulations). Taxes used punitively—for example, sin taxes levied on such things as alcohol and tobacco—force the generator of an externality to internalize the cost of making life worse for everyone else. The alcohol tax forces those who drink alcohol to internalize some of the increased costs and risks of the behavior, such as the need for more police to protect the public from drunk drivers. Laws against littering coerce fast-food patrons to properly dispose of the trash from their meals, and if they are caught breaking the law, they incur a fine that acts like a more painful sin tax and covers the cost of their irresponsibility toward everyone else.

Similarly, emitting excess carbon dioxide may make life easier for you, but it makes it worse for everybody else. Energy and extraction companies have known this for forty years, so, like the tobacco companies, it's likely they will one day be sued for their knowing degradation of the commons and individual health and quality of life while investing billions to mislead the public. Using Friedman's logic, now that we are widely aware, through science, of the costs that carbon emissions shift onto society and future generations, "to overcome neighborhood effects widely regarded as sufficiently important to justify government intervention," it's incumbent upon us to find ways to make excessive carbon emitters internalize those costs in order to maximize freedom and market efficiency.

We can do this by imposing a new form of sin tax—a carbon tax, by implementing cap-and-trade or carbon tax schemes to use the market to punish carbon externalization and reward carbon internalization—or by employing other methods to get excess emitters to internalize the cost of what they are taking from our freedom.

Okay, we might say, that is all well and good. We understand that cap-and-trade is a market-driven, conservative economic idea, as George W. Bush's own chief economic advisor, Douglas Holtz-Eakin, pointed out. We know that market-driven environmental laws and regulations can reduce soiling of the commons and increase individual freedom by incentivizing and rewarding internalization—responsibility and honesty, if you will. But how do we get at Hardin's biggest concern? How do we manage the environment in a limited world when everyone only cares about economic growth and their own prosperity? Are we not rushing toward ruin? Can we have economic growth and a clean environment in a limited world?

It's the Economy *and* the Environment, Stupid

Former NOAA administrator Jane Lubchenco thinks we can. Lubchenco is one of America's leading voices on tackling climate disruption, but she challenges the idea of a limited world, pointing out that science has consistently expanded the economy beyond the zero-sum days of Thomas Hobbes. "By creating new knowledge, science changed economics away from a zero-sum game," she says.

Evolutionary biologist Simon Levin thinks that the way to tackle the problem is to develop an economic model that values the commons as capital. "I see a real split even among my students," he says, "many of whom think economic growth and environmental protection are antagonistic elements and therefore a primary goal must be to reduce consumption and economic growth. I don't believe that's the case, but the interplay is an example of the sorts of problems we have to wrestle with."

Levin got together with 1972 Nobel Prize–winning economist Ken Arrow and ecologist Paul Ehrlich to look at the question of whether we're consuming too much, and he says they found that the answer was "not uniformly—[in] some areas [we] are and [in] some [we] are not." How does one evaluate that? It requires coming up with an acceptable definition of general welfare. If we take a comprehensive wealth measure that includes the ability to eat, to preserve the environment, to maintain our health, and other similar measures, in some areas we're actually underinvesting and underconsuming, Levin argues.

Limitless Growth?

In broad strokes, there is truth to Levin's approach. The iPhone I hold in my hand uses far fewer resources than the mainframe computer I used to program in BASIC over a teleterminal in junior high school. It has far more power and, more important, it adds far more value to my life. Better quality and utility combined with lower resource use and cost can be a formula for economic growth and wealth generation in other sectors as well, particularly as nano-technology becomes more sophisticated.

But look closer. Yes, per-capita resource use is lower now, but the market is vastly larger. Not everyone had a computer back then—I used a teleterminal to access one. But now most of humanity owns a cell phone. The global population is becoming unsustainable, and the road from ENIAC—the first electronic digital computer, completed in 1945—to my iPhone is littered with glass, plastic, steel, gold, silver, platinum, and toxic metals including barium, cadmium, chromium, copper, lead, nickel, and zinc. My iPhone may weigh only six ounces, but I'm responsible for a burden of about 180 pounds of discarded computing technology, on average. And these amounts are now being dwarfed by the e-waste being generated in the rest of the world.

The solution is to be smarter and more, yes, conservative—to find ways to grow the economy without depleting the corpus of natural capital, and thus inducing neighborhood effects. In a business sense, this means not spending one's assets to finance current operations—i.e., don't burn cash to heat your home. The concept is not new. Conservative US president Teddy Roosevelt articulated it in a 1910 speech to the Colorado Livestock Association in Denver. "The Nation behaves well if it treats the natural resources as assets which it must turn over to the next generation increased, and not impaired, in value," Roosevelt told the gathered cattlemen.

This is largely a matter of perspective. Imagine, for example, that we live in a galactic economy and you are the CEO of the corporate planet Earth, Inc. Your job is to maximize the shareholder value of Earth. To keep a healthy balance sheet, you will want to account for, monitor, and restore assets both known and unknown in Earth's minerals and biodiversity, ranging from the sources of new miracle drugs to those of the life-expanding high-tech resources of tomorrow. What makes sense for competition on this scale is different from what makes sense within a smaller frame of reference. If ignorance is tyranny because it removes choices, we need to develop a way to quantify the cost of the unknown wealth—or choices—we may be leaving on the table—in other words, our opportunity cost.

This opportunity cost, and arriving at a present value for the unknown economic bounty our current resource stock may contain, is the most fundamental value equation we need to solve. If we burn our capital, we won't be able to compete. The corporations currently mining these resources may be motivated to become their best protectors if we can find a way to define and value that future economic opportunity. This shifts the business activity from mining resources to farming them. It seeks to maximize productivity and biodiversity to provide maximum biological creativity and the attendant economic potential for finding the next big thing. It seeks, in other words, to maximize both freedom and wealth because, in the end, they are the same thing.

Ecosystem Services and Natural Public Capital

For an economic model to include environmental sustainability, value must be assigned to common commodities, be they biodiversity, a stable climate, clean water, rain forests, parks, mineral resources, topsoil, and so on. Because the economy is a human system that trades in human values, the first principle of valuing the commons is to state it in terms of human values. Some environmentalists argue that this approach is wrong, that the values of these things transcend human purposes. That may be true, but for the purposes of trading in the human economic system, we need to translate things into economic terms. Any value we put on biodiversity has to relate to the way people value it, which comes down to a question of utility and, ultimately, money. This is consistent if the objective of utility is to provide greater freedom. As a means for beginning to think about this, environmental scientists are now talking about ecosystem services—the services that ecosystems provide to humans.

Included in ecosystem services are fuels, foods, and other direct benefits like natural resources, pharmaceutical products, the filtering of surface water by wetlands, and the buffering of coastal cities from storm surges by coastal wetlands. Ecosystem services also encompass indirect benefits in terms of ethical values we choose to assign to the environment, such as the benefits of being able to visit the wilderness or to experience the transcendent and rejuvenating beauty of nature.

These services are the starting point. But to think of them as simply *services* is only part of the financial calculation. The services are provided by working capital—the natural capital owned by the business Earth, Inc. And since the economic system is a human one, we need to relate it to humans as natural public capital. We are the shareholders in Earth, Inc. We have inherited it; we can invest it, even borrow from it, but we should never spend down our capital because it supports our ongoing operations and delimits our future growth.

So how do we arrive at values for natural public capital? How do we deal with uncertainty, inequity in distribution of wealth, intergenerational equity? And how do we put a value on things that don't have a market price, like our love for our children?

A Full Accounting

In 1997, ecological economist Robert Costanza set out to answer those questions. He and a team of fellow researchers embarked on a quest for metrics that would put a value on the ecosystem services that nature provides to our economy.

Published, appropriately, in *Nature*, the paper had a powerful effect because it put a price tag on the commons, showing that the annual value of the world's ecosystem services to the economy was at least $33 trillion, or nearly twice as much as the $18 trillion that made up the world's combined gross national product—the sum total annual value of each nation's traded goods and services—and possibly as much as $54 trillion, depending on certain assumptions, or three times the total world economic activity. For comparison, the US gross domestic product in 1996 was about $6.9 trillion. The economics behind this valuation were hotly debated, but ultimately the paper held up to the scrutiny it received.

Duke environmental scientist Stuart Pimm, who wrote the News and Views column for that issue of *Nature*, describes the moment as a watershed. "Within a decade you could go to meetings at the UN and we had a world that's saying, 'We need to raise $20 billion,' and that would be the price of stopping global deforestation, which is putting 1.5 billion tons of carbon into the atmosphere each year. The cheapest way to slow climate change is to stop tropical deforestation." The argument now had currency in a way that made it discussable in public-policy and economic terms.

Most policymakers outside the United States now understand that we can't treat the global commons as if they have no value. They are exceedingly valuable; in fact, they are the source of our biological existence and the foundation of our economy, and it's worth our while to protect them. "Globally, everybody except the United States now gets this," says Pimm. The political paralysis in the United States on both the science and the economics has become the largest single global impediment to addressing climate disruption.

"When you're dealing with equity issues, it's got to involve conversations with social scientists, humanists, economists, and the public," says Levin. And it's complicated because it's not just about today's people, but also about tomorrow's. How do we value the right to spend down natural public capital that depletes the wealth and freedom of future lives—reducing their liberty—either by polluting it or by using it up? What is ethical, and what is "spending my children's inheritance," as the popular bumper sticker proclaims?

Midnight in the Garden of Good and Evil

Clearly, broadening our frame of reference from Ayn Rand's "rational self-interest" to the broader perspective of "rational self-interest in a shared and limited pasture" held by a farmer—who surveys the field, does a little science, and realizes that if everyone adds one cow all will fall to ruin—relies on political speech, not altruism. Nevertheless, Stuart Pimm sees religious engagement as critical. He's focused much of his career on species extinction and how to prevent it, and says that speaking with Christian groups in particular is important. "You've got very large groups of Eastern Orthodox, Anglican Communion, Methodists, Unitarians, all of whom come down strongly on this notion [that] there needs to be creation care," he says, "versus the group on the Cornwall Declaration saying, 'All of this is deeply evil and we need unfettered access to natural resources.'"

"The Cornwall Declaration on Environmental Stewardship" and its associated "An Evangelical Declaration on Global Warming" are public letters expressing the libertarian economic opinion of the Cornwall Alliance, a coalition of evangelical clergy, theologians, and policy experts, including Charles Colson, James Dobson, Jacob Neusner, and R. C. Sproul. With striking similarity to the indictment of Galileo, the group's position on climate disruption is:

> We deny . . . that Earth's climate system is vulnerable to dangerous alteration because of minuscule changes in atmospheric chemistry. . . . There is no convincing scientific evidence that human contribution to greenhouse gases is causing dangerous global warming. . . . We deny that carbon dioxide—essential to all plant growth—is a pollutant. Reducing greenhouse gases cannot achieve significant reductions in future global temperatures, and the costs of the policies would far exceed the benefits. We deny that such policies, which amount to a regressive tax, comply with the Biblical requirement of protecting the poor from harm and oppression.

Meanwhile, the Cornwall Alliance's position on sustainability is:

> Many are concerned that liberty, science, and technology are more a threat to the environment than a blessing to humanity and nature. . . . While some environmental concerns are well founded and serious, others are without foundation or greatly exaggerated. . . .

> Some unfounded or undue concerns include fears of destructive manmade global warming, overpopulation, and rampant species loss. . . . Public policies to combat exaggerated risks can dangerously delay or reverse the economic development necessary to improve not only human life but also human stewardship of the environment. . . . We aspire to a world in which the relationships between stewardship and private property are fully appreciated, allowing people's natural incentive to care for their own property to reduce the need for collective ownership and control of resources and enterprises, and in which collective action, when deemed necessary, takes place at the most local level possible.

This split view of sustainability—as either a matter of religious stewardship value or something associated with socialism or communism—is uniquely American. And the fact that American libertarians, who are against authoritarianism, are now advocating for tyrannical policies, illustrates the ways the limited pasture is turning our old ideas on their heads. In Europe, for example, few at either end of the political spectrum take any particular issue with the facts of climate change, and European countries have taken the lead in tackling the problem. After Tony Blair was elected British prime minister in 1997, the Labour Party's first conference was on climate change. Chief scientific advisor Lord Robert May had prepared a report that Blair presented to all attendees, and it helped set a national course. "The wheels of government grind slowly," says May, but Britain's Climate Change Act, adopted a few years later, provided for specific carbon targets and established "a Climate Change Committee to recommend the targets and monitor progress towards them. The UK is the only country to have such formal apparatus in place." In a more recent example, Norway committed $1 billion to help combat deforestation, which some, like Pimm, say is the cheapest way to combat climate change. The amount was roughly seventy times more on a per capita basis than the United States promised in the Copenhagen Accord. The list goes on. Pimm sees the disparity between the United States and other nations on climate change as symptomatic.

> It speaks to a very deep division, and one only has to look at the political influence of people like Sarah Palin and James Inhofe. These are not politicians arguing on the edges, these represent very

deep emotional issues where a good half of the US population isn't marching in step either with the science or with the religious, spiritual, and ethical motivations in other countries. So you can't just say we're doing a damn lousy job of explaining science in the media. It's way beyond that. Clearly the US has had a very long and powerful urge to be isolated from the rest of the world. If it was only the Atlantic seaboard that voted in presidential elections, we'd have a very different country. It's not surprising [that] you have Inhofe in the middle of the country. Fewer than 30 percent of Americans even have passports, and roughly half of those are only for travel to Canada and Mexico. The size of the country has people in the middle very isolated from the rest of the world. They sort of admire the fact that Sarah Palin thinks that Africa is a country.

The differences between American perspectives and those prevalent in other Western countries are of course not limited to climate change. University of Pennsylvania bioethicist Jonathan Moreno points out how concern about genetically modified food draws large numbers of students to campus demonstrations in Europe, but, in the United States, "you get maybe eight people. If it's about reproductive choice, the situation is reversed."

Pimm says that the "extraordinary isolation" of Americans is dangerous.

This idea that we can go and tell the rest of the world to go f— themselves with impunity, that's a very stupid thing. We need resources, they cost an extravagant amount of money, and that's getting worse as China comes online competing with us—not just as a producer, but worse, as a consumer, particularly an energy consumer. I live in Florida, where the junior US senator, Marco Rubio, denies global warming, but many of his constituents are losing their homeowners' insurance because insurance companies are worried about costly storms from global warming.

The Insurance Cost of Climate Change

Richard Feely, senior scientist in charge of monitoring ocean chemistry for NOAA, estimates that the United States and other countries could stop climate disruption for an investment roughly equal to 2 percent of their gross domestic product, which in 2015 was about $17.7 trillion.

If this is correct, an annual US investment of roughly $350 billion could get the American economy off carbon. This would not be a cost so much as a tremendous economic investment, generating massive amounts of internal economic activity through domestic businesses that could propel prosperity for decades. The economics of this approach should attract the attention of anyone who has run a business. If you could improve the company's cash position and eliminate a number of long-term ongoing expenses while positioning the business for the future with new technology and strengthening its balance sheet, all for just 2 percent of current value, you'd jump at the chance. In 2015, I put 30 solar panels on my house at a cost of $18,000. I received a 30 percent federal tax credit that lowered my net cost to $12,600. The 8.55 kilowatt system is expected to generate about $1,200 annually worth of electricity, for a return of nearly 10 percent. That's better than the stock market. A few years ago I did a similar thing, replacing our furnace with a geothermal system that is 500 percent efficient. The economics work. It's now all about implementation.

Insurance companies, naturally, are focused on these emerging economics. They are already transferring increased risk from climate-related disasters into policy premiums or dropping coverage altogether, particularly in coastal areas, but they are seeing sharp upward trends in exposure across the board.

Ernst Rauch, head of the Corporate Climate Center at Munich Reinsurance America, a major reinsurer of global insurance companies, reported in the 2014 *Natural Catastrophe Year in Review* webinar that global natural disasters had risen from fewer than four hundred events in 1980 to a new high of 980 in 2014, reflecting an increase, on average, of a factor of more than two and a half. Geophysical events, such as earthquakes, tsunamis, and volcanic eruptions, were relatively unchanged. Most of the increase was in atmospheric perils exacerbated by climate change, he reported, including the heavy rains and flooding in the Balkans, Pakistan, the United Kingdom, and the Eastern and Southwestern United States, and historic droughts in Brazil and the Western United States. It was a case of "both too much precipitation and too little precipitation that occurred in both the US and worldwide," said Carl Hedde, head of risk accumulation for Munich Reinsurance America.

Worldwide catastrophic loss events related to the climate—storms, floods, droughts, and wildfires—in 2014 have more than quadrupled since Munich Re began tracking them in 1980, but have remained unchanged in other categories of loss. Hedde reported that "approximately eighty percent of the (US) insured losses for 2014 were attributable to severe thunderstorms." Globally, more than nine out of ten (92 percent) of the loss-related natural catastrophes were due to weather events.

At the 2014 midyear natural catastrophe webinar, Peter Höppe, head of Munich Re's geo risks research and Corporate Climate Center, informed attendees that "this month has been the warmest May since routine temperature measurements started more than one hundred years ago." He identified how "more frequent storms and severe events will increase losses and variability caused by natural disasters, and that the changing loss patterns will challenge insurance systems to offer affordable coverage and provide more risk-based capital. So this shows the direct affectedness of the insurance industry by global warming."

The growing population is also a driver. Global population is currently roughly 7.4 billion, with roughly eighty million more people being added every year. By 2050, the world's population is expected to reach 9.6 billion. As population increases and science makes society both more interconnected and more high-tech, the costs associated with each disaster increase. A hurricane in Florida causes far more economic damage now than it did in the less populated, less wired, and less wealthy Florida of 1980. Multiply the increasing costs by the increasing population by the increasing number of catastrophic storms—and by the fact that an estimated 44 percent of the world's population now lives in coastal regions—and one can see why insurance premiums are skyrocketing. Sea-level rise is accelerating. As oceans warm, they swell, raising sea levels. Add to that the melting ice from the Antarctic continent and Greenland, both of which are over a mile thick and together contain over 99 percent of all freshwater ice on Earth, and sea levels could rise by as much as 61 meters, or 200 feet, putting most current coastal regions, and even states like Florida, deep under the ocean. If we manage to stop global warming at two degrees centigrade (it's about 1 degree now, with another .5C baked in due to anthropogenic aerosols that will leave the atmosphere as we stop burning fossil fuels), sea levels could rise between 6 and 11 meters, still inundating most of the world's coastal cities, and placing structures and roads further inland at vastly increased risk from storm surge and tsunamis.

From 2001 to 2010, US losses from climate-related catastrophes, most in coastal areas, were greater than the $359 billion Feely's numbers suggest would be the cost of conversion to a carbon-free economy. Considering the growth curve in losses, one can predict that within the next decade economics will totally transform the climate-change debate.

While insurance-industry modeling is limited in scope, it does provide some useful ways for thinking about the interplay between wealth and the environment. What makes it interesting is that insurance turns the tragedy of the commons on its head. Here's how it works:

Like science, wealth affords freedom by expanding the power of choice. Insurance is itself a commons, a shared pasture. It provides financial security by spreading individual risk to the group. In exchange, the individual gives up some small portion of freedom, represented by the monetary cost of the premium (a tax) and by policy terms that limit behavior (regulations). The small limitation on freedom is, generally, viewed as a worthwhile tradeoff for the greater freedom from the risk of financial catastrophe that insurance coverage provides. Insurance is essentially economic regulation that expands freedom by protecting against the brutality—or tyranny—of random catastrophic events, in the same way that property titles and sewage regulation expand freedom, by protecting against the might-makes-right of roving marauders and the ignorance of polluters who allow waste to run off into common waterways.

John Hubble, father of Edwin Hubble, was a staunch Baptist and an insurance underwriter, and he wrote in 1900 about the core human conflict of self versus the collective that was captured by insurance. "The best definition we have found for civilization," he wrote, "is that a civilized man does what is best for all, while the savage does what is best for himself. Civilization is but a huge mutual insurance company against human selfishness."

Insuring the Commons

That is also a useful way to think about protecting global ecosystem services and natural public capital. Once we establish their economic value, we can create an insurance vehicle that protects against the exposure to catastrophe. Costanza suggests that the best way to handle it, at least for accidents such as the Deepwater Horizon oil spill, is by requiring an "assurance bond"—a sum of money large enough to repair the involved ecosystems if an accident occurs. The face value of each bond would be based on the total value of the ecosystems at risk and be set by an independent agency or government-chartered body. Companies posting a $50 billion bond to drill in the Gulf of Mexico would be made aware that they were engaging in a very risky business and that investments like acoustic blowout preventers, at a cost of $500,000, are good deals and thus justifiable to shareholders.

This approach is not new. We regularly ask private parties to purchase insurance to cover the risks they pass on to the public. Purchasing automobile insurance is now mandatory in the United States. Building contractors must carry performance bonds in order to build sizable projects. In the film business, performance bonds protect investors from the risk of the director or producer failing to bring a high-quality picture in on budget.

The risks of such instruments, however, become apparent when looking at the longer time scales often involved in environmental questions, such as climate change and clean-water protection. Consider the nonferrous mining industry. There is a significant deposit of nonferrous, precious metals in Northern Minnesota, near Lake Superior and the Boundary Waters Canoe Area Wilderness. But nonferrous mining excavates minerals found in sulfur-bearing rock, and the waste rock, piled in holding ponds, reacts with water and air to create sulfuric acid, which can be a catastrophic source of water pollution. Mining companies have long sought a means of mining the minerals while meeting environmental regulations, but their submitted environmental-impact statements show the need for continuous treatment of polluted water for at least five hundred years.

The question over such a time scale becomes how to value and structure a financial assurance instrument that can fulfill its intended purpose. A bond is a government instrument, but, in the past, such instruments provided for other purposes have been raided by subsequent legislatures unable to resist spending "free" money, especially in a budget deficit. This actually happened in Minnesota, where a billion-dollar tobacco settlement fund intended to promote public health was raided to fill a budget gap. A bank could hold a deposit, but banks go bankrupt. And, with a massive amount of money paid into a financial instrument unrecoverable for centuries, how can the government incentivize a mining company to clean up the project in twenty years' time, rather than simply declaring bankruptcy when the minerals are gone and leaving the cleanup to the taxpayers, as has frequently occurred? A trust account is similarly vulnerable in the event of political or corporate change. The United States has been a country for less than 250 years, so is it reasonable to consider a financial instrument that would extend twice that time into the future? There is also considerable science to show that humans are not very good at thinking in such time scales, especially when considering consequences.

These kinds of vehicles obviously have some limitations that still need to be worked through, but their purpose at shorter time scales is to cause private interests to internalize the risks they are undertaking and recognize them ahead of the fact, instead of dealing with them afterward—an idea that has been developing since the early 1980s as a market-based, more efficient alternative to the command-and-control method of law and litigation.

The latter system is the one we currently use for mitigating most environmental damage. It's inefficient because it doesn't do a great job of motivating prevention, it requires costly government monitoring of behavior, it is politically vulnerable, it shifts the burden of proof to the public, and it seeks reparations only after the damage is done. The *Exxon Valdez* oil spill is a case in point: litigation took decades to make it through the courts and was ultimately resolved for a fraction of the actual loss. Clearly, this route to accountability is ineffective, unjust, and inefficient. Whether the goal is to preserve the environment or maximize economic efficiency, or simply to solve problems and avoid costly litigation, shifting the economics of risk to the front end by being proactive is better business practice. That's because managers are made aware of the economics of risk in advance, can measure them, can control for them in their business models, and can account for them in their product pricing, thereby protecting the company against otherwise uncontrollable market punishment. Regulation makes it fair for all companies to do the job right. Peer-reviewed scientific research by Costanza, Andrew Balmford, and others has now shown a return on investment of at least $100 for every $1 invested in preserving remaining wild nature—a figure that will make any Wall Street trader's eyes light up if we can find a way to monetize it.

On Leadership

To achieve that, governments—whose purpose is to administer the commons and maximize freedom—will need to be more involved in crafting policy, fair and equitable regulation based on science, and international treaties that level the playing field—not just on climate, but on all evidence-based policy issues. This needs to be done in cooperation with the insurance industry, environmental groups, scientists, and the private sector, so that forward-looking, innovative companies, nations, and individuals—those focused on the future—aren't at a competitive disadvantage. We need, in the words of Pope Francis, "a new dialogue about how we are shaping the future of our planet."

It's also going to take the involvement of Wall Street and agencies that work to develop accountability in public- and private-sector financial reporting, such as the Governmental Accounting Standards Board, the Federal Accounting Standards Advisory Board, and the Financial Accounting Standards Board. Currently, no laws require the government or private industry to use full-cost accounting methods that include the value of ecosystem services. Some of this has begun occurring as pension and other investment funds have begun to realize the fiduciary risk in holding energy-industry stocks and other financial instruments with exposure to evidence from science. But the market is still far behind the curve on this. In addition, many local governments have adopted a limited version of the methods to account for non-cash costs associated with solid waste management. Such programs offer a conceptual base on which to build.

We are living in a time of supranational corporations and a global economy based on feudalism and the false, unexamined assumption that the world is effectively unlimited. This has financial implications for investors. We have globalized the economy but have no global regulatory structure, and so have recreated the Wild West on a global scale, requiring the battles over labor and environmental standards be fought anew. These corporations can no longer be considered constituents of any one nation. This ultimately results in the freedoms of all individuals and all nations being reduced by these corporations' takings from the natural public capital, and that kind of corporate act is beginning to come home to roost in stock valuations.

This lack of a governing system is not the corporations' fault, but there is no reason to assume that this situation will change on its own; these corporations have been freed by the global market that has developed because of poorly thought-out free trade agreements and because of technology and the Internet, and are now battling attempts at regulation. Without concordant global regulation, they have become locked into a pattern of quarterly competition for returns that do not account for the burning of public capital. They chase the cheapest labor, the weakest environmental regulation, and the lowest taxes under the premise that they are creating shareholder value, when in a bounded pasture they are in fact destroying it. Like the herdsmen of Hardin's commons, they are, by the Wild West structure of the supranational global marketplace, pushed to consider only "What is the utility to *me* of adding one more animal to my herd?" But it can be different.

The way to free them from that tyranny of the unregulated global marketplace is to adjust the market, to level the playing field with fair and equitable regulation, as Friedman suggested, and to refuse to let them burn public capital to artificially inflate their financial performance. That is a cheap and dirty way to see quarterly gain, but it is a great externality and thus a tyranny antithetical to free markets and to democracy. Corporations must be asked, on a fair and equitable basis, to internalize their takings. Those who cannot probably should not be in business, because they are either too unresponsive to markets, are abetting bloody physical tyrannies for their gains, or have unrealistic cost structures and are thus bound to fail eventually. This, too, is a fiduciary risk that is actionable by shareholders, regulators, and citizens, and corporations should be held to account.

Instead of the bubble-economy model, based on its false assumption of a limitless world, we need to implement a new economic model based on the reality that the world is finite. This is where leadership comes in. For all the talk of ExxonMobil and the Koch brothers spending billions to scuttle cap-and-trade, there are also new corporations led by younger entrepreneurs who aren't trapped in the discredited anti-government thinking of Ayn Rand. Corporations like Tesla Motors under Elon Musk and Walmart under its new leadership could form a corporate-policy cooperative of dozens of major companies to encourage forward-thinking legislation: cap-and-trade, a carbon tax, and laws favoring the rapid adoption of innovations such as electric cars, geothermal heating and cooling, solar energy, cleaner nuclear technology, and uniform global environmental and labor standards.

We don't lack the ability; clearly. We lack the leadership, the organization, and the courage to invest in a new and different model of the future. The assumption that other economies will not follow the lead of the major Western powers—or the lead of China or India—or the lead of US corporations—is incorrect. Much of the world is well aware of the issues and places value on the environmental commons, but is even more powerless. By exercising leadership, a major economic power would be able to maintain and strengthen its moral prestige, and could do so without damaging its economy. It would be in a position to rein in supranational corporate feudalism and to dictate a new era of foreign policy, via the International Monetary Fund, the United Nations, and its own foreign policy—one that rewards and punishes based on national economies' willingness to fall in line with policies of living wages, human dignity, and environmental regulation that maximize sustained freedom on a global basis. The US State Department is starting to look at issues like these, and at science as a tool of diplomacy, but the thinking is still in its nascent stages. It's a battle worth playing hardball to win.

Ultimately, the war on science comes down to the rise of authoritarianism—corporate authoritarianism, antigovernment authoritarianism, religious authoritarianism, and, in its rejection of objectivity, a rejection of the idea that we could all work together with a common view from both nowhere and everywhere to make life better for everyone while preserving and improving our shared home. These words, however—shared, common, together, everyone—have, to a certain brand of libertarians, the ring of collectivism. But they are the undeniable reality of the bounded pasture, and business and political leadership would be irresponsible and ultimately uncompetitive not to recognize that. For in providing real solutions that address the real situation, one addresses real human needs in an ethical way, and there is a large and growing market for that.

Winning the war on science is this generation's calling. But are we capable of battling back the authoritarian resurgence? Do we have an understanding of science adequate to defend its unique role in human history and policymaking, or even to see the issues clearly—to base our political arguments and our journalistic coverage on knowledge and not just on the confused and endless cacophony of warring opinions from whence the modern era first emerged? Are we able to look up from the grist mill long enough to consider the vast economic and political potential of a new and innovative world economy, circular, decarbonized, reinventing, wealth-building, and sustainable—and to fight with all we have to make it happen? Do we have the vision to even realize we are in such a battle, and that the future goes to those who act? These are the very serious questions by which this generation, and the human race itself, will ultimately be judged, and they remain unanswered.

Chapter 12

Battle Plans

Scientists would see no reason why, just because the individual condition is tragic, so must the social condition be. Each of us is solitary:
each of us dies alone: all right, that's a fate against which we can't
struggle—but there is plenty in our condition which is not fate, and
against which we are less than human unless we *do* struggle.

Most of our fellow human beings, for instance, are underfed and
die before their time. In the crudest terms, that is the social condition.
There is a moral trap which comes through the insight into man's lone-
liness: it tempts one to sit back, complacent in one's unique tragedy,
and let the others go without a meal.

As a group, the scientists fall into that trap less than others. They are
inclined to be impatient to see if something can be done; and inclined
to think that it can be done, until it's proved otherwise. That is their real
optimism, and it's an optimism that the rest of us badly need.

—C. P. Snow, 1959

What Must Be Done

We've now had the briefest of tours of the vast intellectual, ideological, and
economic war on science. We know who's waging it, we know why, we've seen
some of the political changes in Western society that have allowed it to spread
and that it has caused, and we know what the stakes are if it is not won. We've
examined some of the generals on either side of this war, and we've explored
what it is in our own minds that makes us such easy targets for conscription on
one side or the other.

We've also looked at the issues that form the conditions—the intellectual
soil, if you will—of the whole debate: that science has succeeded beyond our
wildest dreams, and that, in so doing, it has torn away some of the spiritual

mysteries of life and disrupted our sense of our place in the cosmos. But more importantly, it has enabled us to increase our population and our environmental impact beyond the capacity of our one small planet to support us. This, above all else, is breaking apart the foundation on which modern society has been built—that individuals, acting in their own self-interest in a free marketplace, can deliver the highest and most efficient good to society, and that such economic activity can expand without limit. Population plus individualism plus technology may be our ultimate undoing.

But the operative word of course is *may*. As we've heard Jane Lubchenco and Simon Levin argue, science has been able to resolve such dilemmas before. Life is not a zero-sum game; the pasture's bounds can be increased. The tool for that increase is the human imagination and its capacity for problem solving and innovation. And while taking a square look at the challenges is daunting and may make it seem as if all hope is lost, all hope is not lost. The human capacity to innovate can always be unleashed, given the right support and circumstances. We know what it takes—a marriage of science and engineering with creativity, artistic design, a vibrant exchange of ideas, freedom of inquiry, investment in and support of basic research, ambition, and cultural support. Those are the elements that have always produced giant leaps forward, and they certainly can again.

But we are running out of time. When policymakers could be encouraging innovation, far too many are fighting against it. When the media could be reporting on the true state of affairs, far too many are making facts up, using false balance, shying away as if they are intimidated, and abdicating their role as democracy's feedback mechanism. And when religious leaders could be parsing great human questions, they are too often flailing in an intellectual quagmire of fundamentalism and free-market libertarianism.

We need to beat back the war on science in order to provide the space, resources, money, and motivation to research, to learn—and, finally, to bring our many fragmented cultures together in common cause. That means drawing some moral lines in the sand. Attacks on science are attacks on democracy and freedom, and we need to start treating them with that level of seriousness and sanction in our public discourse and legal system. As tobacco companies were found liable for misleading the public, which led to increased deaths from cancer, so too should we look at sanctioning other companies and privately funded networks of think tanks who engage in disinformation campaigns to spread "uncertainties" about science they know there is no real uncertainty about.

Because of this quality—that attacking science is attacking the foundation of democracy—democratic forms of government need to take such attacks extremely seriously, and to enact legislation that fights them back. Additionally, the other institutions of civil society should and can do their part to battle back the war on science.

There is no one strategy by which the war may be won, no nuclear option, no shock and awe. But there are many individual battle plans. Some of them can come from government, some of them from within academia, some from business or religious institutions, and some of them from concerned individuals who have had enough and want to do something. Each of these plans deals with different factors that allow antiscience to thrive, and each can contribute in important ways to a new future. In aggregate, they may help to bridge the gaps between science and society, and to push the forces of antiscience back. The list is by no means intended to be complete or inclusive; it is merely offered as a starting point.

Battle Plan 1: Do Something

The summer of 2014 was the hottest that had ever been recorded up until then. By September, a critical mass was being reached. More than three hundred thousand people demonstrated in the People's Climate March in Manhattan as the United Nations gathered for a summit. The march brought together a coalition of more than 1,500 groups and was organized by 350.org, the climate action group founded by environmental writer Bill McKibben. The group Avaaz, another organizer of the march, presented a petition with more than 2.1 million signatures demanding action on climate change. "It's a testament to how powerful this movement is," said Ricken Patel, the executive director of Avaaz. "People are coming in amazing numbers."

What made the march so successful was the size of its coalition, its multipronged approach to communication, and the passionate involvement of individuals and their social networks, according to a study of the event by sociologists Dana Fisher and Anya Galli of the University of Maryland's Program for Society and the Environment. Thirty-three percent of survey respondents found out about the march from an organization or group, 22 percent found out from flyers or posters, 21 percent found out from social media, and 18 percent found out from websites. The findings show that, although the Internet was an important channel for publicizing the event, traditional media were equally important. However, personal networks ruled above all: when asked about how they found out about the event, 42 percent of participants mentioned family or friends. Social networks also played a central role in bringing people out to the streets. Most people (60 percent) came to the march with friends or family, whereas just under 20 percent came with organizations. Personal connections and personal actions by concerned individuals made all the difference.

Fed up by lack of policy action on climate, more and more individual citizens are getting involved in a wide variety of creative ways. In the process, they are becoming one of the major anti-establishment social movements of the twenty-first century. Taking personal action in the face of policy paralysis is both gratifying and gives life meaning, and it takes only a small amount of initiative to take personal, committed action to begin to turn the war on science around.

In June of 2015, by far the hottest June ever recorded to that point, a loose-knit group of "kayaktivists"—led by local Greenpeace activists, but also involving unaffiliated individuals—paddled out to block Royal Dutch Shell's offshore drilling rig, the Polar Pioneer, from departing Seattle for the Chukchi Sea. The protest made international headlines. After weighing the mounting bad press against the price of oil and the costs of exploration, in September Shell announced it would abandon its controversial plans to drill in the Arctic "for the foreseeable future"—a decision climate activists hailed as a victory.

Environmental activism isn't the same thing as science activism, but the two are often related because science creates power and scientists are therefore inevitably political actors. These stories can provide useful models in an age when massive disinformation campaigns are heavily influencing public policy and drowning out scientific knowledge.

One of the most important things a concerned citizen can do is organize, which means taking a public stand against the war on science, staging or participating in events that dramatize their concern, inviting local policymakers and media, and asking friends and family to join in. You don't need to hop in a kayak or march in New York City. You can start wherever you are. Identify a need and do something about it. Use your life as a tool to live your values, and you may find you are coming alive for the very first time. Create a narrative. Invite the press. Use the Internet and social media to reach out. It will give you a chance to give your life new focus and integrity. That's what the six cofounders of ScienceDebate.org found. We began to change the national conversation around science in public policy and influenced the president's selection of top scientists to cabinet-level positions. I have advised similar efforts in other countries. By changing the conversation, we can change the politics.

There are many things one can do, from personal actions like installing solar panels or buying into a community solar garden, to broader ones like trying to change public policy. Consider one of the most successful legacy organizations now tackling the antiscience crisis, the Union of Concerned Scientists. Originally formed by scientists concerned about nuclear proliferation, since the beginning of the twenty-first century the organization has broadened its focus to target antiscience efforts such as the climate battle, to take on the US federal government over lapses in scientific integrity, and to form a new Center for Science and Democracy that focuses on many of the questions discussed in this book. Its then–policy directors Francesca Grifo and Michael Halpern were early and ardent supporters of the first US Science Debate effort, and Halpern kept fighting against the war on science over those years, helping reinvigorate the venerable UCS organization.

Another successful group is the League of Conservation Voters, which, while on the environment rather than science, has nevertheless provided an inspirational model for generating useful metrics and data to shape public opinion about issues related to environmental science.

Taking more militant action to draw attention to many science and environmental issues such as climate disruption, over-logging, and overfishing, Greenpeace has set an international model for combining civic action with creative communication to draw attention to critical issues and cast them in terms of justice. But the group also supports scientists in their work, particularly in politically contentious areas, and tracks and exposes how much money antiscience campaigns funnel into front groups to spread disinformation. Unfortunately, the group also occasionally gets involved in antiscience, such as its work in China against the adoption of GM crops, and must guard against this.

Then there's the National Center for Science Education, which is on the front lines of the battle to prevent efforts to change school textbooks to include intelligent design and climate-change denial, and to stop such efforts within schools. The group provides legal, strategic, media, and scientific resources to parents, students, teachers, and other individuals concerned about the erosion of science for political purposes in classrooms and textbooks, empowering them to do something about it and surrounding them with the support to be successful.

Students are getting fed up and acting on their own as well. Zack Kopplin, then a high-school senior, brought together seventy-eight Nobel laureates in his 2010 campaign against the Louisiana Science Education Act, which allowed science teachers to use supplemental materials that call into question evolution and climate change. While Kopplin's effort did not get the law changed, he was so impassioned and articulate that he became a television spokesperson on these issues, triggering an ongoing national discussion, an outcome not unlike that of Clarence Darrow's. Young people can often have an outsize impact by reinspiring an older generation.

Americans United for the Separation of Church and State was founded in 1947 to work on similar issues in education and other areas where evangelical Christians have tried to insert religion into public policy. The organization increased these efforts with the rise of the religious right, fighting important battles against teaching creationism in science classes, funding religious schools through vouchers, and funding various "faith-based" initiatives, such as abstinence-only sex education, using public tax dollars.

CREDO uses an innovative model to funnel profits from its mobile phone service to fund progressive issues, many of which focus on battling the war on science. They have led the curve of a new trend in business: for-profit companies that serve social good motivations.

If one truly considers the role of business in society, it's not simply to maximize shareholder value. In fact, the *Hobby Lobby* decision expressly states the law does not require such a strict view. The role of business is rather to make life better by addressing a need—i.e., to maximize stakeholder value. That's why a business is started and how it survives in the long run. A good business makes life better for its customers. But it also makes life better for employees by organizing the market and protecting employees from risk. It makes life better for shareholders by providing a return on investment and creating real value. It makes life better for society by being a good citizen in the community. It pays for the public resources it uses with taxes. But it can't do any of these things well if it is unsustainably mining resources from the environment, dumping pollution back in, or externalizing costs that should be rightly borne in the price of the product. Such companies become combative and duplicitous, seeking to battle back all regulation so it is unfettered, when instead they should advocate for fair regulation to level the playing field so all parties can compete sustainably. The opportunities to engage constructively in the business cycle abound. Consider ways your for-profit business could rethink the old model.

A new wave of climate-focused nonprofit organizations has also recently arisen, providing excellent models to tackle issue-specific battles in the war on science. 350.org was one of the first. It focuses on climate activism through local chapters, and has led massive marches and successful efforts to get large investment funds to divest from fossil-fuel corporations. This is an argument that was not taken seriously at first by the financial industry, which is particularly limited in how it defines fiduciary duty and often conscribes its meaning to rule out anything beyond the immediate investment performance. However, with the successful association of climate risk to the risk of fossil-fuel business performance, 350's divestment movement now has a fiduciary argument that even investment advisors are beginning to take seriously: fossil-fuel companies must change or die, and there is significant market risk based on environmental issues, political issues, economic competition from renewables, and the companies' own histories of climate denial, that suggests they have looked at this and are unable to change—and so may no longer be the wise investment they once were. In fact, they are increasingly risky, and when the market moves away from them, it may move very quickly, leaving fiduciaries without enough foresight to bear significant losses.

Climate Progress, an arm of the progressive think tank the Center for American Progress, founded by ScienceDebate supporter and former Bill Clinton chief of staff John Podesta, is edited by science blogger, author, and physicist Joe Romm, who has become one of the most influential thought leaders on climate change.

Climate Nexus works to localize stories of climate change and make the issue personal, concrete, and accessible for journalists and others. ClimateDesk works in similar ways to create broadly accessible mainstream content on climate-science issues.

InsideClimate News is a small nonprofit that has hired excellent science journalists out of the mainstream media and put them to work addressing the opportunities for coverage of climate that mainstream journalism was mostly failing on. The organization won the 2013 Pulitzer Prize for its work.

The relatively new Super PAC Climate Hawks Vote, led by R. L. Miller, Hunter Cutting, and Brad Johnson, has had a powerful influence in focusing public attention on key issues, including Exxon's climate-denial funding, and on supporting candidates who are proscience.

Media Matters for America is dedicated to a similar mission of exposing and correcting (largely conservative, industry) disinformation in the mainstream media on a variety of science- and climate-oriented topics, as well as on progressive policy issues. Its focus is to provide journalists and supporters with resources to debunk false claims in the media.

The Climate Disobedience Center is an example of a new startup dedicated to building a larger movement of civil disobedience to draw attention to climate disruption and its moral, ethical, and legal implications.

The Sallan Foundation works to disseminate information that can produce greener cities, and that exposes the climate-change disinformation battle in the war on science.

NextGen Climate has done some important work debunking the assumption, often argued but unsupported, that tackling climate change will be a hit to the economy. It funded an economic study that shows that, in fact, it will be an economic stimulus. Its associated PAC supports candidates who commit to tackling climate change. Its founder Tom Steyer knows that the best way to leverage change is by intervening in the political process in some form.

Global online networks of concerned scientists, climate bloggers, environmental groups, and energy transition experts are springing up, helping provide resources to journalists and others concerned about attacks on science and about getting the science and engineering right around politically contentious issues.

Indigenous peoples are finding themselves on the moral and environmental front lines of many climate and environmental efforts. Several First Nations in Canada are working to block fracking and the exploitation of the tar sands. American Indian tribes are using hunting, fishing, and ricing treaty rights and civil disobedience to block proposed oil pipelines. Tribes are holding powwows, press conferences, and protests, forging new alliances within the broader environmental movement and hiring legal experts to block pipeline expansion through ecologically sensitive areas.

There are other examples of activists who started where they lived. In Pennsylvania, Darlene Cavalier was a cheerleader for the Philadelphia 76ers basketball team and a Republican, but was alarmed by the erosion of science in the public dialogue. She formed ScienceCheerleader.com, an organization that grew to more than three hundred current and former NFL and NBA cheerleaders, many of them also professional scientists and engineers, who work nationally to promote science and engineering. Cavalier also ran into scientists along the way who needed research assistants, and at the same time noticed how isolated science was becoming from the public, so she helped form a national movement to reinvigorate citizen science, connecting avid nonscientist citizens with researchers.

State-based nonprofits such as Fresh Energy and the Center for Environmental Advocacy in Minnesota and Clean Energy Action in Colorado work on moving policy at the state and local levels, which is a critical strategic approach, since battling antiscientific and anticlimate forces at the federal level has resulted in gridlock. These organizations are representative of hundreds of similar nonprofit startups. Usually involving the marriage of scientists, attorneys, activists, and lobbyists, they do original science to develop sound policy prescriptions and then lobby for legislative solutions. They provide excellent, nonpartisan, evidence-based models for citizen-led groups that focus on the intersection of science, politics, the media, and the public.

Environmental caucuses organized within political parties attempt to exert political pressure on a variety of issues. Joining or forming such a caucus gives one the opportunity to endorse like-minded candidates and recruit volunteers for their campaigns. Examples in the United States include environmental caucuses in the Democratic Party in California, Florida, Minnesota, and Oregon, as well as the National Democratic Party. Each of these was started by concerned citizens obtaining a caucus charter, gathering like-minded people, and starting to organize.

Nationally, in the United States, the National Caucus of Environmental Legislators works to provide legislators with accurate science information, resources, and policy frameworks to enact environmental legislation that is based on knowledge and evidence instead of politics and ideology.

The groups and individuals listed above are just the tip of the iceberg. The takeaway: see a need, take a stand, seek out others, and do something about it. It doesn't matter if you're a CEO or a waitress. You can indeed change the world.

Battle Plan 2:
A National Center for Science and Self-Governance

To be successful, self-governance relies upon the well-informed voter. We cannot take that for granted; instead, we need to introduce certain safeguards to protect it.

The accelerating quantity and complexity of science is producing a depth and breadth of knowledge no longer possible for any one voter to attain. This has opened up an opportunity for antiscience campaigns to gain an unprecedented foothold in the democratic process, undermining the role of science and data in decision making.

At the same time, scientific knowledge now plays a major role in most public policy challenges, and is the main arbiter and protector of individual freedom and social justice. A question arises: how best to bridge the gap between the voter and science so that democracy can be preserved?

A well-endowed, university-based Center for Science and Self-Governance could work to bridge that gap. Such a center could focus academic resources on developing the scientific knowledge and legal strategies necessary to address this growing problem in a nonpartisan way. An interdisciplinary approach to the study and defeat of antiscience, antidemocratic forces would be a rich, multifaceted effort with profound positive impacts for society, and would be well within the charter of most institutions of higher learning.

What would such a center look like, and how would it be guided? Efforts would naturally fall along eight lines of inquiry where the greatest vulnerabilities exist: process, journalism, outreach, education and research, electoral and public policy, foreign policy, religious, and legal.

Process Initiatives

Since we cannot know every issue of import in advance, attempting to bridge the gap must begin with corrections to the process itself. Generally, these efforts should fall along the four paths where the problems arise most often:

1. Improving the quantity and quality of media coverage of science policy issues
2. Improving the quantity and quality of scientist interactions with the public on policy issues
3. Improving the public's engagement with, and understanding of, the scientific process, critical thinking, and high-quality scientific information relative to policy issues
4. Improving lawmakers' use and mastery of science in decision making

The overall strategic approach is to consistently work at the intersection where the four quadrants of science, policymaking, the media, and the public come together. This is true even if the quadrants are only representative; i.e., the intersection is within the mind of an individual student, say, in an education initiative, or voter, in an electoral or public-policy initiative. The important guiding principle as that to be most effective, a strategy should incorporate all four of these elements. Additionally, the approach should always take a pro-science perspective, which means a nonpartisan perspective but never a bipartisan perspective or a multipartisan "stakeholder" perspective, instead letting the chips of partisan interests fall where they may and going with what the knowledge from science suggests.

Journalism Initiatives

1. Create, host, and manage a seal of approval to certify the accuracy of online science information on contentious political issues. The *Tampa Bay Times* did something like this with Politifact.com, which became a national fact-checking site of politicians' claims that won the Pulitzer Prize. There is a need for similar fact checking of public claims about science, not just by politicians but also others who influence public policy.

2. Develop and run an interdisciplinary science-civics-journalism program to train and certify journalists to understand how to work with knowledge from science.

3. Develop a prestigious continuing-education program or fellowship for journalists that teaches them how to incorporate objective scientific knowledge into their reporting and educates them about the core roles of science and journalism in a democracy. Fellows could be certified to report on policymaking.

4. Work with journalism schools to develop curricula that refute the false notion that there is no such thing as objectivity, and to identify where objective and subjective reporting are each appropriate and where each is not.

5. At the same time, show how journalists' own biases, like those of individual scientists, can influence a story, which is why the knowledge gained from replicable science should be given higher authority, and framing a story should take that into account.

6. Create and endow a prestigious international award similar to the Nobel Prize for journalists who consistently incorporate high-quality, science-based, objective knowledge and avoid false balance and antiscience framing in their reporting on public policy and electoral politics.

7. Create a low-cost access point to full-text scientific papers for independent journalists and bloggers, as well as links to related coverage

and papers, so that knowledge is more readily accessible and easily disseminated.

8. Work with public-relations firms to create a Public Relations Code of Ethics governing the use of public deception, the third-party technique, and disinformation campaigns that challenge, skew, or cast "uncertainties" on established science. Work to create a policing mechanism.

9. Develop a method of exposing public-relations firms who act unethically in their handling of scientific knowledge.

Public-Outreach Initiatives

1. Provide and work to require mass-communication training for graduate students in the sciences, so that scientists are able to communicate as successfully as the media-trained shills.

2. Develop tools and models to reform the tenure system to encourage, honor, and reward public outreach and interdisciplinary teaching.

3. Research and develop models to refute postmodernist, fundamentalist, and public-relations ideas about science and objectivity.

4. Create a means for identifying, encouraging, educating, and venerating generalists (scientist-statesmen) who do public outreach and can help the public and scientists themselves put it all together in a big-picture sense, and research ways to develop career paths for generalists.

5. Work to reform granting organizations' public-outreach guidelines and investment to require and fund principal investigators to hire science communicators to do public outreach about their research.

6. Build a CAD-like media modeling toolbox to help scientists and science communicators to quickly integrate advanced visualization technology to illustrate complex concepts and physical interactions in concrete visual terms for the public and decision makers.

Education, Research, and University-Related Initiatives

1. Develop model curricula and provide training for science-civics classes at the secondary and postsecondary level so that non-science students develop an understanding of how science works in decision making and public policy, and how it relates to their daily lives.

2. Create and promote a standard for reproducing science-paper abstracts in plain English and suggesting how conclusions may apply across disciplines.

3. Proselytize the new reality that we are in an era of silo-breaking computational power in which generalists and aggregators are as important as specialists.

4. Work to establish viable, well-paid, and prestigious career paths for science generalists and science communicators (bloggers, journalists, and media creators).

5. Establish multidisciplinary university programs to study science denialism.

Electoral and Public-Policy Initiatives

1. Engage the public in combination with national and international science partners, news outlets, and media networks to develop and host federal and state science debates among candidates for public office, then have scientists recap and rate their answers based on what the best current science indicates, and publicize the ratings.

2. Develop a guide to making good decisions about science-related issues.

3. Using a nonpartisan team of scientists and public-policy experts, develop model bills on contentious public-policy issues based solely on the best science, and make them freely available, along with abstracts and commentary.

4. Using the model bills as a benchmark, rate current and proposed laws by how close they are to what the current science suggests, then use this to highlight differences and create public discussion.

5. Build a database of research into current and past antiscience initiatives.

6. Develop forums for the proactive discussion of the ethical and public-policy issues at hand, assuming the science is not a point of attack or contention. Too often, public-policy discussions become sidetracked by special interests into debates over the science itself. But what if the science was considered settled? Then what would the debate look like? Focusing on ethics and public policy with the science as a given can potentially provide useful models for what productive public policymaking

looks like in a science-dominated age, and may lead to leapfrogged solutions—giant insights that would not be possible to apprehend if the base of knowledge is not already a given.

7. Develop and support model policies and legislation that make it more difficult for antiscience forces to influence the public debate.

8. Develop and test model legislation limiting public-relations efforts that seek to misinform the public or cast "uncertainties" about established science.

9. Develop and test model legislation requiring full disclosure of donors and other financial supporters of front groups, astroturf organizations, and other third-party techniques of propaganda campaigns.

10. Work to restore the Fairness Doctrine in the United States and to codify it in federal law.

Foreign-Policy Initiatives

1. Develop model foreign policies at the SEEP juncture of science, economics, environment, and population control.

2. Hold and promote major foreign-policy dialogues about solving SEEP challenges.

Religious Community Initiatives

1. Initiate a religious-community-outreach program to build collaboration with the faith community to promote the use of science in decision making and policymaking, and to hold faith-oriented community discussions about the moral and ethical issues new knowledge presents, in order to help smooth the process of the social, ethical, and legal integration of new knowledge.

2. Work with faith community leaders to develop guidelines regarding when denial of science is and is not moral and ethical, and how people of faith with strongly held convictions can morally and ethically respond to knowledge from science that offends their beliefs.

Legal Initiatives

1. Develop models for working with the legal community in knowledge transfer: i.e., help formulate models for future public policy about

emerging issues to get ahead of the debate, and strategies for framing the issues in advance to avoid common political pitfalls. For example, as our understanding of the human brain and the various brain systems that can affect perception and free will continues to explode over the next 25 years, how will our legal system need to be adjusted to accommodate this new knowledge and the finer understanding of when we may or may not be able to exercise free will and how free will may be interfered with by drugs, devices or biological processes? What strategies can be developed now to facilitate that knowledge transfer and avoid lengthy public-policy battles and costly legal battles over personal responsibility?

2. Work with the legal community to develop models for legal sanctions against organizations and individuals who engage in science-denial public-relations campaigns. Misrepresentation of known facts is regulated by several federal and state laws.

3. Work with Native constitutional and indigenous law experts to develop models for defense of certain evidence-based environmental-science claims, such as danger from pollution caused by copper-nickel mining or oil pipelines, under assertion of Native treaty rights or sovereignty.

4. Continue to develop legal strategies based on extant law to force regulatory responses, such as the many cases brought by the Center for Biological Diversity on various environmental-science issues.

5. Develop a network of proscience think tanks that explore and promote the relationship between regulation and freedom: when is regulation restrictive and when does it increase freedom? Develop legal theory through these think tanks to clarify this.

Battle Plan 3: Push for Science Debates

Beyond personal activism and the development of policy responses, there is also a need to work at the level of public sentiment. "Public sentiment," Lincoln said, "is everything. With public sentiment, nothing can fail; without it nothing can succeed. Consequently he who moulds public sentiment, goes deeper than he who enacts statutes or pronounces decisions. He makes statutes and decisions possible or impossible to be executed." This is, of course, the description of a politician seeking to mold sentiment in the national dialogue of a democracy. And a key way those who care about the role of evidence in public policy can work to influence public sentiment and counter antiscience efforts is to hold science debates.

Scientific advances now influence every aspect of life on the planet, and play a major role in our most pressing policy challenges. With climate disruption, genetically modified foods, vaccines, the dangers of artificial intelligence and scores of other science-related issues in the public discourse, it is not too much to ask candidates running for national office to address scientific topics during election season.

Most Americans (87 percent) say that candidates should have a basic understanding of the science informing public policy, according to a 2015 US public opinion poll I instigated that was commissioned by Research!America and ScienceDebate.org. This finding holds true across the political spectrum, with a very large majority of Democrats, Republicans, and independents agreeing that candidates should debate key science-based challenges facing the United States, including health care, climate disruption, energy, education, and innovation and the economy. In addition, more than three-quarters of Americans agreed that public policies should be based on the best available science.

While it may be no surprise that a majority of Americans (87 percent) said scientific innovations are improving their standard of living, what is remarkable is that the presidential candidates had up to that point been mostly silent on scientific topics, with the result that less than half of voters (45 percent) said they were well informed about the candidates' views on policies and public funding for science and innovation.

Seemingly, unless the issue is mired in political ideology, such as whether girls should be given HPV vaccinations, science tends to be an afterthought in debates, town-hall meetings, and other campaign activities. This is a missed opportunity, not only for candidates, but also for voters eager to learn their positions. Evidence from science is the great equalizer in a democracy, an objective source of knowledge that can draw us together and create new opportunity. Much of the quashing of science and the passage of antiscience policies occur long after an election and sometimes with little scrutiny. Science debates can bring policy prescription into the light of day, where the public has the greatest leverage in the discussion.

In an age when science drives well over half of all economic activity, what is each candidate's vision for maintaining a competitive edge? How will the candidates tackle climate disruption? What are their thoughts on balancing energy and the environment? How should we manage biosecurity in an age of rapid international travel? Nuclear weapons? Stem-cell research? Freshwater resources? Ocean fisheries? Health care? Science education? The sixth mass extinction? The teaching of evolution? Balancing privacy and freedom on the Internet? Is it acceptable for a president to implement policies that are contradicted by science?

Many candidates do not have science advisors on tap, and the advent of a science debate offers an excellent opportunity to correct this, and one that also provides an opportunity for scientists to be involved in helping inform public policy with objective evidence.

As it stands, it's far too easy for politicians, who largely ran away from science classes after high school and haven't looked back since, to ignore these issues, and that leaves them—and by extension, the rest of us—vulnerable to disinformation campaigns.

Science debates reunite the four fractured elements of society—politicians, scientists, the media, and the public—and bring them back together under scrutiny to discuss important science issues in a way that, if done well, is informed by knowledge instead of disinformation or uneducated opinion. The way it works is quite simple. The debate topics are the big issues of political import of the day, some of which were touched on above. The possible questions could go on for pages. In fact, when ScienceDebate.org develops its questions, they generally come from thousands that have been submitted by supporters and the public.

Beginning with these big policy questions, candidates debate in the regular format, with a skilled journalist as a moderator, accompanied by skilled science communicators to signal to the audience that this is something different, science-based, more serious.

Candidates are incentivized to base their responses on knowledge in the same way Thomas Jefferson did, instead of pandering to antiscientific political forces, because their answers are graded by a nonpartisan panel of scientists who are recognized in their fields, based on how well their policy responses were supported by knowledge from science.

In this way, the public gets a chance to assess not just the policy positions that a candidate espouses, but also their quality of thought and how realistic their responses are from a scientific, evidenciary perspective. Thus, science is re-injected into the discussion and rhetoric is tethered back to objective reality, "till the mind is brought to the source on which it bottoms," as Locke advised—advice Jefferson bore in mind when contemplating his new form of government.

If this sounds like pie in the sky, consider what happened in online forums that ran below the print stories on the candidates' responses to the early

ScienceDebate initiatives. At the time, news directors and newspaper editors told me that the idea of a science debate was a niche topic and the public just wasn't interested. But once we posted the candidates' responses online, their answers—through two presidential cycles—made close to two billion media impressions, and sparked terrific online discussions. The public was hugely interested. Science reporters wrote great stories. The information grabbed the public's attention because science was finally being presented in the form it was first born in, and in which adults are most used to taking in complex information—the ongoing national and international political and policy dialogue. The stories generated more stories, which in turn generated even more. The public, as it turns out, was hungry for this sort of information, for candidates to respect their intelligence and capture their imagination about what we can do to solve our problems.

But science debates have a much larger effect—they capture the imaginations and transform the thinking of the participants, and in so doing they transform the world. When he first ran for office, Barack Obama was not science-friendly. He was an attorney and a community organizer, with little to no interest in the idea of a science debate until we hounded him into it. But as a result of that effort, of becoming convinced of its value, and then forming a team to help him understand and answer the questions, Obama became science-literate, and he eventually came to see science as a central aspect of his administration. He appointed prominent science debate supporters who were scientists to his cabinet, more than any other president in history, and he became the first president to go into office with a science team and a science policy already in place. Just asking these questions has value because it forces the conversation.

But beyond transforming the thinking of politicians, science debates begin to break down the wall between the science community and the public. Today, science is still somewhat walled off from the general population, a subject left to experts, something noted science philosopher Karl Popper frequently warned against. Science has become commoditized. The public is merely presented with the conclusions and not exposed to the process. But watching candidates who know they will be held accountable for the scientific integrity of their remarks and positions, and watching scientists deliver that accountability by grading them and discussing where science and policy do and don't meet, helps the public become familiar with science and knowledge-based argumentation as opposed to rhetoric, to learn or relearn how to distinguish between the two, and to use this process not only in making electoral decisions but also in discussing issues with their kids or over the water cooler at the office. By creating a means to inject the rigorous honesty of science into our political discussions, we have the opportunity to transform our public dialogue for the age in which we currently live—an age dominated by science.

Finally, science debates should not be limited to presidential or prime ministerial races. They should be held at every level of government, especially the congressional and parliamentary levels, but also in state and provincial legislatures and in mayoral and city-council races. If we want to transform the quality of the political leadership, we have to transform the quality of the debate, and science debates are the way to do that.

The fact that science debates are supported by leading figures on the political right and left provides an important means of breaking down identity politics and partisanship. Science is political in that it is a top-wing activity—i.e., it grounds arguments in facts, not in the authority claims of vested interests—but it is not partisan. There are scientists who are progressive and scientists who are conservative, but science itself is both and neither.

Battle Plan 4:
Using Science Advisors More Effectively

One of the keys to winning the policy battle is winning the intellectual battle in the mind of the policymaker. The only way a policymaker can make well-informed votes on critical issues in the age of science is to have an independent, nonpartisan science advisor. Otherwise, too many policymakers are left science-blind, and key policy issues sail over their heads. The role of the science advisor is to bridge the world of science and values-based policymaking. That's not to say there are no values in science—there are values in what scientists choose to study, and in how they apply or use the results. There are values implicit in the process, including integrity, honesty, humility, self-examination, and doubt. But the process of science is values-free on the questions under study, as it was designed to be, and that is how it can claim to create objective knowledge that is independent of our values systems.

The president of the United States has a science advisor, as does the US secretary of state, as do prime ministers of several Western countries. But every legislator and executive at every level of government, from international to national to state to large municipalities, needs science advisors to navigate today's science-driven policy issues intelligently and effectively.

If you are a scientist, reach out. You can help. Contact your mayor, city councilor, governor, or member of Congress or Parliament. Ask them if they have a science advisor. Chances are they will say no, which means they are flying blind when it comes to the complex science information that informs their policymaking, and are probably relying on lobbyists and the Internet, neither of which is a very reliable source of objective knowledge. If they do say no, volunteer for the position, and assemble a network of advisors you can go to to provide them with objective summaries of what the science says on given topics. This is an important and immediate way you can effect better policy outcomes for everyone.

One of the problems with the way the US presidential science advisor's role is structured is that the advisor is appointed by the president. This introduces partisanship in the perceptions of the advice the person provides. It doesn't mean the advice is partisan, just that the other side won't be as willing to trust the advisor and will be more likely to use the fact that she or he is part of the administration as a political excuse for ignoring advice it finds inconvenient. It is additionally problematic in countries like the United States, where Congress has no nonpartisan science advisory body of its own.

In Canada, where the Harper government first sidelined and then abolished its national science advisor position in 2008, the group Evidence for Democracy worked to build support for a number of different approaches to restore and improve scientific integrity in Canadian government offices. "We're making two recommendations," says Katie Gibbs, executive director of the group. "A parliamentary science officer and a chief scientist position. The parliamentary science officer would be an independent overseer of science in government and function somewhat like the parliamentary budget officer, auditing government science policies and reporting to the people. But there's also a role for someone who answers to the government and provides them with science advice."

New Zealand recently created a science advisor position and took a slightly different path toward the same goal. Like Australia, the country has its share of anti-fluoridation and anti-GM activists. But it generally has much less antiscience than the United States, and the hardcore religious science denial around human reproduction—stem-cell research, the morning-after pill, abortion, gay rights, etc.—never really got traction in New Zealand.

That makes it an interesting place to look at how science advice can work in government. In fact, New Zealand recently developed a solution to this problem that can serve as a model for how a successful science-advisor position could be structured. Sir Peter Gluckman, the first science advisor to the prime minister, provides us with a case study of how his role has been structured to provide independence, and the effect that independent approach can have on politics and public policy.

> A science advisor is not just a scientist but also must have a set of skills in understanding the policy process and what we call political intelligence. One of the reasons why science academies haven't usually been effective in advising government is they haven't appreciated those nuances.
>
> If the person is seen as a lobbyist for the science community, for example, the chance of failure is high. Being a lobbyist for the science community or being the public face of science is not the job; then you are simply representing a vested interest.
>
> Another one is that scientists think they know a lot, and therefore when they recommend something government must act on it. The reality is scientists may know a lot but there's also a lot they don't know, and their input into the policy process may be limited. We live in a democracy, and there's more than logic that enters into political decision making. Most areas that cause contention are where the science is not complete and there are considerable values involved and the values are really what's in dispute, and the attacks on science are a proxy for the value discussion.
>
> In my opinion, the science advisor has to be independent from the politics of the day. If I became a political tool of one political party or the other, the trust in what I would say to policymakers or the public would be diminished. And so the prime minister and I agreed that the post had to be independent. I was also available to and could talk to the opposition leader and to the opposition spokesman on science. As a result, I now have a good relationship and a nonpartisan relationship with both major political parties.

So how does this all play out in practice? Consider New Zealand's high teen-pregnancy rate—the same as that of Texas. New Zealanders wanted to do something about it. Looking at how New Zealand's science-advice process helped develop policy to address this issue is instructive of the independence Gluckman was working to achieve.

We have a high rate of teen suicide, a high rate of teen pregnancy, and a high rate of alcohol abuse by young people. And there's been a genuine concern in New Zealand about the number of young people who in the transition from childhood to adulthood fall into mental-health problems that can affect them for life.

The prime minister approached me about setting up a committee to study adolescents. I replied that I didn't think a stakeholder committee would make a substantive difference, because many people will come in with vested interests.

This was a critical insight. Very often, such committees are formed and include "stakeholders"—vested interests from a variety of perspectives—mostly industry, religion, and nonprofit advocacy groups. It is a popular approach among politicians because it gives all the vested interests a seat at the table, but it is very difficult to develop objective knowledge with a stakeholder committee. Sir Peter continues:

I suggested we instead get the academic community to write a substantive review of what we know and don't know about these issues from science. The prime minister fully supported it.

It produced a three-hundred-page report, which was peer-reviewed both nationally and internationally.

The principle behind the report was that it would focus only on the peer-reviewed literature. We had people in the room whose role was to point out when people were going beyond the data and starting to put biases into the discussion. This was the opposite of a stakeholder committee in that sense.

The prime minister then took Gluckman's science committee's report and appointed a second committee, this one made up of senior policymakers, to look at the report and recommend policy options based on the knowledge that had been culled by the scientific committee. They came up with several.

The prime minister then asked Gluckman to chair a third committee to review those options from the scientific perspective, to determine which, based on current knowledge, had the best chances of succeeding, and to make sure areas that needed emphasis weren't being overlooked. Again, Sir Peter:

The whole process led to the government announcing a major investment in twenty-two programs to address youth mental health—in the middle of the global financial crisis. I shared the platform with the prime minister at the press conference, and he said that the scientists were not certain what would and would not work in this raft of programs. However, we're implementing them, as the scientific advice is that this is the best way ahead based on the evidence we have at this point in time. We will also introduce a raft of evaluation processes so we can look back and decide which of these programs help and do not help as we go forward.

So here was a complex area, and science was able to point policymakers to where they might go, but everybody avoided hubris. And the public has accepted that government can't get it right every time but it's making the best attempt with the knowledge we have now to address an issue we all agree is very important.

It's extraordinary in my experience to see a prime minister announce a raft of measures, which was big in money and big in impact, and admit at the same time, they may or may not work. Most politicians want to sell programs in a way that it's absolutely certain that they will work. And so I think this was a very good example of how the science, policy and political communities can work together in a new way.

Battle Plan 5: Preaching in the Age of Science

There is an old Christian joke that was popular in New Orleans after Hurricane Katrina. You've probably heard some variation of it.

A preacher was caught in a hurricane, and as the waters began to rise he decided he'd better climb onto his roof. As he struggled with the ladder, the fast-rising water threatened to sweep him away. A man in a boat came by and said, "Preacher, hop in!" But the preacher was a man of faith and said, "No, no, my son—Jesus will provide." So the man left. After some struggle, the preacher got onto his roof just as the ladder was swept away. One of his neighbors, a scientist, came back in a boat and said, "Preacher, I know you don't want to leave, but the measurements we've got say it's gonna get a lot worse!" And the preacher said, "You can measure all you want, but the Lord will provide," and waved him away. But the waters kept rising. Soon the preacher's legs were engulfed in the flow as it rushed over the peak of his roof, and he had to hang on to his chimney for dear life. A helicopter spotted the poor man and lowered a ladder. A man yelled down and said, "Preacher! Grab the ladder!" But the preacher yelled, "Get away, ye of little faith!" And so the helicopter had no choice but to fly away. The waters rose again, and the preacher was swept away into the rush and drowned. He woke up outside the Pearly Gates. You can imagine how angry he was that the Lord had let him down after a lifetime of faith and devotion. He saw Jesus walking inside and called to him, "Lord, I held firm in my faith, but you betrayed me!" And Jesus said, "What are you talking about? I sent two boats and a helicopter!"

Science and engineering produced those boats and designed that helicopter, and turning away from it means turning away from the natural law of the Puritans—our ability to reason. "What we understand Scripture and God are saying to us is sometimes better understood by knowing what science tells us," says the Reverend Peg Chemberlin. Chemberlin is a past president of the US National Council of Churches, the leading organization for ecumenical cooperation among Christians in the United States, which includes about forty-five million people in more than one hundred thousand congregations.

Chemberlin says she understands facts as things to be interpreted, not ignored, by Scripture. "It's fairly arrogant to think that we know all the ins and outs of Scripture," she says. "So to hang on to a worldview and not allow science to engage that worldview keeps us from opening up to that."

This view is similar to that of Pope Francis. "Science and religion," he has said, "with their distinctive approaches to understanding reality, can enter into an intense dialogue fruitful for both. Given the complexity of the ecological crisis and its multiple causes, we need to realize that the solutions will not emerge from just one way of interpreting and transforming reality."

Galileo's treatment notwithstanding, Francis comes from a long tradition of popes who have appreciated what science offers and sought to incorporate its knowledge into their moral and ethical reflections. "Science and technology are wonderful products of a God-given human creativity," argued Pope John Paul II in 1981. Technology "expresses the inner tension that impels man gradually to overcome material limitations," argued Pope Benedict XVI in 2009.

Chemberlin sees that as a value in both the Protestant and the Catholic churches, which modern preachers have to get back to. She says it hearkens back to natural law and the idea that understanding the physical world is an important part of understanding Scripture and therefore of our relationships with both science and God.

I often ask parishioners, "How many creation stories are there?" They usually say "Two," and I say, "Name them," and they tell me the first two stories in Genesis. But there are more than two: there's also a creation story in Proverbs; there's a creation story in Job; there's a creation story in Corinthians; there's a creation story in the Gospel of John. All of them are different stories. If we're just understanding them on a factual level, they conflict. So we have to do some kind of interpretations of those facts—how do we understand them in a broader way—to understand all those Scripture stories. And I think that's very much the same kind of process we have to bring to understanding conflicts between Scripture and science. With all that science throws at us, it's very exciting to have that invitation in front of us.

People of faith have endless and exciting opportunities to see their faith from fresh new perspectives as science advances and creates new knowledge about the way the physical world—God's creation—exists and functions. In contrast to the views often heard from extreme fundamentalists, this approach allows for a more constructive and responsible exchange on many issues that are critical to the future of the planet. In fact, it was the view that gave birth to science in the first place.

And that is the key to reshaping the relationship between science and people of faith. It is churches, more than any other form of community organization, that are public houses of moral and ethical reflection. If, rather than attacking science in an adversarial relationship, pastors and priests learned the knowledge science has built up, they could use that as a basis to guide parishioners in deeply engaged discussions of the real, current problems of which science is the cause, solution, or both.

They could bring scientists into their houses of worship to discuss the science of major issues like the sixth mass extinction, climate disruption, stem-cell research, and a host of other pressing issues. In short, they could collectively put an end to much of modern antiscience, by refusing it themselves and guiding parishioners to do the same.

Scott Westphal, the head pastor of a Lutheran church in Minnesota's conservative Sixth Congressional Dsistrict, is frustrated with the harshness of certain AM talk-radio personalities, or the unwillingness of certain members of his flock to listen to traditional Lutheran teachings of curiosity and openness.

"Harshness, mockery, and intolerance," he says, "are not a natural part of being either conservative or Christian."

And yet there are still the prominent televangelist supporters of the Cornwall Alliance who argue that we should reduce collective ownership of natural resources, that we should increase our appreciation of private property, and that there is no such thing as anthropogenic global warming. But are these theological arguments, or business arguments masquerading as theology? To what extend have churches been corrupted by industry and industrial PR?

"It's very much the moderate world that has to speak up on this," says Chemberlin. "I recently got an award from the Islamic Resource Group [a US-based educational group seeking to further understanding of Islam]. And I told them, 'Your extremists and our extremists want to fight. And then we're all in trouble.' That's true with faith and science, too. But in a democracy, because I believe that my God, my transcendent value, shares a value about democracy, then I have to live in the democracy and that means compromises on a daily basis."

The magnitude of the current issues suggests that this would be a very good time for church leaders to reach out to scientists. The West is in a moral, intellectual, economic, and ecological crisis, and it matters little whether a preacher is conservative or progressive if he or she is incorporating knowledge into moral reflections. As a new Catholicism may be emerging, so too is it time for a new Protestantism, and for the spirit of Luther's questioning to be reborn, and a new Islam, in which the ideas of equality and science that are implicit in Islam, the one-time protector of science, are embraced anew.

Battle Plan 6: Teachers Should Teach Science Civics

The next battle plan tackles some of the confusion in students' minds and helps prepare them for the science-driven world in which we now live. It uses a few different techniques to create cognitive dissonance and ground science deeply in a student's understanding of the world.

As good teachers often say, the best way to get a student to learn is to first raise their level of concern by showing them why something matters to them personally, and then to create cognitive dissonance by showing them that things aren't the way they thought they were. Then, and only then, is a student truly ready to learn new information.

One of the problems with the way science is often taught in the classroom is its separation from both humanities and the student's personal life. In those circumstances, it's a lot harder for a teacher to raise the student's level of concern, and few science teachers are skilled at doing this. Those who do often use an entertaining whiz-bang approach and get into the fun of things and then move on to cognitive dissonance. On the civics side, ideas about civic participation in a democratic republic are often presented as just that—ideas, divorced from the problem solving and onward striving that gave rise to them. Rarely are these two subjects taught together.

But they should be. Teaching science and civics together can reach and inspire a broader cross-section of students who otherwise may not realize they are interested in science. Science civic classes trace the political and historical contexts of when and how scientists made their discoveries and what they meant in the intellectual and policy dialogue of the day, much as some of the stories in Part 2 of this book do, but adding in more science, lab units, and math. By understanding the *process* of science in its narrative context, versus the *products* of science, and by understanding its relationship to the pressing problems of the day, and to the ongoing drive of progress, curiosity, alleviating suffering, and increasing freedom, students can connect the ideas informing process with the emotions of why those ideas are important, better equipping them both as citizens and as scientists in their future careers.

This is the same thing that made Galileo so powerful: he emphasized the process. It is hard to argue with science when the process is laid bare, because one is arguing with the evidence of one's own senses. It is that personal relationship with science, that personal cognitive dissonance, that makes people understand it, accept it, and value it.

Hold Student Science Debates

Another fun tool is holding high-school or college science debates on politically contentious subjects. The teacher picks a science issue that has political traction at the time, such as climate disruption, vaccines and autism, GM food or the like, and phrases it as an antiscience proposition that students will argue for, using rhetorical arguments, or argue against, using scientific arguments. The trick is that the teacher does not determine who will argue for or against the motion until the day of the debate, by a coin toss.

This way, all students have to research both sides of the debate. In so doing, they quickly learn the difference between the knowledge-based scientific arguments against the antiscience proposition, and the non-scientific, emotionally persuasive rhetorical arguments in favor.

The debate should be conducted in the Oxford style, with the class voting before the debate on whether they agree with the proposition or not, then voting again after the debate, and comparing the two votes to determine which argument was more convincing and thus who won the debate. Depending on the teacher and the community, the audience could include parents as well.

This is then followed up by a discussion of the different forms of argument; the differing motivations of each; the role of motivated reasoning, belief systems, propaganda arguments, statistics, and confirmation bias; and what the students discovered about all of these factors as they researched the arguments in preparation and analyzed the audience votes. Discussion can also happen about which arguments were most persuasive, why that is, how the persuasion worked, whether that's a good or bad thing, media reporting, false balance, subjectivity and objectivity, how to guard one's thinking against propaganda, and the defects in our public-policy process.

Establish Science Literacy Requirements

There is also a bigger issue: we simply have to start requiring science literacy and proficiency in order to graduate from high school, and especially from college, where science classes are often optional. And not just physics-for-poets classes. Kids not only need science to participate in the economy—they need it to be good citizens in a democracy. As discussed above, it's really about opening up the process of science rather than reproducing the right answer. Science is not about reproducing steps to attain predetermined outcomes—quite the opposite—and if we teach it that way we're not teaching science. Science is about exploration. Classes need to emphasize this process-oriented approach, building the skills of observation, hypothesizing, experimental testing, skepticism, and the need to base statements on evidence. That can only be taught in an open-ended approach where the right answer is not a stated goal, but instead exploration and discovery—the process of science.

Set Parental Expectations

Beyond science civic classes, science debates, and process-oriented education, parents need to be made to understand that their attitudes and involvement play a major role. A postmodernist idea that parents in Western countries often have is the belief that science is something one must have an aptitude for in order to be good at. If a child doesn't do well in science classes, parents tend to excuse it and say, "Well, that's okay, not everybody's good at science, but you can be good at something else, like writing, or feminist studies, or law." That's true in the case of students with intellectual disabilities, but it's not true of most students. Parents in China, whose children score at the top of international science rankings, tend to regard science proficiency as a matter of effort, and that if you don't do well in science class, it's because you weren't working hard enough.

There's considerable evidence to show that it's this parental attitude that is perhaps the largest factor in these students' superior performance. Chinese parents also are more likely to help their kids with science homework and to make sure they understand it. This insight has been borne out by other studies of cross-cultural and ethnic differences between low-scoring and high-scoring countries, which consistently find that proactive parental involvement in kids' science homework and studies is important.

This is especially true in highly diverse Western countries. Racial and cultural diversity create challenges. Finland tends to perform near the top of international rankings, in part because of homogeneity in its student bodies. A homogenous student body requires far less management time, which allows for more focus on academics. However, even among diverse American classes, Chinese-American students often outperform their peers, because of their parents' attitude and support.

Battle Plan 7:
Granting Bodies Should Require and Fund
More Outreach

While scientists still rank among the most highly regarded professionals in public-opinion polls, with only teachers and members of the military ranking higher, the majority of Americans can't name a living scientist—unlike their local pastors and doctors. Science is distant, something done by nameless, faceless others in anonymous white lab coats.

We've discussed how scientists in the United States, the world leader of scientific research in the second half of the twentieth century, fell silent in the public dialogue. The National Science Foundation and other government granting agencies did not require public outreach. Additionally, university tenure systems developed that contain strong disincentives to do public outreach, and scientists themselves began to look down on other scientists, like Carl Sagan, who engaged in public outreach.

There is a communication and trust gap that this error created. The taxpayers who were paying for this research didn't often hear from scientists about what their money was buying, and scientists developed a superior attitude toward engaging in civic organizations, giving money, engaging in the political process, or explaining themselves. Imagine if a school board functioned that way. Trust with the community would be broken.

Rebuilding public rapport requires public engagement on a massive scale. But this cannot be just the whiz-bang of classroom or science-museum demonstrations, and not just for children and teenagers. It requires scientists everywhere to come out of their laboratories and talk about what they know in the public square, to adults, so the public understands not just *that* science is important, but *why* it's important. It requires them to provide a constant reminder for all citizens of the process by which we build knowledge, why this process is important, how facts are different from belief and opinion, why citizens should care, and how science intersects with their lives. It requires a reengagement.

That doesn't mean scientists should impose their personal values on society—other than the values of knowledge and evidence and honesty and skepticism and curiosity and humility that are bedrock qualities of the scientific process. "Where science begins to appear political is when scientists, who unfortunately are also people, come to impose or even voice their own values, conflated with the data and their interpretation," cautions Alan Leshner, the former CEO of the American Association for the Advancement of Science. A reengagement simply means presenting their work fairly and publicly. Learning how to do that is another matter.

There is a difference between being *political* and being *partisan* that Leshner's perspective does not fully tease out, but it is an important distinction. Science is always political. It is an antiauthoritarian activity. But it is never partisan, left or right.

Much would be done to reintegrate science into the public if granting agencies required that principal investigators—those in charge of the lab a grant is funding—hire science communicators to explain the importance of the lab's work to the public, to act as liaisons, to promote the work, and to organize public interaction with the lab. The granting agency should also fund such positions as full-time staffers with the same pay and benefits as scientists to emphasize this shift in thinking.

In the past, when granting agencies have instituted "broader impact" components of research grants, it has been in the neighborhood of 1 to 2 percent. Researchers, faculty members, and graduate students from federal research laboratories and research universities became more involved in K–12 and public-outreach activities. Most frequently they gave presentations, tutored, and organized or judged science fairs. This is not the sort of public outreach we are discussing here, because it is preaching to the choir, or at least to the science pipeline. The kind of public outreach that is needed is adult public outreach in broader society: reaching out to and speaking in churches, rotary clubs, local chambers of commerce, and other civic organizations about current science and its implications for society.

By making scientists and science communicators known and more accessible to the public, there will be substantially more opportunities for informal exchanges and education on the key science topics of the day, increasing the role of evidence in the public dialogue, which can go a long way to countering antiscience disinformation campaigns.

Battle Plan 8:
Scientists Should Adopt a Scientific Code of Ethics

A not-insignificant reason the public distrusts science is because of scientists who engage in unethical behavior, such as misconduct for personal gain or overdramatized conclusions to get attention in the media. Considering all the problems that have been caused by these breaches, it is surprising that there is no broad scientific code of ethical conduct, no Hippocratic oath for scientists, beyond the scientific method itself.

Part of the problem, of course, is that science is such a large endeavor, and each branch deals with very, very different ethical issues. Because of this, certain disciplines or even certain journals have developed their own codes. But others have not. Some of the legitimate journals in which Willie Soon published his industry-funded climate-denial papers, for instance, had no code of ethics about funding disclosure and conflicts of interest. That shouldn't happen, and when it does, it can give whole swaths of science a black eye.

What would a broad scientific code of ethics look like? It would lay out the moral and ethical principles under which science should and should not be conducted. It would have some means of enforcement or exposure of scientists who did not meet this standard, so the public could evaluate their claims in that context. It would have guidelines for best practices in research, peer review, independence, and disclosure.

Below is one such possible code of ethics, based on suggestions initially made by UK science advisor Sir David King but developed further:

1. First, do no harm.

    Obviously, this is inspired by the Hippocratic oath. Because the question of what is harm depends upon context, there must be an ethical process for determining what we mean by harm, or what is the lesser harm. But the intent of science should be to advance knowledge for the purpose of improving life. Sir Joseph Rotblat, a nuclear physicist and the only scientist to resign from the Manhattan Project, first suggested a Hippocratic Oath for scientists in his 1995 acceptance speech for the Nobel Peace Prize, and campaigned for such an oath until his death in 2005.

2. Tell the truth.

    Do not mislead; present evidence honestly. Do not withhold information. Do not limit statistical or evidence considerations in order to build an argument to support a desired outcome. Do not extend conclusions into statements that are not supported by the evidence. Present criticism, weakness, and limitations fairly.

3. Practice humility.

    Question yourself and your possible biases, conflicts, and weaknesses. Declare biases, weaknesses, and conflicts of interest. Question peers to ensure their conclusions are based on evidence rather than biases.

4. Act with skill and care, and keep skills up to date.

    Scientists share a responsibility to use current best practices and knowledge, to be as precise as possible, and to use the most powerful measuring tools available, because the conclusions they draw have real-world implications.

5. Respect and acknowledge the work of other scientists.

    Acknowledge intellectual priority and indebtedness. Do not steal or claim credit for work that belongs to others.

6. Ensure that research is justified and lawful.

    Because science creates knowledge, and knowledge is power, a free society has a legitimate right to set ethical and legal bounds for research. Research must be ethically and financially justified in the context of other social needs, and must be lawful.

7. Minimize impacts on people, animals, and the environment.

    Beyond the question of harm is the question of impact, which may not rise to the level of harm when considered individually but may become harmful at scale or cumulatively. Minimize experimental impact on humans without their knowledge and willing participation. Minimize experimental impact on animals and ecosystems. Consider cumulative environmental impacts, impacts compounded by other factors, and impacts that could become harmful at scale, both in experiments and in tech transfer to broad applications.

8. Discuss issues that science and technology raise for society.

Discuss the social implications—both positive and negative—your work could have on society. Dedicate 5 percent of your time to communicating with the public about the process of science, its role in a free society, and the implications of your own work for freedom.

9. Participate in civic society.

Give 5 percent of your income and time to volunteer work and/or civic activities outside of science. Scientists have an obligation to nurture free society as a foundation of science.

10. Defend basic research.

Defend investment in government-funded, curiosity-driven basic research. The purpose of government is to invest in things we cannot do for ourselves or in other ways. Scientists have an obligation to defend basic research as a foundation of science.

Battle Plan 9:
Business Leaders Should Form
a Chamber of Progressive Commerce

There is a structural imbalance in many Western countries in what is allowed in the political dialogue. On the one hand, we tend to allow industry-funded science-denial public-relations efforts, where a powerful vested interest can seek to confuse voters with "uncertainties" and stall unfavorable policy. But we do not do a good job of allowing funding mechanisms for voices that support science in the national public dialogue, especially when it comes to policy advocacy, even though such voices in the long run create the conditions favorable to innovation and the creation of new economies. Scientists who receive government funding can rarely use it to engage in public relations, public outreach, or intervening in the political process, especially on contentious political issues. Nonprofits, because of their tax-exempt status, are forbidden from engaging in partisan politics, and foundations are cautious about funding anyone engaged in such interventions because any statements that could be interpreted as political advocacy may put their nonprofit status at risk. The government itself rarely engages in direct intervention in the political process.

As a result, there is a tremendous power imbalance between the voices of vested interests and those relatively few activist organizations that speak up for science and the democratic process, such as the Union of Concerned Scientists and ScienceDebate.org.

We touched on one of the ways this may be countered earlier. There are a growing number of socially conscious entrepreneurs—owners, CEOs, and chairs of corporations that are both business- and mission-driven. Some of these companies are in the energy space, such as Sun Power and Solar City; some are in transportation, such as Tesla; some, like Google, look for disruptive connectivity opportunities, and some bring ethically produced foods and goods to market.

These companies, as well as emerging clean-energy companies, represent a contingent of corporate citizens that, collectively, have business models at odds with those of the fossil-fuel industry and others engaging in antiscience public-relations campaigns. This divide was made stark by revelations that the US Chamber of Commerce was funding efforts to kill antismoking laws around the world—a stance that caused CVS Pharmacy to quit its membership. It is in the best interests of ethical companies and new-economy companies to form a new Chamber of Progressive Commerce, to work toward advancing their business interests and their ethical concerns in a common forum and supporting nonprofits that share this perspective. Such an alliance could have the resources to counter disinformation campaigns and leverage its power to create meaningful social change, synergistically creating new markets for its superior products and services (because it doesn't need to employ antiscience and obstructionist tactics to prosper). Some alliances like this were forged in the two years leading up to the Paris climate talks, strictly focused on the climate issue, but the effort could go much, much further.

Battle Plan 10:
Diplomats and Elected Leaders Should Use Transformative Foreign Policy

Nations frequently use broad economic, human-rights, and military issues as factors in foreign policy. But rarely do they use proscience policies. And rarely do they demand environmental, economic, and labor policies based on evidence. Consequently, nation-states rarely use their power to intervene on issues that actually help strengthen the health and power of democratic government.

The US National Academies recently looked at part of this when it issued a 2015 report, *Diplomacy for the 21st Century: Embedding a Culture of Science and Technology Throughout the Department of State*:

> The mission of the Department of State (department) is to shape and sustain a peaceful, prosperous, just, and democratic world and foster conditions for stability and progress for the benefit of the American people and people everywhere. The strategy calls for the department to become more efficient, accountable, and effective in a world in which rising powers, growing instability, and technological transformation create new threats but also opportunities.

The report argues convincingly that science and technology advancements are changing the world at an unprecedented pace, altering the way everyone lives and conducts business. The revolution in the biosciences is extending human life, freeing women via contraception, improving agricultural productivity, and protecting essential ecological resources. But unchecked population growth and industrialization are threatening valuable ecosystems, changing climate patterns, and redirecting ocean currents. The same Internet that unites families and businesses also allows drug gangs and cybercriminals to prosper. Violent extremists use the Internet for recruiting and logistical purposes, and increasingly gain access to advanced destructive technologies. Therefore, all nations need to begin retooling how they think about science and technology, and how to use it as a foreign-policy tool.

In particular, nations need to get much more aggressive about requiring compliance with hard-won environmental, labor, technology, and privacy protections that have been established according to the findings of science in the leading industrialized nations. Too often, foreign policies, enacted by diplomats and elected leaders who are scientifically illiterate and technologically uneducated, have abdicated opportunities to influence and strengthen democracy and our shared resources, both economic and environmental—and the basis of future economies. There is not now, nor will there ever be again, a time when the world's major powers can exercise a foreign policy that does not aggressively negotiate on science and technology policy issues and have any illusions of control over their own national destiny.

The National Academies found the US State Department was seriously understaffed in terms of foreign-service officers with science and technology training adequate to understand key issues—a condition that appears to be common throughout developed nations.

This speaks to the Two Cultures divide C. P. Snow was so concerned about. At the same time, countries are intensely focused on the prospects for new opportunities to advance their economies and provide better livelihoods for their populations using science and technology. Globalization has produced rising powers, and no nation is able to establish global or regional security, or political or economic agendas, unilaterally. Europe and Asia, in particular, have important science and technology centers that rival or exceed US capabilities. However, the US opportunity to use its science and technology prowess to support peace and prosperity in other countries remains unrivaled.

"For several decades, and increasingly since 2000," the National Academies noted, "the Department [of State] has addressed S&T as an important appendage to the mainstream of foreign policy formulation and implementation." But this is the problem in a nutshell. It can no longer be an appendage. Science and technology need to be infused in every foreign-policy consideration, not set apart as an appendage. They need to be used as a transformative force in the foreign policies of nations as a whole. This is something the report noted, saying, "The department should provide the S&T Adviser with organizational status equivalent to that of an Assistant Secretary."

Examples of this need abound all over the world. By 2030, the demand for food will increase by 35 percent and the demand for energy by 50 percent, even as we must cut greenhouse gas emissions, deforestation, and pesticide use. Nearly one-half of the global population will be living in areas of severe water stress, a condition that will be exacerbated by climate disruption. In the next few years, the Arctic region will be opened for new maritime routes and for resource exploitation of considerable economic and environmental significance. A series of dams is currently being planned to convert the Congo River basin into the world's largest hydropower complex, with enormous environmental consequences. "The related foreign policy considerations of S&T advances are driving diplomatic agendas throughout the world on a daily basis," the National Academies acknowledges. "The [State] Department needs to upgrade its S&T capabilities and related policies and programs accordingly."

A transformative science and technology foreign policy would seek to get ahead of these major S&T issues, and seek to establish international norms. If free trade is established between countries, so too should a shared regulatory framework, to prevent a race-to-the-bottom exploitation of human and environmental resources simply to provide cheap short-term economic gain at the cost of the commons. Countries that exploit labor or environmental resources should be treated similarly to countries that dump currency or commodities into international markets or that sponsor terrorism; i.e., subject to sanctions and exclusion from key markets.

It is important for conservative readers to note that this is not necessarily a pro-labor or pro-environment agenda, since science is always political but never partisan. This is an evidence-driven agenda against the authoritarian practices we know run counter to the values of democracy—exploitation and dumping onto the commons. By leveling the playing field with a transformative science-driven foreign policy, nations can—and have the moral obligation to—rein in the feudal regulatory race to the bottom at the expense of the global commons.

The National Academies report was clear about the steps that should be taken, but the US State Department, like other federal government departments, is populated with civil-service employees who have an established way of doing things that is not necessarily in line with the nimble needs of modern society. Jane Lubchenco once described changing the culture at the National Oceanographic and Atmospheric Administration as akin to steering the *Titanic* with a canoe paddle. This is not a criticism of the employees, but an acknowledgement that the structure of large enterprises creates a culture with established procedures and precedents that in turn create enormous inertia, that political appointees may have a hard time influencing. The problem is that the rest of society is changing fast, and federal departments and agencies need to change fast as well. Other nations should regard the report as a call to examine their own foreign-affairs departments to make certain they are dealing with the major science, economic, and environmental issues of *this* century, not just the last. Failing to do this is abdicating the ability to influence the future.

Battle Plan 11:
Candidates Should Sign Science Pledges

Why a Pledge

We live in a time when the majority of the unsolved policy challenges facing the United States revolve around science. These challenges have continued to accumulate to the point that many of them are threatening the economic well-being of the nation, and the ongoing health and vitality of its citizens and their environment.

We elect our representatives, senators, governors, members of parliament, and presidents to tackle tough issues. Many in this generation of elected officials have retreated from that level of leadership. They have failed to solve the accumulating challenges, preferring to punt them into the future or, increasingly, deny they even exist.

With the continued delay this failure of leadership imposes, these problems get worse, and solutions become even more difficult and expensive. With each step away from reason and into fraud and denial, democracy moves toward a state of tyranny, in which public-policy decisions come to be based not on knowledge, but on the most loudly voiced opinions.

We need candidates of all major parties who will lead on these tough science questions, and who will reassert objective evidence from science as the best basis for informed, effective, and fair public policies in a diverse nation and world.

In order to reflect candidates' commitments to basing public-policy decisions on knowledge, versus opinion or belief, we need a vehicle. The Contract from America, the Taxpayer Protection Pledge, and the No Climate Tax Pledge have all sought to restrict reasoned debate. We need a pledge that seeks to expand it.

Citizens can print out the Science Pledge and challenge their elected leaders to sign it. The Science Pledge asks candidates to commit to the kind of civic-minded leadership citizens are owed in a democratic republic. It seeks to separate freedom lovers from authoritarians, evidence-based decision makers from those governed by "but faith, or opinion," and independent thinkers from ideologues.

A Science Pledge

A renewed commitment to civic leadership based on the principles of freedom, science, and evidence.

Whereas democratic republics have accomplished great things by setting freedom, knowledge, and science as the bases for sound, effective, and equitable public policy;

Whereas preserving freedom and solving policy challenges requires elected leaders to make a renewed commitment to science and evidence as the foundation of the democratic process;

I hereby pledge that:

PUBLIC DECISIONS MUST BE BASED ON EVIDENCE

I will support public-policy decisions based on well-established evidence produced by science, which may be informed by economic interests and personal values but never superseded by personal opinions or political objectives.

KNOWLEDGE MUST NOT BE SUPPRESSED

I will protect and defend the basis of freedom—the scientific consensus of knowledge—against political forces that seek to deny it, suppress it, or substitute for it rhetoric or opinion.

SCIENTIFIC INTEGRITY MUST BE PROTECTED

I will oppose all efforts to reduce freedom by holding back or altering scientific reports or evidence that conflict with personal opinions, ideological positions, economic interests, or political objectives.

FREEDOM OF INQUIRY MUST BE ENCOURAGED

I will oppose acts that attack, intimidate, prosecute, disparage, or silence scientists and academics whose research or reports conflict with personal opinions, economic interests, or political objectives.

MAJOR SCIENCE ISSUES MUST BE OPENLY DEBATED

To demonstrate my commitment to these principles, I will participate in one or more substantive, nonpartisan, public, independently moderated debates on the science issues I may encounter if elected.

I hereby make this pledge on this ____ day of _____, 20___.

_____
Candidate's Signature

_____
Candidate's Name

_____
Office Sought or Held

Battle Plan 12:
Editors Should Insist on Pro-Evidence Journalism, and Investigative Journalists should Target the Fraud of Science Denial

Restore Objectivity

The fact that journalists have fallen prey to the war on science is now well established. Journalism schools teach there is no such thing as objectivity, new reporter guidelines contain the phrase, and top journalists repeat it in speeches. Journalists then fall prey to skewed perceptions of what balanced journalism really is, and seek to falsely balance the scales by ignoring the weight of evidence and treating knowledge as just one of many opposing opinions, or by countering every quote they have with another from an opposing viewpoint, so-called "he-said she-said" journalism. The split screen between a scientist and someone with an opinion is another, larger example of this approach, which winds up skewing public dialogue toward the views of extremists. These effects are occurring individually and collectively, as editors as a whole run far more stories attacking politically contentious science like climate disruption than is warranted by the evidence.

Then there is the high number of journalists who move to the field of political communications, and from there into public relations, seeking to manipulate the thinking of their former colleagues in the media.

Newspapers and media outlets need to take a long look at the ethical problems created by the belief that there is no such thing as objectivity, to more clearly define the roles of objectivity and subjectivity in reporting, and to develop changes to their ethical guidelines that establish when and how subjective and objective reporting are appropriate and when they are not. Journalism schools need to start reaching out and collaborating with working scientists to help students learn that there is such a thing as objective knowledge, and that, when a public-policy issue has a scientific dimension, they have an ethical obligation to report it as objective knowledge, and to explore the hidden agendas of those who disagree with that knowledge. Journalists should personally research and understand the entwined history of science, democracy, law, and journalism, so they develop a clearer sense of how to change their own professional culture to better serve their readers and viewers, and better equip them to make decisions as informed voters in an age dominated by the objective knowledge that produces increasingly impactful science and technology.

Practice Pro-Evidence Journalism

Understanding that science is never partisan but is always political charts a clear path for journalists. Similar to science, the job of a journalist is to report on political issues not in a bipartisan or multipartisan way, but in a *nonpartisan* way, which means reporting the facts that are supported by the evidence and letting the political chips fall where they may. As MPR News producer Stephanie Curtis pointed out, this allows for an approach to reporting that focuses on the specificity of science-driven questions and not on the top-level politics of support or denial. This makes for much richer, more impactful, and more useful reporting.

This is not a liberal or a conservative position, but it is a position that skews toward both science and democracy—and the interests of the public—in that it is antiauthoritarian. But serving the interests of the public, not the powerful, is what the job of journalists has always been. When journalists have gotten off track is when they have become confused or cowed by authoritarian tactics.

It is within journalists' power to reopen this conversation and guide it in a way that is most helpful to democracy—arguably, it may be their most pressing duty. In Canada, after CBC producer Mary Lynk took the risky step of making the three-hour radio documentary *Science under Siege* based on some of the ideas contained in this book, a national discussion developed among scientists, journalists, and the public that brought politicians from the three opposition parties to the table to sign the Science Pledge of Evidence for Democracy, a nonprofit organization advocating for public policies to be based on evidence. Evidence for Democracy also called for a parliamentary science debate in the October 2015 election that toppled the often-antiscience Harper government and returned Canada's Liberal Party to power under Justin Trudeau. It was, perhaps, the beginning of a new national direction, as antiscience politics was penalized at the ballot box. But at the time, it was a significant risk to stand up for science and criticize the government as a CBC journalist.

"The most surprising thing I found is the reaction, and how many people there are that care deeply about this topic," says Lynk. "I'm still in shock at the number of people that listened and want to hear about this. There was roughly a 90 percent increase in people going to the website and 80 percent of them were new."

Science is as foundational to a functioning democracy as journalism. The first gathers evidence on which policies should be based, and the second disseminates that evidence and reports on those policies to the public. "There is a natural affinity between science and journalism," Lynk says, "because both are steeped in basing their statements on multiple-source evidence they can back up."

But there needs to be more reporting based on that evidence and not on subjective perspectives in politics, and *a lot* more reporting on what is happening to democracy's instruments of self-governance—evidence gathering, evidence-based reporting, and evidence-based policymaking. Journalists have a wealth of stories at their fingertips once they reject the notion that truth is subjective and start asking for evidence and digging into details. They especially have a wealth of stories at their fingertips when they start exploring how science is being intentionally misrepresented by vested interests. Such stories as those uncovered by *InsideClimate News* and the *Los Angeles Times* can give politicians the support they need to take corrective action. Whenever the people are well informed, they have the capacity to change course, but it is journalists who have the calling, mission, and obligation to inform them.

The remarriage of journalism with science may not be as hard as journalists think. Lynk offers this advice to her colleagues:

> Don't be intimidated just because it's science. Realize there's a heartbeat, and look for the heartbeat. And realize that people care—they want to hear this and they're not hearing it, which is why they soaked up *Science under Siege* in such record numbers. They are smarter than you are and they want it in a way that is accessible and still smart. It should be a daily part of journalistic life for everybody. Also, it's easy for journalists because it's highly researched, evidence-based knowledge. A lot of the work has already been done for you, as opposed to when you're interviewing politicians, and you can have confidence in it.

Battle Plan 13:
Scientists Need To Fight Back

One of the biggest things scientists can do is roll up their lab-coat sleeves and fight back. Call a spade a spade, and do it publicly. It is not partisan to call out antiscience activities, even if they are by people in elected office. After the attacks on climate science and scientists following the "Climategate" controversy and the wildly unbalanced way the media covered the event, climate scientists began to realize this. There is a long tradition of scientists becoming more activist in order to defend science and inform the public when the media is doing an inadequate job, including Albert Einstein's vocal concerns about the atomic bomb, Rachel Carson's exposé of pesticides, and Carl Sagan's campaign against science apathy and pseudoscience superstition. When scientists see something that others don't, they have an ethical obligation to bring it to the attention of the public.

In this case, climate scientists realized that they were on their own and that the media was printing and broadcasting disinformation. So they organized themselves into the Climate Science Rapid Response Team (CSRRT), led by John Abraham and Scott Mandia, who recruited more than 150 climate scientists, climate engineers, and psychologists interested in climate science, as well as scores of leading journalists, into an online network and discussion group.

The effort was enormously successful for science, journalism, and the public. The CSRRT engaged hundreds of times in fending off and debunking disinformation campaigns, in alerting journalists to some of the ways they were being manipulated by PR tricks, and in slowing the number of disinformation stories that journalists had been disseminating in the media.

Additionally, they were able to provide quick access to top-level scientists for quotes, background material, and explanations, and to direct journalists' attention to new science about climate change, to the 97 percent consensus of climate scientists that man-made climate change is happening, to information on how to debunk climate-denier myths, and even to new science about climate deniers, their thought processes, and their tendency toward believing conspiracy theories.

By pairing communication expert John Cook with cognitive scientist Stephan Lewandowsky, environmental scientist and blogger Dana Nuccitelli, and top climate scientists, the group was able to create enormously influential new science that has had a powerful impact on the climate conversation and put climate deniers on the defensive.

Battle Plan 14:
Voters Should Support Candidates Who Support Science—And Reject Those Who Don't

The last of these battle plans may seem like the most obvious, but in the roiling, conflicted discussions of democratic politics, it is in many ways the most difficult. Questions about jobs and the economy are often pitted against evidence from science in order to convince voters to vote against their own interests. But voting for the candidates who go with the evidence is, in both the long and short run, the *only* reliable way to protect one's interests. If there is only one issue on which you should base your vote, it is this: whether a candidate supports evidence as the basis for policy decisions.

As we grow in knowledge, we grow in our capacity to solve the problems we perceive around us. There is still time to reverse the systemic problems created by older science, and to elect candidates who will embrace evidence and chart out bold new visions for the future.

Chapter 13

TRUTH AND BEAUTY

Believe in yourself. You gain strength, courage and confidence by every
experience in which you stop to look fear in the face. . . . You must do
that which you think you cannot do. . . . The future belongs to those
who believe in the beauty of their dreams.

—Eleanor Roosevelt

The Unreachable Star

Those fighting against the war on science need to understand what it is they are
fighting for. That question will ultimately determine the outcome. To help us
explore it, let us turn to the story of a scientist who decided to run for Congress
in an antiscience political district.

David Sanders is a Purdue University biologist who ran for Congress three
times as a Democrat in Indiana's Republican-leaning Fourth District. He realized that the district tilted heavily against Democrats, but he ran anyway.

Sanders felt it was part of his civic duty as a scientist. "So many of the
issues that we face as a society have a science or technological basis," he says.
"We've simply got to talk about them. I've had National Institutes of Health
grants and National Science Foundation grants. By taking that public money,
I think it's my responsibility as a scientist to do good science, but also to participate in the public sphere." Win or lose, Sanders puts a face on science and
elevates it in the discussion. Sometimes this process can take years to have an
impact. But eventually, it does.

> I went to the county fairs and I'd meet lots of people. A not-
> infrequent response was, "Oh, since you're a scientist you probably
> believe in global warming." This was from people who didn't. So
> there was this cognitive dissonance. They know that my being a

scientist would make me think that global warming was real, and they both didn't believe it themselves and thought that there was some dispute about it. And they would maintain the position that it's an open scientific question. But in person at least, I didn't feel like they were hostile to me. It's hard to know how sustained you have to be.

At Indiana county fairs there are a lot of church booths. I started introducing myself. Some were like, "It's wonderful to have somebody running for office." They were the good side of religion. Welcoming, open, wanting what is best. Other ones, the first question was, "Are you a Christian?" I said no. Then they said, "Well, are you pro-life?" I asked them, "What do you mean? Do you mean stem cells? Or abortion? Or abortion for women who have been raped?" But they couldn't tell me. So I probed, and asked, "If it's murder, do you want abortion to be banned, or do you want physicians to go to jail, or women who get abortions to go to jail?" I'd never get a clear answer. They had some sort of concept of what they thought was right and wrong, but you couldn't push them to the logical consequences. It wasn't possible.

Sanders says he ran into something similar when he taught microbiology to a class of 250. These students had already had two semesters of biology. Somewhere between 15 and 20 percent of these students, in an advanced biology class, didn't believe in evolution. They were offended by it. They gave him the usual arguments—the flagellum, for example. Flagella are long spiral propellers protruding from many bacteria that are connected to tiny biological motors that drive their rotation. They are one of the ways bacteria move around. Creationists argue that the flagellum couldn't possibly arise through evolution, because, if we take one part away, it doesn't work anymore. It's irreducibly complex and could only arise from design. So Sanders would show them there are intermediates: flagella that turn just one direction; one with two rings; one with three rings. He explained that that's the point of evolution; if there were an extra part, it would go away. He very patiently worked to educate them out of their "vulgar Induction."

The students complained that he worked evolution into every lecture, he says. But biology at this level is essentially evolution in the context of chemistry and physics. Even in these classes, he concluded, there is a certain segment of the population that simply cannot be reached. Sanders says he would have the occasional student come to him upset and say, "Before I believed, and now I am questioning it and I don't know what to do." He'd tell them, "I think that you can continue to practice your religion and still accept evolution. I don't see any conflicts there. If you go with 'What's the message?' I don't think there is any contradiction."

It comes down to the question that Galileo himself had to deal with near the beginning of the scientific revolution, and which we have to deal with still. When they are in conflict, which do we believe: the interpretations of reality made by peasants and translated from an ancient language in an old book, or the evidence we can see, measure, and confirm with our own eyes? "My dear Kepler," Galileo wrote, "what would you say of the learned here, who, replete with the pertinacity of the asp, have steadfastly refused to cast a glance through the telescope? What shall we make of this? Shall we laugh, or shall we cry?"

Sanders's approach is not unique. From Vern Ehlers to Bill Foster to Rush Holt, scientists are, very slowly, beginning to run for Congress. But they don't all have to; they just have to have conversations in the community. Sanders says he began to realize this when he was teaching. The students who couldn't accept evolution, he realized, were simply repeating the arguments coming from a vast and well-organized propaganda network of conservative religious authoritarians. The frame the students would ultimately come to, Sanders says, was, "I can accept microevolution, but not macroevolution"—i.e., evolutionary changes within a species, but not the evolutionary emergence of new species from prior ones. So he'd say, "Give me your best shot, but if that proves not to be true, then you're going to agree to change your position." But they wouldn't take the bet.

If it is this difficult for an educated person in an advanced biology class to accept evidence that contradicts a preconceived belief—if the fear of being cast out by an unforgiving God or an unforgiving family or an unforgiving political party is so profound—imagine how difficult it would be for someone without that education who has little economic stake in a scientific career. This is where conservative authoritarian religions and nonprofit organizations do the world such a great disservice—instead of helping followers understand the moral issues science presents, they slow down humanity in its efforts to solve problems.

"In the end," says Sanders, "some people simply have a different approach to discussion of rational thought. You hope that as a scientist you look at the facts and make a conclusion. That's what we're trained to do." The problem arises when people look for the facts that agree with their preconceptions. In the public and political spheres, that's what happens. As an example, Sanders cites discussions about a common argument of climate-science deniers popularized by Christopher Monckton: that satellite measurements show the troposphere isn't warming. "I bet you most people who said that couldn't tell me what the troposphere was. It turns out those data weren't correct, but even if they had been, that was the outlier in a very large body of evidence saying otherwise. But once you dispel that, they go on to some other argument."

This is, of course, because we use motivated reasoning to make rhetorical arguments in order to navigate life—arguments that we think will convince other people of something even if we don't believe it ourselves. Arguments that are confirmed, like Marxism, no matter where we look. Instead of arguing on the basis of data, we argue to win. At some point, we've made that argument so often or so publicly that it becomes impossible to retreat from it.

Despite his electoral losses, Sanders remains upbeat. "I am hopeful as an educator," he says. "I have to believe that what I do is worthwhile. I think that if people of goodwill with skills—scientists—don't speak out, then we're totally sunk, then the cynicism has seeped into us as well. We need as a society to take pride not only in our rock stars and sports heroes, but also in the beauty of the science we accomplish."

Defending the Beauty of a Dream

In the end, this is what matters most: *the beauty*. It's why scientists do science: to apprehend the great beauty of nature. They are Puritans, four hundred years hence, with more data, but still searching for that direct communion with wonder and the aesthetic. In this way, science is not unlike its progenitors: art and religion. In fact, art often anticipates and reflects the forms science discovers, as does religion. Science is about much more than just solving challenges, as important as they are. It is about who we are as human beings, about our ability to love, to wonder, to imagine, to heal, to care for one another, to create a better future, to dream of things unseen. To figure out the world and our place in it, and to capture the great beauty of the world in representations others can apprehend. They may use the language of math instead of paint or music, but they are artful representations just the same.

The English poet Alfred Noyes was on hand on the night of November 1, 1917, as George Ellery Hale's only invited guest for the dedication of the Hooker one-hundred-inch telescope on the top of Mount Wilson outside of Pasadena, California. "Your Milton's 'optic tube' has grown in power since Galileo," Hale had written his friend. Hale ordered the giant telescope to be trained, as Galileo's had been, on Jupiter and its moons. The experience of being among the first to ever see those moons so clearly inspired Noyes to write his epic poem *Watchers of the Sky*, which charts the long emergence of astronomy from religion, and their close kinship still. The poet took a break shortly after midnight and wandered out onto the mountaintop in a state of near-rapture. Almost all of the tiny sparkling lights of the distant Los Angeles had vanished by then, and "the whole dark mountain seemed to have lost its earth and to be sailing like a ship through heaven." His poem concluded:

> When I consider the heavens, the work of Thy fingers,
> The sun and the moon and the stars which Thou hast ordained,
> Though man be as dust I know Thou art mindful of him;
> And, through Thy law, Thy light still visiteth him.

The evening would not have been possible without the financial support of Andrew Carnegie, who had given Hale a check for $10 million, making an investment in the unknown and building what was, in its time, the greatest scientific laboratory on Earth—all without having any idea of what it might find. Using the Mount Wilson Observatory, Harlow Shapley redefined our relationship to the cosmos, showing that the sun was not at the center of the Milky Way. Using it, Edwin Hubble created the whole field of cosmology, redefining again our place in the universe, our understanding of its vastness, and our ideas about its creation. Those discoveries did not bear any direct financial returns. They did not add to the national defense. But they did something far more important: they changed our lives and the way we think about ourselves. They are among the most profound discoveries of science, and they had no financial justification whatsoever. They were seeking truth and beauty amid chemistry and light.

In 1967, the physicist and sculptor Robert R. Wilson was hired to build a similar dream of truth and beauty: his "fantasy of a utopian laboratory," a particle accelerator that would smash subatomic particles together at close to the speed of light, perhaps yielding new understandings about the underlying nature of the universe. But no one knew what practical applications might come of it, and this time science was being funded not by visionary philanthropists, but by the less-than-visionary senators and representatives of the federal government.

Because of this, Wilson's goal was a hard sell. The nation was already spending unprecedented amounts of money on the risky Apollo program. Science was unpopular with the general public, who saw the spending as wasteful when so many pressing social needs were going unmet. Science was despoiling the environment, and nuclear physics in particular had embroiled us in an arms race that would perhaps end us all in mutually assured destruction.

By 1969 the nation had tumbled into yet another recession, the emerging baby boom was loudly questioning the older generation's priorities, and Congress was looking for ways to cut spending. There was no political capital to be gained by funding a massive new science investment like a particle accelerator unless there was some greater justification. Wilson was called up before the congressional Joint Committee on Atomic Energy and asked to supply just that.

By using a Cold War rationale, there would at least be some chance of getting approval, and Wilson was no fool. Fear of Russia was running at a fever pitch. Science and the military were joined at the hip. Wilson had been a key player in the Manhattan Project and knew these politics well. So when Senator John Pastore (D-RI) asked him to justify allocating $200 million to build the world's largest particle accelerator, the room was stunned by Wilson's reply.

Pastore: *"Is there anything here that projects us in a position of being competitive with the Russians, with regard to this race?"*

Wilson: *"Only from a long-range point of view, of a developing technology. Otherwise, it has to do with: Are we good painters, good sculptors, great poets? I mean all the things that we really venerate and honor in our country and are patriotic about. In that sense, this new knowledge has all to do with honor and country but it has nothing to do directly with defending our country—except to make it worth defending."*

Quite breathtakingly, Wilson had reached not for defense, but for the profound—not fear, but wonder. Miraculously, the funding was approved. Fermilab, as the facility that housed the accelerator came to be called, was built despite offering no quantifiable return to defense or industry, and, when it began operating in 1972, it was the most powerful particle accelerator in the world. Reflecting Wilson's hope, the design of its main building, Wilson Hall, recalled the great French cathedrals of Chartres and Beauvais, bucking the attitude in Washington that Atomic Energy Commission buildings "did not have to be cheap, they just had to look cheap." His sense of the important connection between science and the aesthetic went beyond architecture. "I have always felt," he later wrote, "that science, technology, and art are importantly connected, indeed, science and technology seem to many scholars to have grown out of art. In any case, in designing an accelerator I proceed very much as I do in making a sculpture. I felt that just as a theory is beautiful, so, too, is a scientific instrument—or that it should be." The futuristic superconducting technologies Wilson pushed for helped keep the accelerator vital under the leadership of his successor, the physicist, Nobel laureate, and great science humanitarian Leon Lederman, and it remained the world's most powerful until 2006, when the Large Hadron Collider opened at CERN, the European Organization for Nuclear Research, in Geneva.

To Awaken, Perchance to Dream

As Wilson noted, science, like art, is a cultural expression that makes a nation worth defending. Like great art and great music, its true value lies in exploring the unknown. Today, the opposite argument, the commoditization of science, is virtually the only one heard. It has metastasized from the smaller-minded appeals of the Cold War to all of human learning and higher education. Education and knowledge are no longer values of truth and beauty that make life worth living—they are reduced to an economic means to the ends of greater pay and more consumption.

In 2010, legal and humanities scholar Stanley Fish wrote of this triumph of small-mindedness in an eloquent criticism of *Securing a Sustainable Future for Higher Education*, a set of recommendations made by an independent panel to the British government. It advocated "student choice" in funding higher education. Among the report's palliatives: "Our proposals put students at the heart of the system. . . . Our recommendations . . . are based on giving students the ability to make an informed choice of where and what to study. . . . Students are best placed to make the judgment about what they want to get from participating in higher education." The idea is that the money follows the students. Courses that compete successfully for student attendance survive and prosper; those that do not wither and die. The assumptions of market economics have triumphed: ideas are now considered commodities.

Fish took objection to this, and to the assertion that "students are best placed to make the judgment about what they want to get from participating in higher education."

> The obvious objection to this last declaration is, "No, they aren't; judgment is what education is supposed to produce; if students possessed it at the get-go, there would be nothing for courses and programs to do." But that objection would be entirely beside the point in the context of the assumption informing the report, the assumption that what students want to get from participating in higher education is money.

The point of education isn't to train students with specific job skills they have already identified; it is to show them what they don't already know, and that they don't know they don't know. In other words, to help them build a telescope so they can discover their own talents and interests as scholars and human beings. Fish wrote of listening to:

> . . . a distinguished political philosopher tell those in attendance that he would not be speaking before them had he not been the beneficiary, as a working-class youth in England, of a government policy to provide a free university education to the children of British citizens. He walked into the university with little knowledge of the great texts that inform modern democracy and he walked out an expert in those very same texts.
>
> It goes without saying that he did not know what he was doing at the outset; he did not, that is, think to himself, I would like to become a scholar of Locke, Hobbes and Mill. But that's what he became, not by choice (at least in the beginning) but by opportunity.

The commoditization of education, its reduction to the status of information age vocational training, is yet another place where the pernicious myth of the "marketplace of ideas" has corrupted Western thought with authoritarianism. That it was Fish who made this impassioned defense is all the more meaningful, since it was also Fish who had most prominently criticized the Sokal hoax and defended postmodernism some fifteen years before.

The approach is extending throughout public universities. As public funding has been cut and cut again, universities have come to rely on increased use of patents and control of knowledge for revenue streams, and on corporate branding and influence. This market-driven, privatized approach to funding higher education has in turn created an administrative style catering to commercialization—with presidents making the big dollars and academics and educators being relegated to the role of workers in the corporate university. The heightened commercialism is changing the approach, tenor and focus of these institutions and their faculties, and the experiences of their students, as institutional goals are shifted away from higher education's traditional core commitment to research, teaching, and the production of public knowledge. A century ago, similar threats led to the birth of the American Association of University Professors, an organization dedicated to advancing the scholarly professions across all disciplines and to safeguarding intellectual and academic freedom.

Henry Buchwald, a professor of surgery and biomedical engineering at the University of Minnesota, one of the world's leading medical schools, speaks and writes often about the rise of commercialization in the fields of medicine and university education. "Surgeons today have seen their autonomy reduced by a sort of corporate bondage as the top-down control of money, resources, and opportunities by and for the benefit of the administration has taken over," Buchwald says. "Benjamin Ginsberg [a professor of political science at Johns Hopkins University] calls this 'managerial imperialism' and I think it's an apt description. The surgeon becomes the money-generating technician for the benefit of management. Every skill, particularly the cognitive, outside of business management is treated like technical training. Career choice for a surgeon becomes either a clinician—worker bee or an overseer."

The same triumph of commoditization is plaguing much of Western culture, as corporations have expanded their influence under globalization. Things that cannot be easily quantified in the marketplace or contained within the confines of what software can measure and manage are assumed to have no value. Unfortunately, much of the rest of the world is now coming online in the new global economy following a similar model of commercialization.

Applying such an approach to education, just as in applying it to art or science, or classics or history, or poetry or math or love or joy, defeats the point and turns universities into trade schools whose sole purpose is to supply the skills that enable one to get a job to earn money to buy things.

This is in a sense a prostitution of our greatest human ideals. Our primary asset is not money. It is the quality of our time on Earth. Money can improve that, but only to a limited degree. The rest comes from intangibles. It is a vast misunderstanding by a generation of policymakers that has lost touch with—or perhaps never really fully knew—what a good education can do: open us up to wonder, and to the great meaning, possibility, and beauty of life.

The Shining City upon a Hill

It is wonder that we lost sight of in our move to a commoditized, national-defense model of science funding, which revolved around dispelling fear. It was perhaps important and it was helpful, but it went too far, and by forgetting the real reasons we do science it was ultimately a colossal error that we are paying for now. As Sanders says, "Science is no longer about science. It's about marketing. It's either 'Gee whiz' or 'What is it going to do for me?'"

The commoditization can be seen in how we oversell the practical benefits of science. The war on cancer provides an example. "We're going to cure cancer in fifteen years," we say. It would be less memorable but more helpful to talk about funding the basic research that led, eventually, to the discovery of monoclonal antibodies, which actually have applications now in fighting cancer, among other things. It is more helpful to value and fund the *process*, rather than the goal.

Antiscience politicians often ripped on the apparent non-commercial impracticalities of basic science to score cheap political points, because the titles and cursory descriptions of these studies make for easy targets.

For more than a decade US Senator William Proxmire, a Democrat from Wisconsin with a background in business administration, gave out the "Golden Fleece Award" to ridicule such studies that sounded to him like a waste of taxpayer dollars. His gave awards to the NSF for funding a study on love, to NASA for funding SETI—the Search for Extraterrestrial Intelligence—and to the USDA for funding a study on the sexual behavior of the screw worm fly. Proxmire thought such a study sounded like weird voyeurism. In fact, screw worm flies laid eggs in the open wounds of livestock and the resulting maggots consumed the flesh, leading to painful infections and often death. The study led to insights that eradicated the pest and saved the cattle industry billions of dollars.

Sarah Palin provides another example. In her first policy speech after being named a vice-presidential candidate, Palin spoke of science. "You've heard about some of these pet projects, they really don't make a whole lot of sense and sometimes these dollars go to projects that have little or nothing to do with the public good. Things like fruit fly research in Paris, France. I kid you not!" The research was into the olive fruit fly, which was a widespread problem in Europe and was threatening the US olive industry.

An entire book could be written about the examples of politicians ignorantly ripping on basic, curiosity-driven research that sounded weird but ultimately delivered great social good. If it weren't for people pursuing truth and beauty by working on weird mouse and bird viruses, for example, we'd be nowhere in treating HIV infection. We weren't even aware of human retroviruses until 1979. There were fewer than a dozen people in the world working in this apparently ridiculously impractical, niche field when the AIDS epidemic broke out. Because we were willing to let those scientists explore unexpected corners, we have wrestled more freedom for ourselves, turning infection with HIV from a terrible death sentence into a manageable chronic disease, pulling back years from the gaping maw of death—giving patients the freedom of life, and their families and society the benefits of their contributions. Not because we set a goal to do it and did it, but because as a society we valued the truth and beauty of science.

The one thing we do know about science—the one thing that is predictable about it—is that if we don't value it, or if we become inhospitable to the tolerance, freedom, and open exchange of ideas that stimulates it, if we wall it off and call it a separate culture instead of something we all should do, if we cease funding it, if we try to be overly directive of it, if we insist on certainty, if we elevate ideology in our public policies, if we attack scientists or call entire fields a hoax, if we stomp on it and call it weird, in short if we become authoritarian, we will stifle creativity and science will suffer or disappear. We won't get the big breakthroughs or cures. We won't get the economic boons. We won't get the national-defense advantages. We won't get a clean environment or healthy children.

Germany proved that this can happen with its precipitous fall from arguably being the most scientific nation on Earth to a nation bereft of scientific enterprise within a single decade of Nazi authoritarianism. China proved it under Mao. The Soviets proved it with Lysenko. The Vatican proved it with Galileo. The Romans and the Ottomans proved it. In each of those cases, science, and with it economic and cultural power, moved on.

We can't move toward the world we could create if we don't value tolerance, freedom, and the beauty of wonder. Ronald Reagan described such a world in his hopeful farewell address, which was markedly different from the fearful farewell delivered by Eisenhower some three decades before.

And that's about all I have to say tonight, except for one thing. The past few days when I've been at that window upstairs, I've thought a bit of the "shining city upon a hill." The phrase comes from John Winthrop, who wrote it to describe the America he imagined. What he imagined was important because he was an early Pilgrim, an early freedom man. He journeyed here on what today we'd call a little wooden boat; and like the other Pilgrims, he was looking for a home that would be free. I've spoken of the shining city all my political life, but I don't know if I ever quite communicated what I saw when I said it. But in my mind it was a tall, proud city built on rocks stronger than oceans, windswept, God-blessed, and teeming with people of all kinds living in harmony and peace; a city with free ports that hummed with commerce and creativity. And if there had to be city walls, the walls had doors and the doors were open to anyone with the will and the heart to get here. That's how I saw it, and see it still.

"My colleagues think our science is beautiful," says Sanders, "but we don't put any effort into communicating that to the public. We don't think the public is interested." But the evidence from cutting-edge narrative journalists like CBC's Mary Lynk or ScienceDebate's national polls shows the scientists are wrong—the public is hugely interested.

The Choice

In the end, politics is about story, and so is science. And so we return to the story of the American Dream that has, for more than two centuries, inspired and guided the entwined progress of science and democracy. Robert McKee, Hollywood's master of storytelling, views the world from the top of America's other great cultural export—its movies. "I think that the American ethos is not science-friendly and never has been," he says, and he hastens to point out that the Western model is today very much the American model.

The American model is Thomas Edison and Henry Ford. Guys who never went to college and who were geniuses and invented things, and people like them. The inventor versus the scientist. Somebody who can go West, discover gold mines, and create a lot of money without an education. . . . The American dream is Hollywood.

Sitting in a drugstore and somebody says, 'You ought to be a movie star.' It's an attitude that life is a game and that what you gotta learn is to play that game well, but it's not on a gridiron where you actually have to practice, it's a game of manipulation and most of that game is somehow bullshit.

The challenges will continue. If climate disruption is solved, there will be the sixth mass extinction to stop, there will be the death of the oceans, the end of antibiotics, the robosourcing of jobs, the deployment of killer robots, the editing of human genes, and an ongoing regression of new issues in a world of unregulated economics, environmental externalities, and an exploding population.

So we are faced with a choice. Will we go the way of the ancient Romans or the ancient Muslims or the ancient Chinese, giving up our science and nosing our heads comfortably into the warm sand, obedient, productive, agreeably alike in thought but rigid, paralyzed, no longer able to control our destiny or even to discern the trajectory of our downfall? Lost in authoritarian politics, ideology, public relations, and subjectivism, will we return to a state of miserable serfs ruled by a wealthy elite of religious and corporate royalty?

Or will we take up the mantle of freedom and leadership that science gave us—the commitment to knowledge over the assertions of "but faith, or opinion" that led to the disquieting idea of equality in the face of uncertainty—an idea that became the foundation of democracy and unleashed the modern era? Will we become skeptical of claims that seek to crowd out the space for knowledge in the public dialogue? Will we equip our children with critical thinking skills and continue to embrace the antiauthoritarian power of wonder, tolerance, and imagination to create a new future—a shining city upon a hill? Will we battle back and reprimand corporate public relations that undermine the democratic process? Will we reject ideological conformity, and reward an evidence-based press and an exploration-based science education? Will we set aside the left-right skirmishes of identity politics and focus—as the founders of modern democracy once did—upon the top-bottom battle of freedom against tyranny?

Will we manage to find new ways to stop the destruction of our bounded pasture, to assume responsibility for our collective actions, and to successfully incorporate increasingly complex scientific knowledge into our societies and our legal systems? Will we reject the commercialization and corporatization of our universities and public institutions, and protect and fund the conditions that perfect democracy's unique dream to set the world free, encouraging diversity, creativity, investment and generosity in art and science—not because of what they do for our pocketbooks, but because of what they mean to our values?

In short, will we, the people, remain well enough informed to be trusted with our own government?

Acknowledgments

Thanks to Rebecca and to Jake, of course and always, for your patience with my long days and evenings at the keyboard. Special thanks to Daniel Slager, my fabulous editor and publisher, and to my passionate agent Joy Tutela, whom I am happy to count as a friend. Thanks to the rest of the visionary folks at Milkweed Editions, including Patrick Thomas, Joey McGarvey, Joanna Demkiewicz, Abby Travis, Meagan Bachmayer, Megan Gette, and Drew Nelles, who are working to change the world, and in the meantime have worked to make this book better. Thanks to my crew at ScienceDebate.org, whose core team includes Matthew Chapman, Sheril Kirshenbaum, Lawrence Krauss, Michael Halpern, Nancy Holt, Ryan Johnson, and Darlene Cavalier. Thanks to Jim Jensen of the US National Academies for his, and their, continued support; to DNC vice chair R. T. Rybak, who pushed the idea of a presidential primary science debate to the Democratic National Committee; and to CNN producer Harrison Bohrman, who helped develop a pitch for the candidates with vision and creativity. Thank you to the many scientists, engineers, policy makers, economists, administrators, artists, theologians, ethicists, philosophers, CEOs, and writers who have allowed me to draw on their insights for this book and in the science debate effort, including John Abrahamson, Natty Adams, Gillian Adler, Peter Agre, Rick Anthes, Paula Apsell, Derek Araujo, Chuck Atkins, Randy Atkins, Norm Augustine, Jennifer Ayers, David Baltimore, Craig Barrett, Bill Bates, Erik Beeler, Rosina Bierbaum, Larry Bock, Ben Bova, Chris Brantley, Bob Breck, Douglas Bremner, David Brin, Deborah Byrd, Troy Campbell, Art Caplan, Joanne Carney, Bill Chameides, Peg Chemberlin, Steven Chu, Pat Churchland, Ralph Cicerone, Rita Colwell, Ben Corb, George Crabtree, Stephanie Curtis, Hunter Cutting, Austin Dacey, Brendan DeMelle, Ronald DePinho, Keith Devlin, Calvin DeWitt, Barbie Drillsma, Ann Druyan, Vern Ehlers, Alex Eilts, Priit Ennet, Harold Evans, Dick Feely, Suzanne Ffolkes, Kevin Finneran, Andrew Fire, Dana Fisher, Ira Flatow, John Forde, Al Franken, Gwen Freed, Peter Frumhoff, Richard Gallagher, Jim Gentile, Jack Gibbons, Katie

Gibbs, Caroline Trupp Gil, Newt Gingrich, Linda Glenn, Peter Gluckman, Wolfgang Goede, Barbara Gordon, Bart Gordon, Kurt Gottfried, Francesca Grifo, Rob Gropp, David Guston, Bo Hammer, Russ Harrison, Kathryn Hinsch, Roald Hoffman, John Holdren, Rush Holt, Doug Holtz-Eakin, Al Hurd, Shirley Ann Jackson, Thomas Campbell Jackson, Mariela Jaskelioff, Brad Johnson, Dean Kamen, Don Kennedy, Alex King, Barbara Kline Pope, Sara Kloek, Kevin Knobloch, Greg Laden, Eric Lander, Neal Lane, Phoebe Leboy, Leon Lederman, Russ Lefevre, Alan Leshner, Simon Levin, Jane Lubchenco, Mike Lubell, Mary Lynk, Scott Mandia, Michael Mann, Elizabeth Marincola, Thom Mason, John Mather, Bob May, Angie McAllister, Jim McCarthy, Robert McKee, Bill McKibben, Marcia McNutt, Ann Merchant, Ken Miller, RL Miller, Chris Mooney, Jonathan Moreno, Jan Morrison, Elizabeth Muhlenfeld, Elon Musk, P. Z. Myers, Hajo Neubert, Matt Nisbet, Leslie Nolen, Peter Norvig, Dana Nuccitelli, Bill Nye, Kevin Padian, Bob Park, Ray Pierrehumbert, Stuart Pimm, Steve Pinker, Phil Plait, John Podesta, Greg Pope, Jerry Pope, Julie Pope, John Porter, John Rennie, Alan Robock, Joe Romm, Brian Rosenberg, Eric Rothschild, Linda Rowan, David Sanders, David Sassoon, Eugenie Scott, Seth Shostak, Kassie Siegel, Maxine Singer, Carl Johan Sundberg, Nick Surgery, Joel Surnow, Jill Tarter, Jim Tate, Al Teich, Meg Ury, Harold Varmus, Chuck Vest, Ethan Vishniac, Cynthia Wainwright, Michael Webber, Ty West, Scott Westphal, Frank Wilczek, Robb Willer, Deborah Wince-Smith, Dennis Wint, Susan Wood, Mary Woolley, and many, many more who have helped move the ball forward. To those who I have neglected to thank here, forgive me; you are also important in leading on this effort.

Notes

Chapter 1. The War on Science

3   *"Wherever the people are well informed"* Thomas Jefferson, "From Thomas Jefferson to Richard Price, 8 January 1789," Founders Online, National Archives, accessed October 12, 2015, http://founders.archives.gov/documents/Jefferson/01-14-02-0196. It should be noted that, like Jefferson's notion of science, his notion of democracy limited the right to vote and participate in society in ways we find objectionable today. It was provisional, based on the ideas of the times, but subject to the core science-based principle of later revision as knowledge from science expanded our understanding.

3   *"invisible hand"* Adam Smith, *An Inquiry into the Nature and Causes of the Wealth of Nations* (Dublin, Ireland: Whitestone, 1776).

4   *Harvard entomologist Edward O. Wilson* E. O. Wilson, *Consilience: The Unity of Knowledge* (New York: Knopf, 1998).

4   *science is poised to create more knowledge* Just as there are more people alive now than in all previous generations combined, there are proportionately more scientists alive now, and qualitatively they are much more productive because of Internet collaboration and the power of modern computational and observational tools. This is a game-changing moment for science.

4   *replace millions of truck drivers* For one example of how this is beginning to occur, see Tavia Grant, "Driverless Trucks Could Mean 'Game Over' for Thousands of Jobs," *Globe and Mail* (Toronto), July 26, 2015, http://www.theglobeandmail.com/report-on-business/autonomous-trucks-could-transform-labour-market-eliminate-driver-jobs/article25715184/.

4   *hunt and kill humans* Audie Cornish, "Tech Experts Warn of Artificial Intelligence Arms Race in Open Letter," National Public Radio, July 28, 2015, http://www.npr.org/2015/07/28/427178289/tech-experts-warn-of-artifical-intelligence-arms-race-in-open-letter.

4   *we must get ahead of artificial intelligence* "Autonomous Weapons: an Open Letter from AI & Robotics Researchers," Future of Life Institute, July 28, 2015, accessed July 29, 2015, http://futureoflife.org/AI/open_letter_autonomous_weapons.

7   *The Islamic drive for* al-asala Bassam Tibi, *Islam's Predicament with Modernity: Religious Reform and Cultural Change* (London: Routledge, 2009) 92.

9   *"I don't expect them to understand"* Xi Xiaoxing, quoted in Matt Apuzzo, "U.S. Drops Charges That Professor Shared Technology with China," *New York Times*, September 11,

2015, http://www.nytimes.com/2015/09/12/us/politics/us-drops-charges-that-professor
-shared-technology-with-china.html.

9   *that looks like a bomb*   Quoted in Sarah Kaplan and Abby Phillip, "'They Thought It Was
a Bomb': 9th-Grader Arrested after Bringing a Home-Built Clock to School," *Washington
Post*, September 16, 2015, https://www.washingtonpost.com/news/morning-mix/wp/2015/
09/16/they-thought-it-was-a-bomb-ahmed-mohamed-texas-9th-grader-arrested-after
-bringing-a-home-built-clock-to-school/.

9   *That's who I thought it was*   Quoted in Kaplan and Phillip, "'They Thought It Was a Bomb.'"

9   *Cool clock, Ahmed*   Barack Obama's Twitter page, September 16, 2015, accessed
October 31, 2015, https://twitter.com/POTUS/status/644193755814342656.

10   *more than five times as doubtful*   Kevin Kalhoefer, "ANALYSIS: Notable Opinion Pages
Included Denial in Coverage of Paris Climate Summit," Media Matters for America,
December 16, 2015, http://mediamatters.org/research/2015/12/16/analysis-notable-opinion
-pages-included-denial/207534.

11   *eleven out of 535 members, or 2 percent*   Jennifer E. Manning, *Membership of the 113th
Congress: A Profile* (Washington, DC: Congressional Research Service, November 24, 2014),
accessed July 8, 2015, http://www.fas.org/sgp/crs/misc/R42964.pdf.

11   *about 4 percent*   Mary Lynk, "Science under Siege," *Ideas with Paul Kennedy*, CBC Radio,
June 4, 2015, http://www.cbc.ca/radio/ideas/science-under-siege-part-1-1.3091552.

11   *where it's 4 percent*   Martin Lumb, *The 43rd Parliament: Traits and Trends* (Canberra,
Australia: Parliament of Australia, October 2, 2013), http://www.aph.gov.au/About
_Parliament/Parliamentary_Departments/Parliamentary_Library/pubs/rp/rp1314/
43rdParl#_Toc368474619.

12   *three questions mentioned UFOs*   Shawn Lawrence Otto and Sheril Kirshenbaum,
"Science on the Campaign Trail," *Issues in Science and Technology* 25, no. 2 (2009), http://
www.issues.org/25.2/p_otto.html.

12   *several major topics surrounding science*   "The 14 Top Science Questions Facing America
(2008)," ScienceDebate.org, accessed February 1, 2016, http://www.sciencedebate.org/2012/
questions08.html.

13   *debate the major science policy issues*   S. R. Kirshenbaum et al., "Science and the
Candidates," *Science* 320, no. 5873 (2008): 182.

14   *What little news coverage there was*   Cornelia Dean, "No Democratic Science Debate,
Yet," *New York Times*, April 8, 2008, http://thecaucus.blogs.nytimes.com/2008/04/08/
no-democratic-science-debate-yet.

14   *didn't seem to affect the campaigns*   Brandon Keim, "Clinton and Obama Talk Religion,
Not Science," *Wired*, April 8, 2008, http://www.wired.com/2008/04/clinton-and-oba-4/.

15   *more than half of all US economic growth*   Committee on Prospering in the Global
Economy of the 21st Century: An Agenda for American Science and Technology, National
Academy of Sciences, National Academy of Engineering, and Institute of Medicine, *Rising
above the Gathering Storm: Energizing and Employing America for a Brighter Economic
Future* (Washington, DC: National Academies Press, 2007), http://www.nap.edu/catalog/
11463/rising-above-the-gathering-storm-energizing-and-employing-america-for.

17   *"Science, like any field"*   George H. W. Bush, "Remarks to the National Academy of Sciences, April 23, 1990," American Presidency Project, accessed February 1, 2016, http:// www.presidency.ucsb.edu/ws/?pid=18393.

18   *"This administration looks at the facts"*   Scott McClellan, quoted in Christopher Marquis, "Bush Misuses Science Data, Report Says," *New York Times*, August 8, 2003, http://www .nytimes.com/2003/08/08/politics/08REPO.html.

18   *"the most liberal"*   Leslie McCarthy, quoted in Andrew C. Revkin, "Climate Expert Says NASA Tried to Silence Him," *New York Times*, January 29, 2006, http://www.nytimes. com/2006/01/29/science/earth/29climate.html.

18   *preconceived ideological agenda*   Andrew C. Revkin, "Bush Aide Softened Greenhouse Gas Links to Global Warming," *New York Times*, June 8, 2005, http://www.nytimes.com/ 2005/06/08/politics/bush-aide-softened-greenhouse-gas-links-to-global-warming.html.

18   *angering many scientists*   Juliet Eilperin, "Censorship Is Alleged at NOAA," *Washington Post*, February 11, 2006, http://www.washingtonpost.com/wp-dyn/content/article/2006/ 02/10/AR2006021001766.html. For a more detailed account, see Union of Concerned Scientists, "Agencies Control Scientists' Contacts with the Media," accessed February 1, 2016, http://www.ucsusa.org/center-for-science-and-democracy/scientific_integrity/ abuses_of_science/a-to-z/agencies-control-scientists.html.

18   *hearings investigating the distortions*   US House of Representatives Committee on Oversight and Government Reform, "Committee Holds Hearings on Political Influence on Government Climate Change Scientists," January 30, 2007, accessed February 1, 2016, http://democrats.oversight.house.gov/legislation/hearings/committee-holds-hearing-on -political-influence-on-government-climate-change.

19   *none of which were abstinence-based*   *Scientific Integrity in Policy Making: An Investigation into the Bush Administration's Misuse of Science* (Cambridge, MA: Union of Concerned Scientists, March 2004), http://www.ucsusa.org/our-work/center-science-and-democracy/ promoting-scientific-integrity/reports-scientific-integrity.html.

19   *breast cancer was falsely linked to abortion*   "Abortion and Breast Cancer," *New York Times*, January 6, 2003, http://www.nytimes.com/2003/01/06/opinion/06MON1Y.html. For a more detailed examination of this issue, see Karen Malek, "The Abortion-Breast Cancer Link: How Politics Trumped Science and Informed Consent," *Journal of American Physicians and Surgeons* 8, no. 2 (2003): 41–45, http://abortionno.org/pdf/breastcancer.pdf.

19   *prevents ovulation after unprotected sex*   Population Council, "Emergency Contraception's Mode of Action Clarified," *Population Briefs* 11, no. 2 (2005): 3, accessed February 1, 2016, http://www.popcouncil.org/uploads/pdfs/pbmay05.pdf. See also A. L. Müller et al., "Postcoital Treatment with Levonorgestrel Does Not Disrupt Postfertilization Events in the Rat," *Contraception* 67, no. 5 (2003): 415–419.

19   *likely to* reduce *the number of abortions*   US Food and Drug Administration Center for Drug Evaluation and Research, "Transcript of Nonprescription Drugs Advisory Committee in Joint Session with the Advisory Committee for Reproductive Health Drugs," December 16, 2003, accessed February 1, 2016, http://www.fda.gov/ohrms/ dockets/ac/03/transcripts/4015T1.doc. The briefing document collection, accessed

February 1, 2016, is posted at http://www.fda.gov/ohrms/dockets/ac/03/briefing/4015b1 .htm.

19   *it was causing heart attacks*   Gardiner Harris, "F.D.A. Failing in Drug Safety, Official Asserts," *New York Times*, November 19, 2004, http://www.nytimes.com/2004/11/19/ business/19fda.html.

19   *more than fifty thousand American deaths*   David Graham, "Testimony of David J. Graham, MD, MPH, November 18, 2004," ConsumersUnion, accessed February 1, 2016, http://consumersunion.org/news/testimony-of-david-j-graham-md-mph-nov-18-2004/.

19   *scientist who brought the problem to light*   Manette Loudon, "The FDA Exposed: An Interview with Dr. David Graham, the Vioxx Whistleblower," *Natural News*, August 30, 2005, http://www.naturalnews.com/011401.html. For an overview and timeline of the Vioxx catastrophe, see Rita Rubin, "How Did Vioxx Debacle Happen?," *USA Today*, October 11, 2004, http://www.usatoday.com/news/health/2004-10-11-vioxx-main_x.htm.

19   *"Doubt is our product"*   "Smoking and Health Proposal," Brown & Williamson document no. 690010954, 1969, Ness Motley Law Firm, Truth Tobacco Industry Documents, University of California, San Francisco, accessed February 1, 2016, https://industrydocuments.library .ucsf.edu/tobacco/docs/#id=xhvf0040.

19   *"For what a man"*   Francis Bacon, *Novum Organum* (London: Joannem Billium, 1620).

20   *"Teach both"*   Sarah Palin, quoted in Brandon Keim, "McCain's VP Wants Creationism Taught in School," *Wired*, August 29, 2008, http://www.wired.com/wiredscience/2008/08/ mccains-vp-want.

21   *host of other science policy topics*   Some of the coverage is archived at the ScienceDebate 2008 news page, accessed February 1, 2016, at www.sciencedebate.org/news08.html.

22   *many newspaper reporters' guidelines*   For example, "There is no such thing as objectivity" in *Voice of San Diego*'s "Guidelines for Reporters," May 29, 2014, http://www.voiceofsandiego .org/all-narratives/vosd/voice-of-san-diegos-guidelines-for-reporters/.

22   *"Some journalists don't even attempt"*   Don Shelby, interview by author, April 3, 2014.

23   *"I think there are a lot"*   David Gregory, quoted in Glenn Greenwald, "David Gregory Shows Why He's the Perfect Replacement for Tim Russert," *Salon*, December 29, 2008, http://www.salon.com/news/opinion/glenn_greenwald/2008/12/29/gregory.

25   *outlets that reached the general public*   Cristine Russell, "Covering Controversial Science: Improving Reporting on Science and Public Policy," Joan Shorenstein Center on the Press, Politics, and Public Policy, 2006, http://shorensteincenter.org/wp-content/uploads/2012/ 03/2006_04_russell.pdf.

25   *focused on global warming*   Andrew Freedman, "NBC Fires Weather Channel Environmental Unit," *Washington Post*, November 21, 2008, http://voices.washingtonpost. com/capitalweathergang/2008/11/nbc_fires_twc_environmental_un.html.

25   *science, technology, and environment news unit*   Curtis Brainard, "CNN Cuts Entire Science, Tech Team," *Columbia Journalism Review*, December 4, 2008, http://www.cjr.org/ the_observatory/cnn_cuts_entire_science_tech_t.php.

25   *renowned science and health section*   Cristine Russell, "*Globe* Kills Health/Science

Section, Keeps Staff," *Columbia Journalism Review*, March 4, 2009, http://www.cjr.org/
the_observatory/globe_kills_healthscience_sect.php.

25 *gravest environmental issues in human history* Katherine Bagley, "About a Dozen
Environment Reporters Left at Top 5 U.S. Papers," *InsideClimate News*, January 17, 2013,
http://insideclimatenews.org/print/23503.

25 *elevenfold* increase *since 1989* Bienvenido León, "Science Related Information in
European Television: A Study of Prime-Time News," *Public Understanding of Science* 17,
no. 4 (2008): 443–460.

26 *between science and the public* Elias Aggelopoulos, "Science View Participates in Major
Horizon 2020 Project 'NUCLEUS,'" *Science View Newsletter*, July 8, 2015, accessed July 29,
2015, http://www.scienceview.gr/wordpress/wp-content/uploads/2015/07/Newsletter
_SV_2015_07-08.pdf.

26 *"flourishing"* Aisling Irwin, "Science Journalism 'Flourishing' in Developing World,"
SciDev.Net, February 18, 2009, http://www.scidev.net/en/news/science-journalism
-flourishing-in-developing-world.html.

26 *problems are beginning to emerge* Discussions with the author at the World Conference
of Science Journalists, Helsinki, Finland, June 30, 2013.

26 *"Only Christianity offers"* Charles Colson and Nancy Pearcey, *How Now Shall We Live?*
(Wheaton, IL: Tyndale House, 1999). Colson, called the "evil genius" by colleagues in the
Nixon administration, pleaded guilty to obstruction of justice for his role in that administration's attempt to retaliate against Daniel Ellsberg, the leaker of the Pentagon Papers.
("The Watergate Story: Key Players: Charles Colson," *Washington Post*, n.d., http://www
.washingtonpost.com/wp-srv/onpolitics/watergate/charles.html.) Pearcey is an evangelical author, editor-at-large for her husband's evangelical mouthpiece, the *Pearcey Report*,
and chapter contributor to the creationist high school–level biology book *Of Pandas and
People*. ("About," *Pearcey Report*, accessed February 1, 2016, http://www.pearceyreport
.com/about.php.)

26 *"a biblical worldview"* Tom DeLay, quoted in Alan Cooperman, "DeLay Criticized for 'Only
Christianity' Remarks," *Washington Post*, April 20, 2002, https://www.washingtonpost
.com/archive/politics/2002/04/20/delay-criticized-for-only-christianity-remarks/
21016a6c-9733-4518-961b-1c4df9312fc6/.

26 *"Our entire system"* Tom DeLay, quoted in Peter Perl, "Absolute Truth," *Washington Post*,
May 13, 2001, http://www.washingtonpost.com/wp-dyn/content/article/2006/11/28/
AR2006112800700.html.

26 *"couldn't have been because"* Tom DeLay, quoted in Ron Nissimov, "DeLay's College
Advice: Don't Send Your Kids to Baylor or A&M," *Houston Chronicle*, April 18, 2002,
http://www.chron.com/news/houston-texas/article/DeLay-s-advice-Don-t-send-kids-to-
Baylor-or-A-M-2060338.php.

27 *"It's important that the implementation"* John A. Boehner and Steve Chabot to Jennifer L.
Sheets and Cyrus B. Richardson Jr., March 15, 2002, Discovery Institute, accessed
February 1, 2016, http://www.discovery.org/articleFiles/PDFs/Boehner-OhioLetter.pdf.

28 *"regarded as settled facts"* National Research Council, *Advancing the Science of Climate Change* (Washington, DC: National Academies Press, 2010), 22, http://books.nap.edu/openbook.php?record_id=12782&page=22.

29 *"Listen [. . .] when you make"* Jon Huntsman, "Huntsman: Republicans Can't Run from Science," video, RealClearPolitics, September 7, 2011, http://www.realclearpolitics.com/video/2011/09/07/huntsman_republicans_cant_run_from_science.html.

30 *pregnancies from rape* Felicia H. Stewart, "Prevention of Pregnancy Resulting from Rape: A Neglected Preventive Health Measure," *PubMed*, US National Library of Medicine, National Institutes of Health, November 2000, http://www.ncbi.nlm.nih.gov/pubmed/11064225.

30 *pregnancy rate from consensual sex* Matt Walker, "Rape—An Evolutionary Strategy?," *New Scientist*, June 23, 2001, http://www.eurekalert.org/pub_releases/2001-06/NS-Raes-1906101.php.

30 *"We are sliding back"* Nina Fedoroff, quoted in Robin McKie, "Attacks Paid for by Big Business Are 'Driving Science into a Dark Era,'" *Guardian* (Manchester), February 18, 2012, http://www.theguardian.com/science/2012/feb/19/science-scepticism-usdomesticpolicy.

30 *efforts to muzzle scientists* Professional Institute of the Public Service of Canada, *The Big Chill: Silencing Public Interest Science, A Survey* (Ottawa, Canada: 2013), accessed February 1, 2016, http://www.pipsc.ca/portal/page/portal/website/issues/science/bigchill.

31 *anti-tax activist Grover Norquist* Brooke Jeffrey, *Dismantling Canada: Stephen Harper's New Conservative Agenda* (Montreal: McGill-Queen's University Press, 2015), 53.

31 *reducing their engagement on climate change* Mike De Souza, "Climate-Change Scientists Feel 'Muzzled' by Ottawa: Documents," *CanWest News Service*, March 15, 2010, http://www.montrealgazette.com/news/Climate+change+scientists+feel+muzzled+Ottawa+Documents/2684065/story.html.

31 *feeling the thumb* Emily Chung, "Unmuzzle Scientists, Federal Leaders Urged," *CBC News*, April 26, 2011, http://www.cbc.ca/news/technology/unmuzzle-scientists-federal-leaders-urged-1.1065080.

31 *needed to do more to raise public awareness* Paul McLeod, "Union: Scientists Scared to Talk about DFO Cuts," *Chronicle Herald* (Halifax), December 14, 2011, http://thechronicleherald.ca/canada/42676-union-scientists-scared-talk-about-dfo-cuts.

31 *they held a mock funeral* Scott Findlay, Katie Gibbs, and Beth Wood, "Marking the Death of Evidence: A Casualty in the War on Science," The Death of Evidence, accessed February 2, 2016, http://www.deathofevidence.ca/media.

31 *"You can't have a functioning democracy"* Katie Gibbs, interview by author, July 28, 2015.

31 *without censure or retaliation* Professional Institute of the Public Service of Canada, "The Big Chill."

31 *were discarded in Dumpsters* Margaret Munro, "Last Chapter for Many Environment Canada Libraries," Canada.com, January 9, 2014, http://o.canada.com/news/last-chapter-for-many-environment-canada-libraries.

32 *making it impossible to track* Lynk, "Science under Siege."

32 *governments to ban Wi-Fi* "Ongoing Efforts Against EMF Pseudoscience," Bad Science

Watch, accessed February 1, 2016, http://www.badsciencewatch.ca/projects/ongoing
-efforts-against-emf-pseudoscience/.

32 *"the media has been"* "Investigation of Anti-WiFi Activism in Canada," Bad Science
Watch, accessed February 1, 2016, http://www.badsciencewatch.ca/projects/investigation
-of-anti-wifi-activism-in-canada/.

32 *"The climate change argument is absolute crap"* "Tony Abbott Joins the 7.30 Report," by
Kerry O'Brien, Australian Broadcasting Corporation, February 2, 2010, http://www.abc
.net.au/7.30/content/2010/s2808321.htm.

32 *a post that had existed since 1931* "Australia: Science Taking a Hit Down Under from
New Administration," *ASLO Aquatic Science Policy Report: November 2013*, accessed
February 1, 2016, http://aslo.org/pipermail/pan/2013/000066.html.

32 *"There will be no carbon tax"* Julia Gillard, quoted in Julia Baird, "A Carbon Tax's Ignoble
End," *New York Times*, July 24, 2014, http://www.nytimes.com/2014/07/25/opinion/julia
-baird-why-tony-abbott-axed-australias-carbon-tax.html.

33 *whose articles ran 82 percent negative* Wendy Bacon, *A Sceptical Climate: Media Coverage
of Climate Change in Australia 2011* (Sydney: Australian Centre for Independent Journalism,
2011), http://www.abc.net.au/mediawatch/transcripts/acij_011211.pdf.

33 *a model for other nations* Rob Taylor and Rhiannon Hoyle, "Australia Becomes First
Developed Nation to Repeal Carbon Tax," *Wall Street Journal*, July 17, 2014, http://www
.wsj.com/articles/australia-repeals-carbon-tax-1405560964.

33 *secular mental-health workers were excluded* Matthew Knott and James Massola, "Tony
Abbott to Keep Secular Workers out of School Chaplaincy Program," *Sydney Morning Herald*,
August 27, 2014, http://www.smh.com.au/federal-politics/political-news/tony-abbott-to
-keep-secular-workers-out-of-school-chaplaincy-program-20140827-1091u0.html.

33 *"It's like sending in clowns"* Lawrence Krauss, interview by author, June 25, 2015.

33 *"one of 10 great public health achievements"* "Ten Great Public Health Achievements in
the 20th Century," US Centers for Disease Control and Prevention, accessed June 28, 2015,
http://www.cdc.gov/about/history/tengpha.htm.

34 *"We understand people want to make"* Lawrence Springborg, quoted in Matt Wordsworth,
"To Fluoride or Not to Fluoride: The Water Treatment Question," Australian Broadcasting
Corporation, February 27, 2013, http://www.abc.net.au/7.30/content/2013/s3699925.htm.

34 *"concerns over the localized impacts"* Greg Hunt, quoted in Jason Scott, "Never Mind
the Pope, Australia Plans Czar to Police Windfarms," *Bloomberg Business*, June 18, 2015,
http://www.bloomberg.com/news/articles/2015-06-19/never-mind-the-pope-australia
-plans-czar-to-police-windfarms.

34 *called the mineral a "toxin"* Ken Perrott, "Fluoridation—The IQ Myth," *Open Parachute*
(blog), September 18, 2013, accessed October 20, 2015, https://openparachute.wordpress.
com/2013/09/18/fluoridation-the-iq-myth/.

35 *propaganda piece that linked fluoride* Steven Novella, "Anti-Fluoride Propaganda as
News," *NeuroLogica Blog*, July 27, 2012, accessed October 20, 2015, http://theness.com/
neurologicablog/index.php/anti-fluoride-propaganda-as-news/.

35 *the conservative cities of Calgary, Alberta* "Dental Decay Rampant in Calgary Children,

Pediatric Dentist Says," *CBC News*, December 8, 2014, http://www.cbc.ca/news/canada/calgary/dental-decay-rampant-in-calgary-children-pediatric-dentist-says-1.2864413.

35 *and Wichita, Kansas* Dion Lefler and Annie Calovich, "Wichita Voters Reject Fluoridated Water," *Wichita Eagle*, November 7, 2012, http://www.kansas.com/news/article1102401.html.

35 *dozens of other, smaller cities* "Communities That Have Rejected Fluoridation Since 2010," Fluoride Action Network, accessed October 20, 2015, http://fluoridealert.org/content/communities_2010/.

35 *the city council in Dublin, Ireland* Aoife Barry, "Dublin City Councillors Vote against Fluoride," *TheJournal.ie*, October 7, 2014, http://www.thejournal.ie/dublin-city-council-vote-fluoride-1710194-Oct2014/.

35 *criticisms of medical associations* Justin Jalil, "Israel to Discontinue Fluoridation of Tap Water," *Times of Israel*, August 25, 2014, http://www.timesofisrael.com/israel-to-discontinue-fluoridation-of-tap-water/.

35 *"There is a danger"* Tony Blair, quoted in "UK 'Developing Anti-science Culture,'" *BBC News*, November 17, 2000, http://news.bbc.co.uk/2/hi/science/nature/1028911.stm.

35 *a process called mutagenesis* Per Sikora et al., "Mutagenesis as a Tool in Plant Genetics, Functional Genomics, and Breeding," *International Journal of Plant Genomics*, December 15, 2011, http://www.hindawi.com/journals/ijpg/2011/314829/.

36 *mutagenesis can, however, be labeled* "GMOs vs. Mutagenesis vs. Conventional Breeding: Which Wins?," *Genetic Literacy Project*, December 3, 2013, accessed October 20, 2015, http://www.geneticliteracyproject.org/2013/12/03/gmos-vs-mutagenesis-vs-conventional-breeding-which-wins/.

36 *"condemning Europe to a new Dark Age"* Andrew Ward, "Brussels Rejects 'Anti Science' Criticism from UK Conservatives," *Financial Times*, January 14, 2014, http://www.ft.com/cms/s/0/8eeb54fe-7d34-11e3-81dd-00144feabdc0.html#axzz3d8z0BtVH.

36 *"We must be bold"* Xi Jinping, quoted in "Xi's Remarks on GMO Signal Caution," *Wall Street Journal China*, October 9, 2014, http://blogs.wsj.com/chinarealtime/2014/10/09/xis-remarks-on-gmo-signal-caution/.

36 *"dodgy GM food"* "Food Fight: A Fierce Public Debate over GM Food Exposes Concerns about America," *Economist*, December 14, 2013, http://www.economist.com/news/china/21591577-fierce-public-debate-over-gm-food-exposes-concerns-about-america-food-fight.

36 *blindness due to vitamin A deficiency* Jane Qui, "China Sacks Officials over Golden Rice Controversy," *Nature*, December 10, 2012, http://www.nature.com/news/china-sacks-officials-over-golden-rice-controversy-1.11998.

36 *"we are heading for an environmental disaster"* "Global Trends 2014: Environment," Ipsos MORI, accessed October 21, 2015, http://www.ipsosglobaltrends.com/environment.html.

37 *Chinese officials tried to cover up* Richard Spencer and Peter Foster, "Chinese Ordered Cover-Up of Tainted Milk Scandal," *Telegraph*, September 24, 2008, http://www.telegraph.co.uk/news/worldnews/asia/china/3074986/Chinese-ordered-cover-up-of-tainted-milk-scandal.html.

37 *lead paint on wooden baby toys* Eric S. Lipton and David Barboza, "As More Toys Are

Recalled, Trail Ends in China," *New York Times*, June 19, 2007, http://www.nytimes.com/2007/06/19/business/worldbusiness/19toys.html.

37 *baking powder that contained the metal* Peter Foster, "Top 10 Chinese Food Scandals," *Telegraph*, April 27, 2011, http://www.telegraph.co.uk/news/worldnews/asia/china/8476080/Top-10-Chinese-Food-Scandals.html.

38 *often used by journalists to describe the pills* Hannah Groch-Begley, "This Conservative Myth about Birth Control Could Sway a Supreme Court Case," Media Matters for America, March 26, 2014, http://mediamatters.org/blog/2014/03/26/this-conservative-myth-about-birth-control-coul/198623.

39 *"It's troubling that"* Steven Salzberg, "Supreme Court Gets Decision Right, Science Wrong, on Gene Patents," *Forbes*, June 13, 2013, http://www.forbes.com/sites/stevensalzberg/2013/06/13/supreme-court-gets-decision-right-science-wrong/.

39 *climate has always been changing* Chris Christie, "Morning Joe Mix: Tuesday, December 1," *Morning Joe Chow*, MSNBC, December 1, 2015. http://www.msnbc.com/morning-joe/watch/morning-joe-mix-tuesday-december-1-577268291522.

39 *"For those of us"* Bernie Sanders's Twitter page, July 10, 2015, accessed October 20, 2015, https://twitter.com/berniesanders/status/619640256371929088. A 2015 Pew Research Center and Association for the Advancement of Science (AAAS) study on science literacy found that 88 percent of AAAS scientists believe genetically modified foods are safe, while 87 percent of AAAS scientists believe climate change is mostly due to human activity. Ninety-eight percent of AAAS scientists believe humans have evolved over time, making evolution the strongest consensus in science. (Cary Funk and Lee Raine, "Public and Scientists' Views on Science and Society," Pew Research Center, January 29, 2015, http://www.pewinternet.org/2015/01/29/chapter-3-attitudes-and-beliefs-on-science-and-technology-topics/.) Among the subset of climate scientists, 97 percent of published and peer-reviewed papers that expressed an opinion say that humans are causing global warming. The most influential study on this is John Cook et al., "Quantifying the Consensus on Anthropogenic Global Warming in the Scientific Literature," *Environmental Research Letters* 8, no. 2 (2013), http://iopscience.iop.org/article/10.1088/1748-9326/8/2/024024.

40 *fight any international agreement* Andrew Restuccia, "GOP to Attack Climate Pact at Home and Abroad," *Politico*, September 7, 2015, http://www.politico.com/story/2015/09/gop-congress-climate-pact-paris-213382.

40 *"our international partners"* Mitch McConnell, "McConnell Statement on Obama Administration International Climate Plan," Office of Senator Mitch McConnell, March 31, 2015, accessed October 31, 2015, http://www.mcconnell.senate.gov/public/index.cfm?p=PressReleases&ContentRecord_id=9d042b43-cbc5-4ffb-8cca-2e4125f769d2&ContentType_id=c19bc7a5-2bb9-4a73-b2ab-3c1b5191a72b&Group_id=0fd6ddca-6a05-4b26-8710-a0b7b59a8f1f.

40 *"The speed at which Republicans"* Jonathan Chait, "The Republican Plot to Destroy an International Climate Agreement," *New York*, September 10, 2015, http://nymag.com/daily/intelligencer/2015/09/gop-plot-to-destroy-climate-agreement.html#.

40 *"In one fell swoop"* Eddie Bernice Johnson, quoted in Katherine Bagley, "Congresswoman

Defends NOAA Scientists from Lamar Smith 'Witch Hunt,'" *InsideClimate News*, November 23, 2015, http://insideclimatenews.org/news/22112015/ranking-congresswoman -defends-NOAA-climate-change-scientists-from-witch-hunt-lamar-smith.

41   *investigated the National Science Foundation*   Jeffrey Mervis, "Battle between NSF and House Science Committee Escalates: How Did It Get This Bad?," *Science*, October 2, 2014, http://news.sciencemag.org/policy/2014/10/battle-between-nsf-and-house-science -committee-escalates-how-did-it-get-bad.

41   *and the Environmental Protection Agency*   Lamar Smith, "Committee Investigation into EPA's Secret Science," US House Committee on Science, Space, and Technology, August 1, 2013, https://science.house.gov/issues/committee-investigation-epas-secret-science.

41   *fossil fuel industry orchestrated a cover-up*   Warren Cornwall, "Climate Scientist Requesting Federal Investigation Feels Heat from House Republicans," *Science*, October 5, 2015, http://news.sciencemag.org/policy/2015/10/turnabout-house-republicans-say-they-ll -investigate-climate-scientist-requesting.

41   *"on the ongoing debate"*   Ted Cruz's Facebook page, December 8, 2015, accessed February 1, 2016, https://www.facebook.com/SenatorTedCruz/posts/825569564222012. See also "Data or Dogma? Promoting Open Inquiry in the Debate over the Magnitude of Human Impact on Earth's Climate," US Senate Subcommittee on Space, Science, and Competitiveness, December 8, 2015, http://www.commerce.senate.gov/public/index.cfm/hearings?ID =CA2ABC55-B1E8-4B7A-AF38-34821F6468F7.

Chapter 2. The Politics of Science

43   *"There is nothing"*   George Washington, "First Annual Message to Congress (January 8, 1790)," Miller Center of Public Affairs, University of Virginia, accessed February 1, 2016, http://millercenter.org/scripps/archive/speeches/detail/3448.

43   *"Knowledge and power go hand in hand"*   Bacon, *Novum Organum*.

44   *"The proposition that the sun"*   "The Crime of Galileo: Indictment and Abjuration of 1633," *Internet Modern History Sourcebook*, Fordham University, January 1999, accessed February 1, 2016, http://legacy.fordham.edu/halsall/mod/1630galileo.asp.

45   *we have been able to double our life spans*   Average life expectancy for an American born in 1789 was 36.5 years. For those born in 2013, it is 78.8 years. See J. David Hacker, "Decennial Life Tables for the White Population of the United States, 1790–1900," *Historical Methods* 43, no. 2 (April 2010): 45–79, http://www.ncbi.nlm.nih.gov/pmc/ articles/PMC2885717/, compared to "Deaths: Final Data for 2013," *National Vital Statistics Reports* 64, no. 2 (February 16, 2016), http://www.cdc.gov/nchs/data/nvsr/ nvsr64/nvsr64_02.pdf.

45   *boost the productivity of our farms*   USDA Economic Research Service, "Farming's Role in the Rural Economy," *Agricultural Outlook* (June–July 2000): 19–22. In 1820, 70 percent of the US population lived on farms. By 1990, that number had fallen to 2 percent, meaning that productivity is thirty-five times greater.

45   *"a war ... of every man"*   Thomas Hobbes, *Leviathan; or, the Matter, Form, and Power of a Commonwealth Ecclesiastical and Civil*, 3rd ed. (London: George Routledge and Sons, 1887).

47   *Earth is about 4.54 billion years old*   William L. Newman, *Geologic Time*, (Reston, VA: USGS Information Services, United States Geological Survey, 1997), accessed February 1, 2016, http://pubs.usgs.gov/gip/geotime/contents.html.

48   *"My dear Kepler"*   Galileo Galilei, quoted in Giorgio de Santillana, *The Crime of Galileo* (Chicago: University of Chicago Press, 1955).

48   *"The anatomist showed"*   Galileo Galilei, *Dialogue Concerning the Two Chief World Systems*, trans. Stillman Drake (New York: Modern Library, 2001), 108.

49   *is it perhaps a life at quickening*   William Blackstone, *Commentaries on the Laws of England: A Facsimile of the First Edition of 1765–1769* (Chicago: University of Chicago Press, 1979), 388, http://press-pubs.uchicago.edu/founders/documents/amendIXs1.html.

49   *one-third to one-half of fertilized eggs*   Rachel Benson Gold, "The Implications of Defining When a Woman Is Pregnant," *Guttmacher Report on Public Policy* 8, no. 2 (May 2005), https://www.guttmacher.org/pubs/tgr/08/2/gr080207.html. The foundational science on this was done by Allen Wilcox, et al, in 1988, which showed a 22 percent implantation failure rate followed by a 31 percent rate of loss after implantation. (Allen Wilcox et al., "Incidence of Early Loss of Pregnancy," *New England Journal of Medicine* 319 (July 28, 1988): 189-194, http://www.nejm.org/doi/full/10.1056/NEJM198807283190401.) Further research by Wilcox et al. in 1999 showed that there is a broad range of days between ovulation and implantation, from six to eighteen days, with the optimal day for implantation being the ninth day. (Allen Wilcox et al., "Time of Implantation of the Conceptus and Loss of Pregnancy," *New England Journal of Medicine* 340 (June 10, 1999): 1796-1799, http://www.nejm.org/doi/full/10.1056/NEJM199906103402304.) More recent research from IVF showed that half of fertilized embryos failed to implant, and half of implanted embryos failed to establish an ongoing pregnancy. (Y. E. M. Koot et al., "Molecular Aspects of Implantation Failure," *Biochimica et Biophysica Acta (BBA): Molecular Basis of Disease* 1822, no. 12 (2012): 1943–1950, http://www.sciencedirect.com/science/article/pii/S0925443912001330.) Thus, the failure rate for fertilized eggs is very high, on the order of 75 percent. Most of these failures occur without a woman's knowledge.

50   *as scientists at the J. Craig Venter Institute*   "First Self-Replicating Synthetic Bacterial Cell," news release, J. Craig Venter Institute, May 20, 2010, accessed February 1, 2016, http://www.jcvi.org/cms/press/press-releases/full-text/article/first-self-replicating-synthetic-bacterial-cell-constructed-by-j-craig-venter-institute-researcher/home/.

51   *In fact, there are four*   Timothy Ferris, *The Science of Liberty* (New York: Harper, 2010).

53   *"And say, finally"*   Thomas Jefferson, "Letter to James Madison, December 20, 1787," in *The Writings of Thomas Jefferson: Correspondence*, ed. Henry Augustine Washington (New York: Derby and Jackson, 1859), 332.

Chapter 3. Religion, Meet Science

57   *"The value of science"*   Thomas Jefferson, "Memorial on the Book Duty," in *Thomas Jefferson: Writings*, ed. M. D. Peterson (New York: Library of America, 1984).

58   *rather, a natural outgrowth of it*   Ferris, *The Science of Liberty*.

58   *and that it provided a plan for living*   I. Bernard Cohen, ed., *Puritanism and the Rise of Modern Science: The Merton Thesis* (New Brunswick, NJ: Rutgers University Press, 1990).

58   *Science was the "handmaiden" to theology*   George Becker, "Pietism's Confrontation with Enlightenment Rationalism: An Examination of the Relation Between Ascetic Protestantism and Science," *Journal for the Scientific Study of Religion* 30, no. 2 (1991): 139–158.

58   *"the vast library of creation"*   Charles Raven, *John Ray, Naturalist: His Life and Works* (Cambridge, UK: Cambridge University Press, 1942).

59   *The Mu'tazilites' primary ethos*   Jim al-Khalili, *The House of Wisdom: How Arabic Science Saved Ancient Knowledge and Gave Us the Renaissance* (New York: Penguin Press, 2011). ·

59   *"what be the very grounds"*   Christopher St. Germain, *The Doctor and Student*, ed. William Muchall (Cincinnati: R. Clarke, 1874).

60   *Coke was a Puritan sympathizer*   Allen Boyer, *Sir Edward Coke and the Elizabethan Age* (Stanford, CA: Stanford University Press, 2003), 65.

60   *"The law of nature is"*   Sir Edward Coke, "Calvin's Case, or the Case of the Postnati," in *The Selected Writings and Speeches of Sir Edward Coke*, ed. Steve Sheppard (Indianapolis, IN: Liberty Fund, 2003).

60   *"royal prerogative"*   Sir Edward Coke, "The First Part of the Institutes of the Lawes of England; or, a Commentary upon Littleton, Not the Name of the Author Only, but of the Law It Selfe," in *The Selected Writings and Speeches of Sir Edward Coke*, ed. Steve Sheppard (Indianapolis, IN: Liberty Fund, 2003).

61   *"I could recount what I have seen"*   John Milton, *Areopagitica: A Speech for the Liberty of Unlicensed Printing to the Parliament of England, 1644* (Salt Lake City: Project Gutenberg Literary Archive Foundation, 2013), accessed February 1, 2016, http://www.gutenberg .org/files/608/608-h/608-h.htm.

61   *"that there had been men"*   Margaret C. Jacob, *Scientific Culture and the Making of the Industrial West* (New York: Oxford University Press, 1997), 28.

62   *Arabic for "completion"*   The word "algebra" derives from the title of the book *Kitab al muhtasar fi hisab al gabr w'al muqubalah* by the Persian mathematician and astronomer Muhammad ibn Musa al-Khwārizmī, who taught in Baghdad. The book's title has been translated as "*A Compact Introduction to Calculation Using Rules of Completion and Reduction.*" *Al-gabr* or *al-jabr* ("completion") referred to one of the two operations he used in the book to solve quadratic equations (the other being "reduction"). The Latin form of the name al-Khwārizmī is Algoritmi, the root of the word "algorithm."

62   *geography, mathematics, medicine, and zoology*   Al-Khalili, *The House of Wisdom.*

62   *in the late eleventh century*   Ibid.

63   *"[We should begin] our investigation"*   Ibn al-Haytham, *Optics*, trans. Abdelhamid Ibrahim Sabra (London: Warburg Institute, 1989), 6.

64   *"The one who enacts the burning"*   Michael E. Marmura, trans., *The Incoherence of the Philosophers by Imam Al-Ghazzali* (Provo, UT: Brigham Young University Press, 2000).

65   *his Ninety-Five Theses*   This is the common name for the document, officially titled *Disputation of Doctor Martin Luther on the Power and Efficacy of Indulgences.*

65 *"Why does not the pope"* Martin Luther, quoted in Hans J. Hillerbrand, *The Protestant Reformation*, rev. ed. (New York: Harper Perennial, 2009).

66 *he began by embracing skepticism* René Descartes, *Discourse on the Method of Rightly Conducting One's Reason and of Seeking Truth in the Sciences* (Salt Lake City: Project Gutenberg Literary Archive Foundation, 2008), accessed February 1, 2016, http://www .gutenberg.org/ebooks/59.

67 *All men are mortal* Aristotle, *Prior Analytics* (a part of *Organon*), trans. A. J. Jenkinson (Internet Classics Archive), accessed February 1, 2016, http://classics.mit.edu/Aristotle/ prior.mb.txt.

67 *disdain for experimentation* Bacon, *Novum Organum.*

69 *they were called "natural philosophers"* Sydney Ross, "Scientist: The Story of a Word," *Annals of Science* 18, no. 2 (1962): 65–85, accessed February 1, 2016, http://www.scribd .com/doc/42338381/Ross-1964-Scientist-the-Story-of-a-Word.

69 *sometimes he used it to refer* Thomas Jefferson, "Letter to Peter Carr, September 7, 1814," in *68 Letters to and from Jefferson, 1805–1817*, University of Virginia Library, accessed February 1, 2016, http://etext.lib.virginia.edu/toc/modeng/public/Jef1Gri.html.

69 *62 percent of the natural philosophers* Robert K. Merton, *Science, Technology and Society in Seventeenth-Century England* (New York: Howard Fertig, 1970).

69 *Royal Society of London were Puritans* Cohen, *Puritanism and the Rise of Modern Science: The Merton Thesis.*

69 *work on light and refraction* Abdelhamid Ibrahim Sabra, "Explanation of Optical Reflection and Refraction: Ibn-al-Haytham, Descartes, Newton," *Proceedings of the 10th International Congress of the History of Science (Ithaca, 1962)*, vol. 1 (Paris: Hermann, 1964), 551–554.

69 *wrote far more on religion and alchemy* Michael White, *Isaac Newton: The Last Sorcerer* (Reading, MA: Addison-Wesley, 1997).

69 *apocalypse would come in 2060* "The World Will End in 2060, According to Newton," *London Evening Standard*, June 19, 2007, http://www.standard.co.uk/news/the-world-will -end-in-2060-according-to-newton-7254673.html.

69 *"not the first of the age of reason"* John Maynard Keynes, quoted in White, *Isaac Newton: The Last Sorcerer.*

69 *dwarfing his scientific work* White, *Isaac Newton: The Last Sorcerer.*

69 *Mathematical Principles of Natural Philosophy* Isaac Newton, *Isaac Newton's Philosophiae Naturalis Principia Mathematica*, 3rd ed., eds. I. Bernard Cohen and Alexandre Koyré (Cambridge, UK: Cambridge University Press, 1972).

70 *barometric pressure and wind speed* Ferris, *The Science of Liberty.*

70 *archeology, paleontology, and civil engineering* Silvio A. Bedini, *Jefferson and Science* (Charlottesville, VA: Thomas Jefferson Foundation, 2002).

70 *frustrated by the cloudy conditions* Kevin J. Hayes, *The Road to Monticello: The Life and Mind of Thomas Jefferson* (Oxford, UK: Oxford University Press, 2008).

70 *conduct it as a scientific expedition* Thomas Jefferson, "Letter to Meriwether Lewis,

June 20, 1803," Monticello.org, accessed February 1, 2016, http://www.monticello.org/site/jefferson/jeffersons-instructions-to-meriwether-lewis.

70 *"Science is my passion, politics my duty"* Thomas Jefferson, quoted in Silvio A. Bedini, *Thomas Jefferson, Statesman of Science* (New York: Macmillan, 1990).

70 *"Never did a prisoner"* Thomas Jefferson, "Letter to Pierre Samuel du Pont de Nemours, 1809," in *Correspondence between Thomas Jefferson and Pierre Samuel du Pont de Nemours 1798–1817*, ed. Dumas Malone (Cambridge, MA: Da Capo Press, 1970).

71 *he had studied as a law student* Thomas Jefferson, "Letter to James Madison, February 17, 1826," in *The Writings of Thomas Jefferson*, vol. 16, ed. Albert Ellery Bergh (Washington, DC: Thomas Jefferson Memorial Association of the United States, 1907), 155.

71 *the College of William and Mary* Thomas Jefferson, *Thomas Jefferson: Writings*, ed. Merrill D. Peterson (New York: Library of America, 1984).

71 *"my trinity of the three greatest men"* Thomas Jefferson to Benjamin Rush, January 16, 1811, Library of Congress, https://www.loc.gov/item/mtjbib020491/.

71 *"in all sorts of reasoning"* John Locke, *An Essay Concerning Human Understanding* (Oxford, UK: Clarendon Press, 1964).

72 *strove to answer that question* Ibid.

73 *"Whatever comes short"* Ibid.

74 *Drawing on Locke* Benjamin Franklin, *The Writings*, vol. 10, ed. Albert H. Smyth (New York: Macmillan, 1907), 148.

74 *antiauthoritarian words "self-evident"* Thomas Jefferson et al., "Declaration of Independence Rough Draft with Edits by Franklin and Adams," *Declaring Independence: Drafting the Documents*, Library of Congress, July 23, 2010, accessed February 1, 2016, http://www.loc.gov/exhibits/declara/images/draft1.jpg.

74 *"The idea of a supreme being"* Locke, *An Essay Concerning Human Understanding*.

75 *"By liberty, then"* David Hume, "An Enquiry Concerning Human Understanding," in *English Philosophers of the Seventeenth and Eighteenth Centuries: Locke, Berkeley, Hume, with Introductions, Notes and Illustrations* (New York: P. F. Collier and Son, 1910).

75 *"A man may imagine"* Isaac Newton, *Sir Isaac Newton: Theological Manuscripts*, ed. H. McLachlan (Liverpool, UK: Liverpool University Press, 1950).

75 *"Whatever definition we may give"* Hume, "An Enquiry Concerning Human Understanding."

Chapter 4. Science, Meet Freedom

78 *"Individuals in reality"* Simon Levin, interview by author, November 21, 2010.

78 *"The law of nature"* Coke, "Calvin's Case, or the Case of the Postnati."

79 *"by analogy with artist"* Laura J. Snyder, *The Philosophical Breakfast Club: Four Remarkable Friends Who Transformed Science and Changed the World* (New York: Broadway Books, 2011), 165.

79 *Ferris shows how powerfully* Ferris, *The Science of Liberty*.

81 *"contempt for tradition and for forms"* Alexis de Tocqueville, *Democracy in America*, ed. Francis Bowen, trans. Henry Reeve (Cambridge, MA: Sever and Francis, 1864).

83   *the arguments by economist Mariana Mazzucato*   Mariana Mazzucato, *The Entrepreneurial State: Debunking Public vs. Private Sector Myths* (New York: Anthem Press, 2013).

84   *"reality has a well-known liberal bias"*   Stephen Colbert, quoted in Daniel Kurtzman, "Stephen Colbert at the White House Correspondents' Dinner," About.com, April 30, 2006, http://politicalhumor.about.com/od/stephencolbert/a/colbertbush.htm.

84   *55 percent of US scientists*   Scott Keeter, "Scientific Achievements Less Prominent Than a Decade Ago," news release, Pew Research Center, July 9, 2009, http://people-press.org/files/legacy-pdf/528.pdf.

85   *"We have many people even here"*   R. A. Millikan, "Science and Society," *Science* 58, no. 1503 (October 19, 1923): 293–298, http://science.sciencemag.org/content/58/1503/293.

85   *"it is absurd for a scientist"*   A. S. Hitchcock, "Remarks on the Scientific Attitude," *Science* 59, no. 1535 (May 30, 1924): 476–477, http://science.sciencemag.org/content/59/1535/476.

86   *famous for his athletic prowess*   Gale E. Christianson, *Edwin Hubble: Mariner of the Nebulae* (Chicago: University of Chicago Press, 1995).

86   *"gentlemen scientists" than professionals*   Ibid.

86   *steel magnate Andrew Carnegie*   David Nasaw, *Andrew Carnegie* (New York: Penguin Press, 2006).

86   *John D. Rockefeller Sr.*   Ron Chernow, *Titan: The Life of John D. Rockefeller, Sr.* (New York: Vintage Books, 1998).

86   *on the side while at Oxford*   Christianson, *Edwin Hubble.*

87   *only to suffer more personal attacks*   Walter Isaacson, *Einstein: His Life and Universe* (New York: Simon & Schuster, 2007).

87   *he decided to leave Berlin*   "Einstein to Leave Berlin: Is Aroused by Unfair Attacks on Relativity Theory," *New York Times*, August 29, 1920.

87   *"This world is a strange madhouse"*   Albert Einstein, quoted in Jeroen van Dongen, "On Einstein's Opponents, and Other Crackpots," *Studies in History and Philosophy of Science Part B: Studies in History and Philosophy of Modern Physics* 41, no. 1 (2010): 78–80.

88   *he worked behind the scenes*   Isaacson, *Einstein.*

89   *fire departments across the country*   Carey McWilliams, "Sunlight in My Soul," in *The Aspirin Age, 1919–1941*, ed. Isabel Leighton (New York: Simon & Schuster, 1963).

89   *against the teaching of evolution*   William Jennings Bryan, "Mr. Bryan on Evolution," Institute for Creation Research, accessed February 1, 2016, http://www.icr.org/article/mr-bryan-evolution.

89   *"It seems to me absurd"*   Charles Darwin, "To John Fordyce, 7 May 1879," Darwin Correspondence Project, University of Cambridge, accessed February 1, 2016, http://www.darwinproject.ac.uk/entry-12041.

89   *"I am sorry"*   Charles Darwin, quoted in Michael Casey, "Darwin Letter Reveals, 'I Do Not Believe in the Bible,'" *CBS News*, September 21, 2015, accessed October 21, 2015, http://www.cbsnews.com/news/darwin-letter-i-do-not-believe-in-the-bible/.

89   *The most notable of these laws*   Tennessee State Legislature, *Tennessee Anti-Evolution Statutes, Public Acts of the State of Tennessee Passed by the 64th General Assembly 1925,*

Chapter No. 27, House Bill No. 185, March 21, 1925, accessed February 1, 2016, http://law2.umkc.edu/faculty/projects/ftrials/scopes/tennstat.htm.

89   *social Darwinism had corrupted students' morality*   Matthew Avery Sutton, *Aimee Semple McPherson and the Resurrection of Christian America* (Cambridge, MA: Harvard University Press, 2009).

89   *"the greatest triumph of satanic intelligence"*   Aimee Semple McPherson, quoted in Sutton, *Aimee Semple McPherson and the Resurrection of Christian America*.

90   *"If you want your kids to learn"*   Kenneth Miller, interview by author, August 18, 2010.

90   *being called "Major Hubble"*   Christianson, *Edwin Hubble*.

91   *Leavitt was paid*   George Johnson, *Miss Leavitt's Stars: The Untold Story of the Woman Who Discovered How to Measure the Universe* (New York: W. W. Norton, 2005).

92   *"overthrow" of the heliocentric universe*   Harlow Shapley, *Through Rugged Ways to the Stars* (New York: Scribner, 1969).

92   *He took out his handkerchief*   Christianson, *Edwin Hubble*.

94   *it is recognized*   Clifford Pickover, *Archimedes to Hawking: Laws of Science and the Great Minds Behind Them* (New York: Oxford University Press, 2008).

95   *crowds of five thousand people*   Christianson, *Edwin Hubble*.

95   *one of the most famous men*   Allan Sandage, interview by author, August 3, 2004.

95   *"didn't talk to other astronomers"*   Ibid.

96   *Humason's special attentions*   Ibid.

96   *colluded with other Republican scientists*   Ibid.

96   *"I am used to seeing you earn"*   Elmer Davis to Edwin Hubble, November 23, 1951, Huntington Library Collection, HUB Box 10, f. 238, Henry Huntington Library, San Marino, CA.

Chapter 5. Gimme Shelter

97   *"Turning and turning in the widening gyre"*   William Butler Yeats, "The Second Coming," in *Michael Robartes and the Dancer* (Churchtown, Ireland: Cuala Press, 1921).

97   *campaign called* "Hitler über Deutschland"   Heinrich Hoffmann and Josef Berchtold, *Hitler über Deutschland* (Munich: Franz Eher Nachfolger GmbH, 1932).

98   *"Hitler's dictatorship differed"*   Albert Speer, quoted in *Nuremberg: The War Crimes Trial: Transcript*, ed. Richard Norton Taylor (London: Nick Hern Books, 1997), 55.

99   *military's adoption of new technologies*   Jerome B. Wiesner, *Vannevar Bush: 1890–1974* (Washington, DC: National Academy of Sciences, 1979), http://www.nasonline.org/publications/biographical-memoirs/memoir-pdfs/bush-vannevar.pdf.

99   *US inventors who collaborated*   Petra Moser et al., "German Jewish Émigrés and U.S. Invention," *American Economic Review* 104, no. 10 (2014), accessed February 1, 2016, http://news.stanford.edu/news/2014/august/german-jewish-inventors-081114.html.

100   *including some of his own top students*   Hans Bethe, *J. Robert Oppenheimer: 1904–1967* (Washington, DC: National Academy of Sciences, 1997).

100   *"without doubt the most concentrated"*   William L. Laurence, "Atomic Bombing of Nagasaki Told by Flight Member," *New York Times*, September 9, 1945.

101 *at first officially denied* William L. Laurence, "U.S. Atom Bomb Site Belies Tokyo Tales," *New York Times*, September 12, 1945.

101 *"The Atomic Age began"* William L. Laurence, "Drama of the Atomic Bomb Found Climax in July 16 Test," *New York Times*, September 26, 1945.

102 *"I am become Death"* J. Robert Oppenheimer, quoted in "The Decision to Drop the Bomb," *NBC White Paper*, produced by Fred Freed, aired January 5, 1965.

102 *"[W]e have made a thing"* J. Robert Oppenheimer, quoted in Kai Bird and Martin J. Sherwin, *American Prometheus* (New York: Knopf, 2005).

102 *"to let the people know"* Albert Einstein, quoted in "Atomic Education Urged by Einstein; Scientist in Plea for $200,000 to Promote New Type of Essential Thinking," *New York Times*, May 25, 1946.

103 *"The double horror of two Japanese city names"* Erwin Chargaff, *Heraclitean Fire: Sketches from a Life Before Nature* (New York: Rockefeller University Press, 1978).

103 *"Our knowledge of science"* Omar N. Bradley, "Armistice Day Speech, Boston, Massachusetts, November 10, 1948," in *The Collected Writings of General Omar N. Bradley*, vol. 1 (Washington, DC: US Government Printing Office, 1977), 584–589.

103 *Roosevelt had asked* Vannevar Bush, *Science the Endless Frontier: A Report to the President* (Washington, DC: US Government Printing Office, 1945). http://www.nsf.gov/od/lpa/nsf50/vbush1945.htm#letter.

104 *the creation of knowledge is boundless* Ibid.

104 *"a bucket full of sunshine"* Sean M. Maloney, *Learning to Love the Bomb: Canada's Nuclear Weapons during the Cold War* (Washington, DC: Potomac Books, 2007).

105 *"This is the narrative"* Robert Jacobs, "Atomic Kids: Duck and Cover and Atomic Alert Teach American Children How to Survive Atomic Attack," *Film & History: An Interdisciplinary Journal of Film and Television Studies* 40, no. 1 (Spring 2010): 25-44, https://www.academia.edu/1468387/Atomic_Kids_Duck_and_Cover_and_Atomic_Alert_Teach_American_Children_How_to_Survive_Atomic_Attack.

106 *mock Soviet atomic-bomb attacks* Ralph Lapp, "An Interview with Governor Val Peterson," *Bulletin of the Atomic Scientists* 9, no. 7 (1953): 237–242.

106 *"Our best life insurance"* *If the Bomb Falls: A Recorded Guide to Survival*, 1961, record album, narrated by David Wiley, Tops Records, accessed October 22, 2015, https://www.youtube.com/watch?v=qxtA7qp9208.

107 *"This generation had America's only"* Michael J. Carey, "Psychological Fallout," *The Bulletin of the Atomic Scientists* 38, no. 1 (1982).

107 *"We may be seeing that growing up"* John E. Mack, "Psychosocial Effects of the Nuclear Arms Race," *Bulletin of the Atomic Scientists* 37, no. 4 (1981).

108 *She urged adults* Joanna Santa Barbara, "Living in the Shadow: The Effects of Continual Fear," in *Nuclear War: The Search for Solutions*, ed. Thomas L. Perry and Diane DeMille (Vancouver: Physicians for Social Responsibility, 1985), 93–104.

108 *"While children were supposedly being trained"* Jacobs, *Atomic Kids*.

108 *lead to post-traumatic stress disorder* R. M. Birn et al., "Childhood Maltreatment and Combat Posttraumatic Stress Differentially Predict Fear-Related Fronto-Subcortical

Connectivity," *Depress Anxiety* 31, no. 10 (2014): 880–92, http://www.ncbi.nlm.nih.gov/pubmed/25132653.

109 *effectively rewire brain pathways* Jonathan E. Sherin and Charles B. Nemeroff, "Post-Traumatic Stress Disorder: The Neurobiological Impact of Psychological Trauma," *Dialogues in Clinical Neuroscience* 13, no. 3 (2011): 263–278, http://www.ncbi.nlm.nih.gov/pmc/articles/PMC3182008/.

109 *Later studies showed an increased incidence* Kari Poikolainen et al., "Fear of Nuclear War Increases the Risk of Common Mental Disorders Among Young Adults: A Five-Year Follow-Up Study," *BMC Public Health* 4, no. 42 (2004), http://www.biomedcentral.com/1471-2458/4/42.

109 *a sharp increase in the adolescent suicide rate* US Centers for Disease Control and Prevention, "Suicide—United States, 1970–1980," *Morbidity and Mortality Weekly Report* 34, no. 24 (1985): 353–7.

109 *30 percent higher than other age cohorts* Julie A. Phillips et al., "Understanding Recent Changes in Suicide Rates Among the Middle-aged: Period or Cohort Effects?," *Public Health Reports* 125, no. 5 (2010): 680–688, http://www.ncbi.nlm.nih.gov/pmc/articles/PMC2925004/.

109 *twice the rate of the prior generation* Susan L. Brown and I-Fen Li, "The Gray Divorce Revolution: Rising Divorce among Middle-aged and Older Adults, 1990-2010," National Center for Family & Marriage Research, March 2013, accessed October 22, 2015, https://www.bgsu.edu/content/dam/BGSU/college-of-arts-and-sciences/NCFMR/documents/Lin/The-Gray-Divorce.pdf.

110 *"On all express city thoroughfares"* General Motors, *To New Horizons*, 1940, film, accessed October 29, 2015, https://www.youtube.com/watch?v=1cRoaPLvQx0.

110 *"dispersal"* Donald Monson, "Is Dispersal Obsolete?," *Bulletin of the Atomic Scientists* 10, no. 10 (1954): 378–383.

111 *concrete, oil, and construction companies* Robert Cantwell, *When We Were Good: The Folk Revival* (Cambridge, MA: Harvard University Press, 1996).

111 *"what was good for our country"* Charles Wilson, quoted in Justin Hyde, "GM's 'Engine Charlie' Wilson Learned to Live with a Misquote," *Detroit Free Press*, September 14, 2008, http://archive.freep.com/article/20080914/BUSINESS01/809140308/GM-s-Engine-Charlie-Wilson-learned-live-misquote.

111 *"the present national dispersion policy"* Allan M. Winkler, *Life under a Cloud: American Anxiety about the Atom* (Urbana, IL: University of Illinois Press, 1999), 117.

111 *"matching a sleeping tortoise"* Hornell Hart, "The Remedies Versus the Menace," *Bulletin of the Atomic Scientists* 10, no. 6 (1954): 197–205.

111 *"from 'Duck and Cover'"* Stephen I. Schwartz et al., "Excerpts from Atomic Audit," *Bulletin of the Atomic Scientists* 54, no. 5 (1998): 36–43.

111 *practicing clearing hundreds of city blocks* Lapp, "An Interview with Governor Val Peterson."

112 *"The struggle for federal aid"* JoAnne Brown, "'A Is for Atom, B Is for Bomb': Civil Defense in American Public Education, 1948–1963," *Journal of American History* 75, no. 1 (1998), http://www.jstor.org/pss/1889655.

113   *to protect themselves by reinvesting*   John W. Finney, "Sputnik Acts as a Spur to U.S. Science and Research," *New York Times*, November 3, 1957.

115   *"Public sentiment"*   Abraham Lincoln, quoted in John Nicolay and John Hay, *Abraham Lincoln: A History*, vol. 2 (New York: Cosimo Classics, 1917).

116   *"A good many times"*   C. P. Snow, *The Two Cultures* (Cambridge, UK: Cambridge University Press, 1960).

117   *one hundred most influential Western books*   "The Hundred Most Influential Books Since the War," *Times Literary Supplement*, December 30, 2008, http://entertainment.timesonline .co.uk/tol/arts_and_entertainment/the_tls/article5418361.ece.

118   *"[I]n holding scientific research"*   Dwight D. Eisenhower, "Farewell Address to the American People," January 17, 1961, Dwight D. Eisenhower Speeches, Eisenhower Presidential Library and Museum, https://www.eisenhower.archives.gov/all_about_ike/speeches/wav_files /farewell_address.mp3.

Chapter 6. Science, Drugs, and Rock 'n' Roll

119   *"Doubt has replaced hopefulness"*   Students for a Democratic Society, "Port Huron Statement of the Students for a Democratic Society," June 15, 1962, http://coursesa.matrix .msu.edu/~hst306/documents/huron.html.

119   *about half of the spending*   "Historical Trends in Federal R&D," American Association for the Advancement of Science, accessed July 5, 2015, http://www.aaas.org/page/ historical-trends-federal-rd.

120   *has historically been about improving health*   Ibid., comparing US basic and applied research spending by character and agency, 1976–2016.

121   *"As we grew . . . our comfort"*   Students for a Democratic Society, "Port Huron Statement of the Students for a Democratic Society."

121   *"The human race may well become extinct"*   "Bertrand Russell: Playboy Interview," *Playboy*, March 1963.

122   *"I am not that interested"*   John F. Kennedy, "Tape Recording of a Meeting between President John F. Kennedy and NASA Administrator James E. Webb," November 21, 1962, White House Meeting Tape 63, John F. Kennedy Presidential Library and Museum.

123   *"urgent national needs"*   John F. Kennedy, "Excerpt from an Address Before a Joint Session of Congress, 25 May 1961," John F. Kennedy Presidential Library and Museum, accessed February 1, 2016, http://www.jfklibrary.org/Asset-Viewer/xzw1gaeeTES6khED14P1Iw.aspx.

123   *some four hundred thousand people*   "Washington Goes to the Moon," radio broadcast, WAMU, May 24, 2001, https://www.prx.org/series/8719-washington-goes-to-the -moon.

123   *he thought he could get*   Robert Dallek, *An Unfinished Life: John F. Kennedy, 1917–1963* (Boston: Little, Brown, 2003).

124   *changed his mind again*   Roger Launius, interview by Howard McCurdy in "Washington Goes to the Moon," https://www.prx.org/series/8719-washington-goes-to-the-moon.

124   *started jumping ship*   H. McCurdy, interview in "Washington Goes to the Moon," https://www.prx.org/series/8719-washington-goes-to-the-moon.

124 *proposing to cut NASA's funding* Launius, "Washington Goes to the Moon."

124 *poverty, race relations, and the economy* Roger Launius, "Exploding the Myth of Popular Support for Project Apollo," personal blog, August 16, 2010, accessed February 1, 2016, https://launiusr.wordpress.com/2010/8/16/exploding-the-myth-of-popular-support-for-project-apollo/.

124 *The cost of funding* Norman R. Augustine et al., "Seeking a Human Spaceflight Program Worthy of a Great Nation," Review of US Human Spaceflight Plans Committee, NASA, October 22, 2009, www.nasa.gov/offices/hsf/meetings/10_22_pressconference.html.

125 *captured this growing fear* Joel N. Franklin, "The New Priesthood—The Scientific Elite and the Uses of Power," book review, *Engineering and Science* 28, no. 9 (1965): 4, http://calteches.library.caltech.edu/2383/1/books.pdf.

125 *"democracy faces its most severe test"* Ralph Lapp, *The New Priesthood: The Scientific Elite and the Uses of Power* (New York: Harper and Row, 1965).

125 *"Today there is great concern"* Ronald Reagan, "Address by Governor Ronald Reagan, Installation of President Robert Hill, Chico State College," May 20, 1967, Ronald Reagan Presidential Library, accessed February 1, 2016, http://www.reagan.utexas.edu//archives/speeches/govspeech/05201967a.htm.

125 *"poet to a generation"* Ralph J. Gleason, "Bob Dylan: Poet to a Generation," *Jazz & Pop*, December 1968, 36–37.

126 *"Everything is going to become"* Kurt Vonnegut, "American Notes: Vonnegut's Gospel," *Time*, June 29, 1970.

128 *an inhuman priority* T. A. Heppenheimer, *The Space Shuttle Decision: NASA's Search for a Reusable Space Vehicle* (Washington, DC: National Aeronautics and Space Administration, 1999).

128 *"One-fifth of the population lacks"* Ralph Abernathy, quoted by Thomas O. Paine in Roger Launius, "Managing the Unmanageable: Apollo, Space Age Management and American Social Problems," *Space Policy* 24, no. 3 (2008): 158–165, http://si-pddr.si.edu/jspui/bitstream/10088/8213/1/Launius_2008_Managing_the_unmanageable.pdf.

128 *"A society that can resolve"* Ralph Abernathy, quoted in Lucas Laursen, "@ApolloPlus40 —A Colossal Perversion," *Nature*, July 7, 2009, http://blogs.nature.com/inthefield/2009/07/apolloplus40_a_colossal_perver.html.

128 *"child's play"* Thomas O. Paine, quoted in Heppenheimer, *The Space Shuttle Decision*.

131 *A 1979 NBC/AP poll showed* Frank Newport, "Landing a Man on the Moon: The Public's View," Gallup, July 20, 1999, www.gallup.com/poll/3712/landing-man-moon-publics-view.aspx.

131 *increasing public skepticism toward science* Georgine M. Pion and Mark W. Lipsey, "Public Attitudes toward Science and Technology: What Have the Surveys Told Us?," *Public Opinion Quarterly* 45, no. 3 (1981): 303–316.

131 *"I was a child"* Carl Sagan, *The Demon-Haunted World: Science as a Candle in the Dark* (New York: Random House, 1995).

132 *"I have a foreboding of an America"* Ibid.

132 *"the greatest media work"* Stephen Jay Gould, quoted in Chris Mooney and Sheril

Kirshenbaum, *Unscientific America: How Scientific Illiteracy Threatens Our Future* (New York: Basic Books, 2009).

133   *not strong enough to justify admission*   Faye Flam, "What Should It Take to Join Science's Most Exclusive Club?," *Science* 256, no. 5059 (May 15, 1992): 960–961, http://science .sciencemag.org/content/256/5059/960.

133   *"I can just see them"*   Stanley Miller, quoted in Beth Luey, "Are Fame and Fortune the Kiss of Death?," *Publishing Research Quarterly* 19, no. 3 (2007): 35–44.

133   *was considered inversely proportional*   Jim Hartz and Rick Chappell, *Worlds Apart: How the Distance between Science and Journalism Threatens America's Future* (Nashville, TN: Freedom Forum First Amendment Center, 1997).

133   *a total of five hundred scientific papers*   Michael B. Shermer, "Stephen Jay Gould as Historian of Science and Scientific Historian, Popular Scientist and Scientific Popularizer," *Social Studies of Science* 32, no. 4 (2002): 489–525.

133   *scientists who engage the public*   Pablo Jensen et al., "Scientists Who Engage with Society Perform Better Academically," *Science and Public Policy* 35, no. 7 (2008): 527–541.

133   *that number had fallen to 27 percent*   Keeter, "Scientific Achievements Less Prominent Than a Decade Ago."

134   *the broad use of the herbicide glyphosate*   Carey Gillam, "CORRECTED—Herbicide Scrutiny Mounts as Resistant Weeds Spread in U.S.," Reuters, September 28, 2015, http:// www.reuters.com/article/2015/09/28/agriculture-glyphosate-idUSL1N11S1Y220150928.

134   *70 percent of rainfall samples*   W. A Battaglin et al., "Glyphosate and Its Degradation Product AMPA Occur Frequently and Widely in U.S. Soils, Surface Water, Groundwater, and Precipitation," *Journal of the American Water Resources Association* 50, no. 2 (2014): 275–290, http://onlinelibrary.wiley.com/doi/10.1111/jawr.12159/abstract.

135   *the largest opinion difference*   Funk and Rainie, "Public and Scientists' Views on Science and Society."

136   *a series of stories about electromagnetic pollution*   Paul Brodeur, "Annals of Radiation," *New Yorker*, June 12, 1989–December 7, 1992, http://www.newyorker.com/magazine/ annals-of-radiation.

136   *Brodeur's popular 1993 book*   Paul Brodeur, *The Great Power-Line Cover-Up: How the Utilities and the Government Are Trying to Hide the Cancer Hazard Posed by Electromagnetic Fields* (Boston: Little, Brown, 1993).

136   *broad epidemiological studies*   J. A. Dennis et al., "Epidemiological Studies of Exposures to Electromagnetic Fields: II. Cancer," *Journal of Radiological Protection* 11, no. 1 (1991): 13–25.

136   *"She held it against her head"*   David Reynard, quoted in Robert L. Park, "Cellular Telephones and Cancer: How Should Science Respond?," *Journal of the National Cancer Institute* 93, no. 3 (2001): 166–167, http://jnci.oxfordjournals.org/content/93/3/166.full.

136   *no links at all*   Christopher Johansen et al., "Cellular Telephones and Cancer—A Nationwide Cohort Study in Denmark," *Journal of the National Cancer Institute* 93, no. 3 (2001): 203–207, http://jnci.oxfordjournals.org/content/93/3/203.full.

136   *70 percent of the world's population*   National Research Council, *Diplomacy for the 21st*

Century: Embedding a Culture of Science and Technology throughout the Department of State (Washington, DC: National Academies Press, 2015), http://www.nap.edu/catalog/21730/diplomacy-for-the-21st-century-embedding-a-culture-of-science.

138 *"Today when the public thinks"*   Leo P. Kadanoff, "Hard Times," *Physics Today* 45, no. 10 (1992): 9–11.

139 *"It was penny-wise"*   Michael Halpern, interview by author, March 28, 2010.

140 *In 1998, Wakefield published*   A. J. Wakefield et al., "Retracted: Ileal-Lymphoid-Nodular Hyperplasia, Non-Specific Colitis, and Pervasive Developmental Disorder in Children," *Lancet* 351, no. 9103 (1998): 637–641, http://www.thelancet.com/journals/lancet/article/PIIS0140-6736(97)11096-0/abstract.

140 *the injection of syphilis*   Sana Loue, *Textbook of Research Ethics: Theory and Practice* (New York: Kluwer Academic/Plenum Publishers, 2000).

140 *The paper has since been discredited*   Fiona Godlee et al., "Wakefield's Article Linking MMR Vaccine and Autism Was Fraudulent," *BMJ* 342, no. 7452 (2011), http://www.bmj.com/content/342/bmj.c7452.full.

140 *"conduct in this regard was dishonest"*   Madison Park, "Medical Journal Retracts Study Linking Autism to Vaccine," CNN, February 2, 2010, http://www.cnn.com/2010/HEALTH/02/02/lancet.retraction.autism/index.html.

140 *outbreaks that raged across the UK*   Sy Mukherjee, "Thanks to Debunked Anti-Vaccine Study, U.K. Sees Dramatic Surge in Measles Cases," *ThinkProgress*, May 21, 2013, http://thinkprogress.org/health/2013/05/21/2040051/uk-measles-outbreak-anti-vaccine-study/.

141 *a horrible 1928 tragedy*   US Food and Drug Administration, "Thimerosal in Vaccines," March 31, 2010, http://www.fda.gov/biologicsbloodvaccines/safetyavailability/vaccinesafety/ucm096228.htm.

141 *"convinced that the link"*   Robert F. Kennedy Jr., "Deadly Immunity," *Rolling Stone*, July 14, 2005.

141 *unvaccinated children tend to be*   Philip J. Smith et al., "Children Who Have Received No Vaccines: Who Are They and Where Do They Live?," *Pediatrics* 114, no. 1 (2004): 187–195, http://pediatrics.aappublications.org/cgi/content/full/114/1/187.

142 *A 2011 study found similar results*   Philip J. Smith et al., "Parental Delay or Refusal of Vaccine Doses, Childhood Vaccination Coverage at 24 Months of Age, and the Health Belief Model," *Public Health Reports* 126, no. 2 (2011): 135–146, http://www.ncbi.nlm.nih.gov/pmc/articles/PMC3113438/.

142 *"Best US City for Hippies,"*   Ryan Nickum, "17 Best U.S. Cities for Hippies," *Estately Blog*, July 31, 2013, http://blog.estately.com/2013/07/17-best-u-s-cities-for-hippies/.

142 *vaccination exemption rates over 60 percent*   Kristian Foden-Vencil, "Oregon Has Highest Vaccine Exemption Rate in US," Oregon Public Broadcasting, February 4, 2015, http://www.opb.org/news/article/oregon-has-highest-vaccine-exemption-rate-in-us/#FindSchool.

142 *"Additionally [. . .] some parents"*   Karlen E. Luthy et al., "Reasons Parents Exempt Children from Receiving Immunizations," *Journal of School Nursing* 28, no. 2 (November 2011): 153–160, http://www.researchgate.net/publication/51790612_Reasons_Parents_Exempt_Children_From_Receiving_Immunizations.

142   *By 2014, the number had skyrocketed*   US Centers for Disease Control and Prevention, "Measles Cases and Outbreaks," accessed July 29, 2015, http://www.cdc.gov/measles/cases-outbreaks.html.

142   *the state with the highest vaccine exemption rate*   Ranee Seither et al., "Vaccination Coverage Among Children in Kindergarten—United States, 2013–14 School Year," *Morbidity and Mortality Weekly Report* 63, no. 41 (October 27, 2014): 913–920, http://www.cdc.gov/mmwr/preview/mmwrhtml/mm6341a1.htm.

142   *"California Gov says yes"*   Jim Carrey's Twitter page, June 30, 2015, accessed July 2, 2015, https://twitter.com/JimCarrey/status/616049450243338240.

143   *"get vaccinated, but do your homework"*   Wayne Rohde, quoted in Steve Karnowski, "Autism Fears, Measles Spike among Minn. Somalis," Minnesota Public Radio, April 2, 2011, http://minnesota.publicradio.org/display/web/2011/04/02/somali-autism-vaccines.

143   *MMR vaccination rates among the Minnesota Somali population*   Amy Norton, "How One Unvaccinated Child Sparked Minnesota Measles Outbreak," *CBS News*, June 9, 2014, http://www.cbsnews.com/news/how-one-unvaccinated-child-sparked-minnesota-measles-outbreak/.

143   *After this was debunked*   Jeffrey S. Gerber and Paula A. Offit, "Vaccines and Autism: A Tale of Shifting Hypotheses," *Clinical Infectious Diseases* 48, no. 4 (2009): 456–461, http://www.ncbi.nlm.nih.gov/pmc/articles/PMC2908388/.

144   *but much easier on the immune system*   Mark Crislip and Stephen Barrett, "Do Children Get Too Many Immunizations? The Answer Is No," Quackwatch, October 3, 2014, http://www.quackwatch.com/03HealthPromotion/immu/too_many.html.

144   *That, too, was debunked*   Anjali Jain et al., "Autism Occurrence by MMR Vaccine Status among US Children with Older Siblings with and without Autism," *Journal of the American Medical Association* 313, no. 15 (2015): 1534–1540, http://jama.jamanetwork.com/article.aspx?articleid=2275444.

144   *A large majority*   World Health Organization, "Human papillomavirus (HPV)," accessed October 25, 2015, http://www.who.int/immunization/diseases/hpv/en/.

144   *and allowing the legislature to overturn it*   "Iowa Conservatives Quickly Vetting Perry," CBS Dallas Fort-Worth, August 15, 2011, http://dfw.cbslocal.com/2011/08/15/iowa-conservatives-quickly-vetting-perry/. Includes audio file of Perry on Jan Mickelson's show at the Iowa State Fair earlier in the day.

144   *"All adult citizens"*   Republican Party of Texas, *Report of Permanent Committee on Platform and Resolutions* (Austin: 2014), accessed July 11, 2015, http://www.texasgop.org/wp-content/uploads/2014/06/2014-Platform-Final.pdf.

145   *"I understand that there are families"*   Barack Obama, quoted in Eun Kyung Kim, "President Obama on Measles: 'You Should Get Your Kids Vaccinated,'" *The Today Show*, February 2, 2015, accessed July 11, 2015, http://news.health.com/2015/02/03/president-obama-on-measles-you-should-get-your-kids-vaccinated/.

145   *"The concern would be measuring"*   Chris Christie, quoted in Heather Haddon, "N.J. Gov. Christie Calls for 'Balance' on Vaccinations amid Measles Outbreak," *Wall Street Journal*,

February 2, 2015, http://blogs.wsj.com/metropolis/2015/02/02/chris-christie-there -should-be-balance-on-whether-to-require-vaccinations/.

145 *"I'm not anti-vaccine at all"* "Rand Paul: Most Vaccines Should Be 'Voluntary,'" video, DNC_Clips, accessed July 11, 2015, https://www.youtube.com/watch?v=CGvBB_nqZWI.

145 *Paul has long been a member* Joseph Gerth, "Rand Paul Part of AAPS Doctors' Group Airing Unusual Views," *Louisville Courier-Journal*, September 24, 2010, http:// freedemocracy.blogspot.com/2010/09/rand-paul-part-of-aaps-doctors-group.html.

145 *"I think there's a big difference"* Carly Fiorina, quoted in McKay Coppins, "Carly Fiorina on Vaccinations: 'Parents Have to Make Choices for Their Family,'" Buzzfeed, February 2, 2015, http://www.buzzfeed.com/mckaycoppins/carly-fiorina-on-vaccinations-parents -have-to-make-choices-f.

145 *"Although I strongly believe"* Ben Carson, quoted in Jonathan Easley, "Ben Carson Backs Vaccinations as 'Safe,'" *The Hill*, February 2, 2015, http://thehill.com/blogs/ballot-box/ presidential-races/231545-ben-carson-backs-vaccines.

146 *"in 1 massive dose"* Donald Trump's Twitter page, September 4, 2014, accessed October 25, 2015, https://twitter.com/realDonaldTrump/status/507546307620528129.

146 *and in debates over the coming year* Michael E. Miller, "The GOP's Dangerous 'Debate' on Vaccines and Autism," *Washington Post*, September 17, 2015, http://www.washingtonpost .com/news/morning-mix/wp/2015/09/17/the-gops-dangerous-debate-on-vaccines-and -autism/.

146 *"Tiny children are not horses"* Donald Trump's Twitter page, September 3, 2014, accessed October 25, 2015, https://twitter.com/realDonaldTrump/status/507158396051927041.

146 *"Spread them out"* Donald Trump's Twitter page, September 4, 2014, accessed October 25, 2015, https://twitter.com/realDonaldTrump/status/507546307620528129.

146 *"We in the educated, advanced"* Stephen Harper, quoted in Laura Payton, "Stephen Harper Tells Parents to Listen to Scientists about Vaccines," *CBC News*, February 25, 2015, http://www.cbc.ca/news/politics/stephen-harper-tells-parents-to-listen-to-scientists -about-vaccines-1.2972483.

146 *a homeopathic alternative to the flu vaccine* Michael Kruse et al., "Review of Canadian Sellers of Homeopathic Flu 'Vaccines' (Nosodes)," Bad Science Watch, November 28, 2014, http://www.badsciencewatch.ca/wp-content/uploads/2014/01/BSW-Homeomark-2014 -Homeopathy-Flu-Vaccine-Report.pdf.

147 *"Parents who vaccinate their children"* Tony Abbott, quoted in Stephanie Peatling, "Abbott Government Withdraws Childcare Payments for Anti-Vaccination Parents," *Sydney Morning Herald*, April 12, 2015, http://www.smh.com.au/federal-politics/political-news/abbott -government-withdraws-childcare-payments-for-antivaccination-parents-20150412 -1mj837.html.

147 *"When we compare"* Arnica website, accessed July 11, 2015, http://www.arnica.org.uk/.

147 *the First International Anti-Vaccination Congress* M. R. Leverson, "Memoir—Mr. William Tebb," *Homoeopathic Physician: A Monthly Journal of Medical Science* 19, (1899): 316.

148   *The group appointed a committee*   "The International Anti-Vaccination Congress,"
       Vaccination Inquirer and Health Review 3 (1882): 110.

148   *Vegetarian Society of Great Britain*   Nadja Durbach, *Bodily Matters: The Anti-Vaccination
       Movement in England, 1853–1907* (Durham, NC: Duke University Press, 2005), 123.

148   *to escape a malaria outbreak*   Molly Whittington-Egan, *Mrs Guppy Takes A Flight* (Castle
       Douglas, UK: Neil Wilson Publishing, 2014), 92.

148   *compulsory laws to protect public health*   Wendy K. Mariner et al., "*Jacobson v
       Massachusetts*: It's Not Your Great-Great-Grandfather's Public Health Law," *American
       Journal of Public Health* 95, no. 4 (2005), http://www.ncbi.nlm.nih.gov/pmc/articles/
       PMC1449224/.

148   *"That is not what I would call"*   John Porter, interview by author, August 31, 2009.

149   *By 1957, two years after*   "Summary of Notifiable Diseases, United States, 1994," *Morbidity
       and Mortality Weekly Report* 43, no. 53 (1995): tables 11–12, www.cdc.gov/mmwr/
       preview/mmwrhtml/00039679.htm.

149   *low as 9 percent in some districts*   Jeffrey P. Baker, "The Pertussis Vaccine Controversy
       in Great Britain, 1974–1986," *Vaccine* 21 (2003): 4003–4010, https://online.manchester.
       ac.uk/bbcswebdav/orgs/I3075-COMMUNITY-MEDN-1/DO%20NOT%20DELETE%20
       -%20PEP%20Quality%20and%20Evidence/QE-PEP-HTML5/media/F8430185-03E3
       -C538-8362-DE46812E97BE.pdf.

149   *"What members of Congress need"*   Porter, interview.

Chapter 7. The Rise of the Antiscience News Media

151   *"Every time we proceed"*   Karl R. Popper, *Objective Knowledge: An Evolutionary Approach*,
       rev. ed. (Oxford: Oxford University Press, 1972).

152   *consider a bit of educational psychology*   Pat Burke Guild and Stephen Garger, *Marching
       to Different Drummers* (Alexandria, VA: Association for Supervision and Curriculum
       Development, 1985).

155   *a concept first developed*   Vannevar Bush, "As We May Think," *Atlantic*, July 1945, www
       .theatlantic.com/magazine/archive/1945/07/as-we-may-think/3881.

156   *"There doesn't seem to be much accountability"*   Marcia McNutt, interview by author,
       August 27, 2010.

156   *"Story is not about facts"*   Robert McKee, interview by author, October 7, 2010.

157   *Soldiers took to social media*   Travis J. Tritten, "NBC's Brian Williams Recants Iraq Story
       after Soldiers Protest," *Stars and Stripes*, February 4, 2015, http://www.stripes.com/news/
       us/nbc-s-brian-williams-recants-iraq-story-after-soldiers-protest-1.327792.

157   *"You are absolutely right"* Brian Williams, quoted in Tritten, "NBC'S Brian Williams Recants
       Iraq Story after Soldiers Protest."

157   *"NBC's Brian William [sic] recounts being shot at"*   Brian Williams, "Target Iraq:
       Helicopter NBC's Brian Williams Was Riding in Comes under Fire," *NBC Nightly News*,
       March 26, 2003, http://www.nbcuniversalarchives.com/nbcuni/clip/5114582427_s09.do.

158   *"My week—two weeks—there"*   Brian Williams, *The duPont Talks: Tom Brokaw & Brian*

Williams on Covering Katrina Pt. 1 of 3, June 26, 2014, video, Columbia Journalism School, accessed July 2, 2015, https://www.youtube.com/watch?v=TBoyOGu6Gt8.

158 *"When you look out"* Brian Williams, quoted in Noah Rothman, "Did Brian Williams Lie about His Katrina Experience, Too?", *Hot Air* (blog), February 5, 2015, http://hotair.com/archives/2015/02/05/did-brian-williams-lie-about-his-katrina-experience-too/.

158 *"I can tell you"* Myra DeGersdorff, quoted in Terrence McCoy, "Brian Williams Perhaps 'Misremembered' Dangers of Katrina, Hotel Manager Says," *Washington Post*, February 10, 2015, http://www.washingtonpost.com/news/morning-mix/wp/2015/02/10/brian-williams -perhaps-misremembered-floating-dead-body-and-gangs-during-katrina-hotel-manager -says/.

158 *"We were never wet"* Brobson Lutz, quoted in John Simerman, "NBC News Anchor Brian Williams' Comments about Dead Bodies, Hurricane Katrina Starting to Gain Attention, Draw Scrutiny," *New Orleans Advocate*, February 9, 2015, http://www .theneworleansadvocate.com/news/11526453-148/nbc-news-anchor-brian-williams.

158 *"I've covered wars, okay?"* Bill O'Reilly, quoted in Tucker Carlson, *Politicians, Partisans, and Parasites: My Adventures in Cable News* (New York: Grand Central Publishing, 2003), 51.

158 *"Having survived a combat situation"* Bill O'Reilly, "Semper Fi," BillOReilly.com, November 18, 2004, https://www.billoreilly.com/site/product?printerFriendly=true&pid =18827.

159 *Several other journalists made similar comments* Brian Stelter, "CBS Staffers Dispute Bill O'Reilly's 'War Zone' Story," *CNN Money*, February 23, 2015, http://money.cnn.com/2015/02/22/media/cbs-staffers-oreilly-argentina/.

159 *Publications' reporter guidelines contain it* Voice of San Diego, "Guidelines for Reporters."

159 *"There is no such thing"* Linda Ellerbee, quoted in Gary DeMar, "The Myth of Objectivity," American Vision, December 4, 2006, http://americanvision.org/1360/myth-of-objectivity/.

160 *repeats on the speaking circuit* Mark Judge, "Anderson Cooper and Objectivity," *Daily Caller*, July 3, 2012, http://dailycaller.com/2012/07/03/anderson-cooper-and-objectivity.

160 *"Why is this the case?"* Lawrence Krauss, "Equal Time for Nonsense," *New York Times*, July 30, 1996.

161 *"shockingly inaccurate"* Hal Arkowitz and Scott O. Lillienfeld, "Why Science Tells Us Not to Rely on Eyewitness Accounts," *Scientific American*, January 8, 2009, http://www .scientificamerican.com/article/do-the-eyes-have-it/.

161 *"For what a man"* Bacon, *Novum Organum*.

162 *"something get dislodged"* Brian Williams, interview by Michael Eisner, "BREAKING: NBC Anchor Brian Williams Lied about Body Floating by Hotel Room in Katrina?," February 5, 2015, video, GotNews, February 5, 2015, accessed January 31, 2016, https:// www.youtube.com/watch?v=TVXAxHFjiOk.

163 *Jay Rosen points out quite eloquently* Jay Rosen, "Objectivity as a Form of Persuasion: A Few Notes for Marcus Brauchli," in "The View From Nowhere at Twilight Hour: A PressThink Series," PressThink blog, July 7, 2010. http://archive.pressthink.org/2010/07/07/obj_persuasion.html

163 *news media coverage can cause cynicism* Kathleen Cross et al., *News Media and Climate*

Politics: Civic Engagement and Political Efficacy in a Climate of Reluctant Cynicism (Ottawa, Canada: Canadian Centre for Policy Alternatives, September 2015), https://www.policyalternatives.ca/publications/reports/news-media-and-climate-politics.

163 *A study by Rutgers University communications researcher* Lauren Feldman et al., "Polarizing News? Representations of Threat and Efficacy in Leading US Newspapers' Coverage of Climate Change," *Public Understanding of Science* (July 30, 2015), http://pus.sagepub.com/content/early/2015/07/29/0963662515595348.

165 *"I'm not aware"* Michael Mann, email to author, January 5, 2016.

166 *"When you're taking it seriously enough"* Stephanie Curtis, interview by author, January 5, 2016.

Chapter 8. The Identity Politics War on Science

171 *"The human understanding is no dry light"* Bacon, *Novum Organum*.

173 *Any claims of scientific objectivity* Michel Foucault, *Madness and Civilization: A History of Insanity in the Age of Reason* (New York: Vintage, 2013), 276.

173 *"Simplifying to the extreme"* Jean-Francois Lyotard, *The Postmodern Condition: A Report on Knowledge* (Minneapolis: University of Minnesota Press, 1984), xxiv.

173 *"the dismantling of conceptual oppositions"* Christopher Norris, *Derrida* (Cambridge, MA: Harvard University Press, 1987).

173 *borrowed significantly from Sigmund Freud* Alfred I. Tauber, "Freud's Social Theory: Modernist and Postmodernist Revisions," *History of the Human Sciences* 25, no. 4 (2012), http://blogs.bu.edu/ait/files/2013/08/Freuds-social-theory-copy.pdf.

173 *"operate a kind of strategic reversal"* Norris, *Derrida*.

174 *"What differs? Who differs?"* Jacques Derrida, *A Derrida Reader: Between the Blinds* (New York: Columbia University Press, 1991), 66.

174 *"to make enigmatic"* Jacques Derrida, *Of Grammatology* (Maryland: John Hopkins University Press, 1997), 16.

174 *"out of you who have chosen"* Friedrich Nietzsche, *Thus Spake Zarathustra: A Book for All and None* (London: Macmillan & Co, 1896).

175 *"Man is a rope"* Ibid.

175 *"all anti-Semites shot"* Friedrich Nietzsche, "Letter to Franz Overbeck," January 7, 1889, in *Selected Letters of Friedrich Nietzsche* (Chicago: University of Chicago Press, 1969), 348.

175 *Ayn Rand in her philosophy of objectivism* Stephen Newman, *Liberalism at Wits' End: The Libertarian Revolt Against the Modern State* (Ithaca, NY: Cornell University Press, 1984), 26.

175 *knowledge claims must be evaluated* John W. Creswell, *Qualitative Inquiry and Research Design: Choosing among Five Approaches*, 3rd ed. (Los Angeles: SAGE Publications, 2013), 27.

175 *"view from nowhere"* Tsarina Doyle, "Nietzsche on the Possibility of Truth and Knowledge," *Minerva: An Internet Journal of Philosophy* 9 (2005), accessed July 5, 2015, http://www.minerva.mic.ul.ie//vol9/Nietzsche.html.

175 *pointed to these scientific roots* Derrida, *Of Grammatology*.

176 *a virtue under postmodernist critique* Ibid.

176 *all systems of thought, or metanarratives* Lyotard, *The Postmodern Condition*, 24.

176 *"I will use the term modern"* Ibid.

178 *"The referent (the path of the planets)"* Ibid.

179 *"It would of course be easy enough"* Karl Popper, *Conjectures and Refutations: The Growth of Scientific Knowledge* (London: Routledge, 1963), 338.

181 *the last medieval institutions* Donald Downs, "Minutes of the Faculty Forum on 'Cornell 1969: Key Issues Then and Now,'" Cornell University, May 3, 1999, http://theuniversity faculty.cornell.edu/forums/pdfs/1969min.pdf.

181 *Science was not the gradual* Thomas S. Kuhn, *The Structure of Scientific Revolutions* (Chicago: University of Chicago Press, 1962).

181 *These shifts were akin* Ibid.

182 *"A new scientific truth"* Max Planck, *Scientific Autobiography and Other Papers* (New York: Philosophical Library, 1949), 33–34.

182 *"numberless"* Bacon, *Novum Organum*.

183 *"We had something"* Patricia Churchland, interview by author, August 24, 2010.

184 *"I do not doubt"* Kuhn, *The Structure of Scientific Revolutions*.

187 *"The world, including the world of science"* Paul Feyerabend, *Killing Time: The Autobiography of Paul Feyerabend* (Chicago: University of Chicago Press, 1995), 142–143.

187 *"The church at the time of Galileo"* Paul Feyerabend, quoted in Lauren Green, "The Messy Relationship between Religion and Science: Revisiting Galileo's Inquisition," *Fox News*, January 16, 2008, http://www.foxnews.com/story/0,2933,323327,00.html.

188 *"When the Catholic Church"* Peter Harrison, "The Territories of Science and Religion" (lecture, Universities of Edinburgh, Glasgow, Aberdeen, and St. Andrews, February 14, 2011), accessed July 27, 2015, http://www.ed.ac.uk/schools-departments/humanities-soc -sci/news-events/lectures/gifford-lectures/archive/2010-2011/prof-harrison/lecture-1 -territories.

188 *"While such views"* "Science Wars and the Need for Respect and Rigour," *Nature* 385, no. 6615 (January 30, 1997): 373, http://www.nature.com/nature/journal/v385/n6615/ pdf/385373a0.pdf.

190 *As Islamic scholar Bassam Tibi writes* Bassam Tibi, *Islam's Predicament with Modernity: Religious Reform and Cultural Change* (London: Routledge, 2009), 92–93.

190 *"Why should we suppose"* Francis E. Dart and Panna Lal Pradhan, "Cross-Cultural Teaching of Science," *Science* 155, no. 3763 (February 10, 1967): 649–656, http://science .sciencemag.org/content/155/3763/649.

190 *"a growing awareness"* Bryan Wilson, "The Cultural Contexts of Science and Mathematics Education: Preparation of a Bibliographic Guide," *Studies in Science Education* 8, no. 1 (1981): 27–44.

191 *"Educators have long viewed"* William W. Cobern, "Alternative Constructions of Science and Science Education" (lecture, South Africa Association for Mathematics and Science

Education Research, University of Durban-Westville, January 27–30, 1994), http://www
.wmich.edu/slcsp/SLCSP122/SLCSP122.pdf.

191 *"there is no representation of reality"* Virginia Richardson, ed., *Constructivist Teacher Education: Building a World of New Understandings* (New York: Routledge, 1997), 8.

192 *"Because reality is in part"* Clive Beck, "Postmodernism, Pedagogy, and Philosophy of Education," *Philosophy of Education Society Yearbook* (1993): 3–6.

193 *"As Richard Dawkins likes to put it"* William W. Cobern and Cathleen C. Loving, "Defining 'Science' in a Multicultural World: Implications for Science Education," *Science Education* 85, no. 1 (2001): 50–67.

194 *"The relativity of truth"* Allan Bloom, *The Closing of the American Mind: How Higher Education Has Failed Democracy and Impoverished the Souls of Today's Students* (Chicago: University of Chicago Press, 1987).

195 *"Openness—and the relativism that makes it"* Ibid.

196 *a polemic attacking* Paul R. Gross and Norman Levitt, *Higher Superstition: The Academic Left and Its Quarrels with Science* (Baltimore: Johns Hopkins University Press, 1994).

196 *"multiculturalism equals relativism"* E. O. Wilson, quoted in Andrew Ross, ed., *Science Wars* (Durham, NC: Duke University Press, 1996), 152.

196 *"Deep conceptual shifts"* Alan Sokal, "Transgressing the Boundaries: Towards a Transformative Hermeneutics of Quantum Gravity," *Social Text* 46/47 (1996): 217–252, http://www.physics.nyu.edu/sokal/transgress_v2/transgress_v2_singlefile.html.

197 *"What concerns me"* Alan Sokal, "A Physicist Experiments with Cultural Studies," *Lingua Franca*, May/June 1996, http://linguafranca.mirror.theinfo.org/9605/sokal.html.

197 *"effete, elitist academic"* Rush Limbaugh, quoted in John Switzer, "Unofficial Summary of the *Rush Limbaugh Show*," May 22, 1996, accessed February 1, 2016, http://jwalsh.net/projects/sokal/articles/rlimbaugh.html.

197 *"He [Sokal] says we're epistemic relativists"* Stanley Aronowitz, quoted in Janny Scott, "Postmodern Gravity Deconstructed, Slyly," *New York Times*, May 18, 1996, http://www.nytimes.com/1996/05/18/nyregion/postmodern-gravity-deconstructed-slyly.html.

197 *"Conservatives have argued"* Scott, "Postmodern Gravity Deconstructed, Slyly."

198 *"It took a New York University physicist"* Ruth Rosen, "A Physics Prof Drops a Bomb on the Faux Left," *Los Angeles Times*, May 23, 1996, http://articles.latimes.com/1996-05-23/local/me-7174_1_academic-left.

198 *"who believed that class oppression"* Michael Bérubé, "The Science Wars Redux," *Democracy* 19 (2011), http://www.democracyjournal.org/19/6789.php.

198 *"a pseudo-politics"* Katha Pollitt, "Pomolotov Cocktail," *Nation*, June 10, 1996.

201 *"There's no such thing as objective truth"* Theodor Schick Jr. and Lewis Vaughn, *How to Think about Weird Things: Critical Thinking for a New Age* (Mountain View, CA: Mayfield, 1995).

203 *"sciences as one would"* Bacon, *Novum Organum.*

203 *"These days, when I talk"* Bérubé, "The Science Wars Redux."

Chapter 9. The Ideological War on Science

205   *"Enlightenment is man's emergence"*   Immanuel Kant, "An Answer to the Question: What Is Enlightenment?," in *Kant: Political Writings*, ed. H. S. Reiss, trans. H. B. Nisbet (Cambridge, UK: Cambridge University Press, 1970).

205   *watched four hours of the "boob tube"*   Debra J. Holt et al., *Children's Exposure to TV Advertising in 1977 and 2004: Information for the Obesity Debate* (Washington, DC: Bureau of Economics, Federal Trade Commission, 2007), https://www.ftc.gov/sites/default/files/documents/reports/childrens-exposure-television-advertising-1977-and-2004-information-obesity-debate-bureau-economics/cabebw.pdf.

206   *three hundred and fifty thousand people*   Cecilia Rasmussen, "Billy Graham's Star Was Born at His 1949 Revival in Los Angeles," *Los Angeles Times*, September 2, 2007, http://articles.latimes.com/2007/sep/02/local/me-then2.

206   *"All across Europe"*   Billy Graham, quoted in Rasmussen, "Billy Graham's Star Was Born at His 1949 Revival in Los Angeles."

206   *"First of all, I want you"*   Billy Graham, "Why a Revival?," September 25, 1949, audio recording, *Billy Graham & the 1949 Christ for Greater Los Angeles Campaign*, Billy Graham Center, Wheaton College, accessed February 1, 2016, http://espace.wheaton.edu/bgc/audio/cn026t5702a.mp3.

207   *"cowardice . . . of lifelong immaturity"*   Kant, "An Answer to the Question: What Is Enlightenment?"

207   *antiscience, anti-intellectual theme*   Billy Graham, "The Solution to Modern Problems," Madison Square Garden, June 1, 1957, audiovisual footage, *Billy Graham New York Crusade*, Billy Graham Center, Wheaton College, accessed February 1, 2016, http://www.wheaton.edu/bgc/archives/exhibits/NYC57/12sample65.htm.

207   *"When Sir Walter Raleigh"*   Billy Graham, "Heart Trouble," Madison Square Garden, May 23, 1957, audiovisual footage, *Billy Graham New York Crusade*, Billy Graham Center, Wheaton College, accessed February 1, 2016, http://www.wheaton.edu/bgc/archives/exhibits/NYC57/18sample97-1.htm.

207   *the ten most admired men*   Lydia Saad, "Barack Obama, Hillary Clinton Are 2010's Most Admired," Gallup, December 27, 2010, www.gallup.com/poll/145394/barack-obama-hillary-clinton-2010-admired.aspx.

208   *twenty-two million viewers per week*   Jeffrey K. Hadden and Charles E. Swann, *Prime Time Preachers: The Rising Power of Televangelism* (Reading, MA: Addison-Wesley, 1981), 47–55, accessed February 1, 2016, http://web.archive.org/web/20080509191119/http://etext.lib.virginia.edu/toc/modeng/public/HadPrim.html.

208   *"It is a political movement"*   Sarah Diamond, *Not by Politics Alone: The Enduring Influence of the Christian Right* (New York: Guilford Press, 2000).

209   *"civil war of values"*   James Dobson, quoted in Diamond, *Not by Politics Alone.*

209   *a November 1989 article*   Marcia Barinaga, "California Backs Evolution Education," *Science* 246, no. 4932 (November 17, 1989): 881, http://science.sciencemag.org/content/246/4932/881.1.

209 *had been chronicled in the magazine* Arthur Miller, "The New Catastrophism and Its Defender," *Science* 55, no. 1435 (June 30, 1922): 701–703, http://science.sciencemag.org/content/55/1435/701.2.

210 *"Now, remember, those bones"* An administrator at the Science Museum of Minnesota relayed this story to the author. Similar stories have also been relayed to the author by staff at other science museums.

210 *Carson has said he believes* Ben Carson, "Celebration of Creationism (2011)," June 15, 2015, video recording, Three Angels Messages, accessed October 28, 2015, https://www.youtube.com/watch?v=YPqq6fr2CF4.

211 *A 2006 panel discussion* "Richard Dawkins & Daniel Dennett vs. Francis Collins & Benjamin Carson," March 10, 2006, audio recording, Internet Archive, accessed October 28, 2015, https://archive.org/details/RichardDawkinsDanielDennettVs.FrancisCollinsBenjaminCarson.

212 *all point to the same conclusion* Jon Perry et al., "What Is the Evidence for Evolution?," October 10, 2014, video, StatedClearly.com, accessed October 28, 2015, https://www.youtube.com/watch?v=lIEoO5KdPvg.

213 *"a self-sustaining chemical system"* Gerald F. Joyce, David W. Deamer, and Gail R. Fleischaker, *Origins of Life: The Central Concepts* (Boston: Jones and Bartlett, 1994), xi.

214 *Over 50 percent of antibiotics* Tom Frieden, *Antibiotic Resistance Threats in the United States, 2013* (Washington, DC: US Department of Health and Human Services, Centers for Disease Control and Prevention, 2013), accessed July 4, 2015, http://www.cdc.gov/drugresistance/pdf/ar-threats-2013-508.pdf.

214 *"drug choices for the treatment"* Interagency Task Force on Antimicrobial Resistance, *A Public Health Action Plan to Combat Antimicrobial Resistance* (Washington, DC: US Department of Health and Human Services, Centers for Disease Control and Prevention), accessed July 4, 2015, www.cdc.gov/drugresistance/actionplan/aractionplan.pdf.

214 *"This will be a post-antibiotic era"* Margaret Chan, "Antimicrobial Resistance in the European Union and the World," keynote address, World Health Organization, March 14, 2012, http://www.who.int/dg/speeches/2012/amr_20120314/en/.

215 *"Then there is the danger"* Alexander Fleming, "Penicillin: Nobel Lecture," December 11, 1945, Nobelprize.org, accessed July 27, 2015, http://www.nobelprize.org/nobel_prizes/medicine/laureates/1945/fleming-lecture.pdf.

215 *are not systematically collected* Interagency Task Force on Antimicrobial Resistance, *A Public Health Action Plan.*

215 *a Union of Concerned Scientists report* Margaret Mellon, Charles Benbrook, and Karen Lutz Benbrook, *Hogging It: Estimates of Antimicrobial Abuse in Livestock* (Cambridge, MA: Union of Concerned Scientists, 2001), accessed July 4, 2015, http://www.ucsusa.org/food_and_agriculture/our-failing-food-system/industrial-agriculture/hogging-it-estimates-of.html#.VZiJ6hNVhBc.

215 *The use of antibiotics* European Commission, "Ban on Antibiotics as Growth Promoters in Animal Feed Enters into Effect," European Union, December 22, 2005, http://europa.eu/rapid/press-release_IP-05-1687_en.htm.

215   *But data collection varies by country*   Øistein Thorsen, "Food, Farming and Antibiotics: A Health Challenge for Business," *Guardian*, August 7, 2014, http://www.theguardian .com/sustainable-business/food-farming-antibiotics-health-challenge-business.

215   *"the lack of progress"*   Ad-Hoc Committee for Antimicrobial Stewardship in Canadian Agriculture and Veterinary Medicine, "Stewardship of Antimicrobial Drugs in Animals in Canada: How Are We Doing?," *Canadian Veterinary Journal* 55 (March 2014), https:// shawglobalnews.files.wordpress.com/2015/06/stewardship-of-antimicrobial-drugs-in -animals-in-canada.pdf.

216   *Consider this cartoon*   This and similar "castle illustration" cartoons have been created and promoted by the leader of the creationist group Answers in Genesis since 1986. See Ken Ham, "Maturing the Message: Creationism and Biblical Authority in the Church," *Answers*, January 1, 2010, https://answersingenesis.org/apologetics/maturing-the -message/.

216   *has been used in public school science classes*   See Garrett Haley, "Evolutionists Infuriated by Creation Cartoon Shown in Public School," *Christian News Network*, July 4, 2014, http:// christiannews.net/2014/07/04/evolutionists-infuriated-by-creation-cartoon-shown-in -public-school/.

217   *"I have no problem with teaching"*   Michele Bachmann, quoted in Greg C. Huff, "Schools Should Not Limit Origins-of-Life Discussions to Evolution, Republican Legislators Say," *Stillwater Gazette*, September 27, 2005.

219   *The experiment is ongoing today*   Lyudmila Trut, Irina Oskina, and Anastasiya Kharlamova, "Animal Evolution during Domestication: The Domesticated Fox as a Model," *Bioessays* 31, no. 3 (2009): 349–360, http://www.ncbi.nlm.nih.gov/pmc/articles/PMC2763232/.

219   *Part of the problem seems to be educational*   *Can You Tell the Difference between Evolution and Natural Selection?* (Powder Springs, GA: Creation Ministries International), accessed February 1, 2016, http://www.creation.com/images/pdfs/flyers/can-you-tell-the-difference -between-evolution-and-natural-selection-p.pdf.

219   *"Natural selection is not the same thing"*   Bachmann, quoted in Huff, "Schools Should Not Limit Origins-of-Life Discussions to Evolution, Republican Legislators Say."

220   *"I'm not going to answer"*   Gary Goodyear, quoted in Phil Plait, "Is the Canadian Science Minister a Creationist?," *Discover*, March 17, 2009, http://blogs.discovermagazine.com/ badastronomy/2009/03/17/is-canadas-science-minister-a-creationist/#.VZrxHxNVhBc.

220   *"Well, of course, I do"*   Gary Goodyear, quoted in "Scientists Still Wary after Science Minister Says He Believes in Evolution," *CBC News*, March 18, 2009, http://www.cbc.ca/ news/technology/scientists-still-wary-after-science-minister-says-he-believes-in-evolution -1.847502.

221   *"intelligent design," a more recent version*   Michael J. Behe, *Darwin's Black Box: The Biochemical Challenge to Evolution* (New York: Touchstone, 1996).

221   *exactly how the eye evolved*   Carl Zimmer, "How the Eye Evolved," *New York Academy of Sciences Magazine*, October 9, 2009, http://www.nyas.org/publications/detail.aspx?cid =93b487b2-153a-4630-9fb2-5679a061fff7.

221   *"While we respect Prof. Behe's right"*   Department of Biological Sciences, "Department

Position on Evolution and 'Intelligent Design,'" Lehigh University, accessed February 1, 2016, http://www.lehigh.edu/bio/News/evolution.html.

221 *as Locke demonstrated, from observation* Locke, *An Essay Concerning Human Understanding.*

222 *In 2012, a commotion occurred* Soo Bin Park, "South Korea Surrenders to Creationist Demands," *Nature* 486, no. 14 (June 7, 2012), http://www.nature.com/news/south-korea -surrenders-to-creationist-demands-1.10773.

222 *Definitive belief in God* "Ipsos Global @dvisory: Supreme Being(s), the Afterlife and Evolution," Ipsos/Reuters, April 25, 2011, accessed October 28, 2015, http://www.ipsos-na .com/news-polls/pressrelease.aspx?id=5217.

222 *"until the apes stand up"* Pervez Hoodbhoy, quoted in Kenneth Chang, "Creationism, Minus a Young Earth, Emerges in the Islamic World," *New York Times*, November 2, 2009, http://www.nytimes.com/2009/11/03/science/03islam.html.

223 *"stimulate an academic debate"* Maria van der Hoeven, quoted in Martin Enserink, "Is Holland Becoming the Kansas of Europe?", *Science* 308, no. 5727 (June 3, 2005), http:// www.sciencemag.org/content/308/5727/1394.2.summary.

223 *"in every culture"* "And Finally . . . ," *Warsaw Business Journal*, December 18, 2006, http:// wbj.pl/and-finally-2/.

223 *"gaps in the theory of evolution"* Quoted in Stefaan Blancke, Hans Henrik Hjermitslev, and Peter C. Kjærgaard, *Creationism in Europe* (Baltimore: Johns Hopkins University Press, 2014).

223 *"straight connection between"* Blancke, Hjermitslev, and Kjærgaard, *Creationism in Europe.*

223 *"some people call for"* Committee on Culture, Science, and Education, *The Dangers of Creationism in Education* (Strasbourg, France: Council of Europe, 2007), http://assembly .coe.int/nw/xml/XRef/X2H-Xref-ViewHTML.asp?FileID=11751&lang=en.

224 *it did reject a bill* Andrew Denholm, "Formal Ban on Teaching of Creationism Rejected," *Herald* (Scotland), May 12, 2015, http://www.heraldscotland.com/news/education/formal -ban-on-teaching-of-creationism-rejected.125864971.

224 *At the beginning of the millennium* National Science Board, *Science and Engineering Indicators 2002* (Arlington, VA: National Science Foundation, 2001), accessed February 1, 2016, http://www.nsf.gov/statistics/seind02/c7/c7s1.htm#c7s1l4a.

224 *But by 2006* National Science Board, *Science and Engineering Indicators 2008* (Arlington, VA: National Science Foundation, 2008), accessed February 1, 2016, http://www.nsf.gov/ statistics/seind08/c7/c7s2.htm.

225 *whistling "Onward, Christian Soldiers"* Bill D. Moyers, *Moyers On America: A Journalist and His Times* (New York: Anchor Books, 2005), 9.

225 *voter registration drives by pastors* David D. Kirkpatrick, "Bush Allies Till Fertile Soil, among Baptists, for Votes," *New York Times*, June 18, 2004, http://www.nytimes.com/2004/06/18/ us/2004-campaign-strategy-bush-allies-till-fertile-soil-among-baptists-for-votes.html.

225 *mobilization of church congregations* David D. Kirkpatrick, "Churches See an Election Role and Spread the Word on Bush," *New York Times*, August 9, 2004, http://www.nytimes .com/2004/08/09/us/churches-see-an-election-role-and-spread-the-word-on-bush.html.

225 *Other studies have shown* William G. Mayer, ed., *The Swing Voter in American Politics* (Washington, DC: Brookings Institution, 2008).

225 *broke sharply for Bush in 2004* Ronald Brownstein, "The Hidden History of the American Electorate," *National Journal*, October 18, 2008, http://www.nationaljournal.com/magazine /the-hidden-history-of-the-american-electorate-20081018.

225 *"You are going to do"* Ralph Reed, "Growing Grassroots," panel, *PBS NewsHour*, March 2, 2004, http://www.pbs.org/newshour/bb/media-jan-june04-grassroots_03-02/.

226 *"a spiritual battle"* Kirkpatrick, "Churches See an Election Role."

226 *could cause them to violate the law* Kirkpatrick, "Bush Allies Till Fertile Soil."

226 *Rove had told an audience* Michael Tackett, "Laying Claim to the Nation," *Chicago Tribune*, November 7, 2004, http://www.chicagotribune.com/news/opinion/chi-0411070166nov07 ,0,595361.story.

226 *"Christians should not be treated"* Ralph Reed, quoted in Kirkpatrick, "Bush Allies Till Fertile Soil."

226 *require its ninth-grade science teachers* John E. Jones, *Memorandum and Order, Kitzmiller, et al. v. Dover School District, et al.*, United States District Court, Middle District of Pennsylvania, December 20, 2005, http://www.pamd.uscourts.gov/kitzmiller/ decision.htm.

226 *"Because Darwin's Theory"* Matthew Chapman, *40 Days and 40 Nights: Darwin, Intelligent Design, God, OxyContin and Other Oddities on Trial in Pennsylvania* (New York: Collins, 2007).

226 *"to provide them with"* Michael B. Berkman et al., "Evolution and Creationism in America's Classrooms: A National Portrait," *PLoS Biology* 6, no. 5 (2008): e124, http://www.plosbiology .org/article/info:doi/10.1371/journal.pbio.0060124.

226 *John Jones, a Republican federal judge* Faye Flam, "The Difference between Science and Religion," *Philadelphia Inquirer*, April 18, 2011, http://articles.philly.com/2011-04-18/ news/29443540.

226 *intelligent design in public-school science* Jones, *Memorandum and Order.*

226 *"stuck the knife in the backs"* Phyllis Schafly, quoted in Lauri Lebo, "Judge in Dover Case Reports Hostile E-Mails: Jones and His Family Were under Marshals' Protection in December," *York Daily Record*, March 24, 2006.

226 *"in their almost infinite wisdom"* John E. Jones, "The Myth of 'Activist Judges,'" *College News*, November 16, 2006, http://collegenews.org/editorials/2006/the-myth-of-activist -judges.html.

227 *pastors are warned* David Karp, "IRS Warns Churches: No Politics Allowed," *St. Petersburg Times*, September 15, 2004, http://www.sptimes.com/2004/09/15/State/IRS_warns _churches__n.shtml.

227 *campaign in churches* Bill Berkowitz, "The Christian Right's Compassion Deficit," *Dissident Voice*, December 30, 2004, http://dissidentvoice.org/Dec2004/Berkowitz1230.htm.

227 *"moral values" are often cited* Katharine Q. Seeyle, "Moral Values Cited as a Defining Issue of the Election," *New York Times*, November 4, 2004, http://query.nytimes.com/gst/ fullpage.html?res=990DE7D7173CF937A35752C1A9629C8B63&pagewanted=all.

227 *millions of voter guides* "Voter's Guide for Serious Catholics," Priests for Life, 2004, accessed February 1, 2016, http://www.priestsforlife.org/elections/voterguide.htm.

227 *by congregation members in the broader community* Melanie Hunter, "Religious Watchdog Group Warns Clergy against Passing out Voter Guides," Christian Headlines, October 20, 2004, http://www.christianheadlines.com/news/religious-watchdog-group -warns-clergy-against-passing-out-voter-guides-1291673.html.

227 *"the lowest levels of belief"* Frank Newport, "Third of Americans Say Evidence Has Supported Darwin's Evolution Theory," Gallup, November 19, 2004, http://www.gallup .com/poll/14107/Third-Americans-Say-Evidence-Has-Supported-Darwins-Evolution -Theory.aspx.

227 *voting blocs that went for Bush* Seeyle, "Moral Values Cited as a Defining Issue."

227 *easily won the 2004 creationist vote* Bootie Cosgrove-Mather, "Poll: Creationism Trumps Evolution," *CBS News*, November 22, 2004, http://www.cbsnews.com/stories/2004/11/22/ opinion/polls/main657083.shtml.

227 *some studies suggest that the brain* Sam Harris et al., "The Neural Correlates of Religious and Nonreligious Belief," *PLoS One* 4, no. 10 (2009): e7272, http://www.plosone.org/article/ info:doi%2F10.1371%2Fjournal.pone.0007272.

228 *That conflict came to a new head* The deleted section of the draft can be viewed here, accessed February 1, 2016: http://www.shawnotto.com/downloads/seind2010deletion.pdf.

228 *"Board members say the answers"* Yudhijit Bhattacharjee, "NSF Board Draws Flak for Dropping Evolution from Indicators," *Science* 328, no. 5975 (April 9, 2010): 150–151, http:// science.sciencemag.org/content/328/5975/150.

229 *"True or false"* National Science Board, *Science and Engineering Indicators 2006* (Arlington, VA: National Science Foundation, 2007), 19, accessed February 1, 2016, http://www.nsf .gov/statistics/seind06/pdf/c07.pdf.

229 *"We're turning that steady march"* M. E. Webber, "Don't Dumb down Texas," *Austin American-Statesman*, September 15, 2009, http://www.statesman.com/opinion/content/ editorial/stories/2009/09/16/0916webber_edit.html.

229 *The cooperative was formed* Michael Webber, interview by author, July 30, 2010.

229 *According to a 2009 report* David Wiley and Kelly Wilson, *Just Say Don't Know: Sexuality Education in Texas Public Schools* (Austin: Texas Freedom Network Education Fund, 2009), accessed February 1, 2016, http://www.tfn.org/site/DocServer/SexEdRort09_web .pdf?docID=981.

230 *Teen pregnancy in Texas went* up Ibid.

230 *60 percent higher than the national average* Joyce A. Martin et al., "Births: Final Data for 2013," *National Vital Statistics Reports* 64, no. 1 (January 15, 2015), http://www.cdc.gov/ nchs/data/nvsr/nvsr64/nvsr64_01.pdf.

230 *"If birth control doesn't work, why use it?"* Webber, interview.

230 *"If you are a kid"* Susan Tortolero, quoted in Gail Collins, "Mrs. Bush, Abstinence and Texas," *New York Times*, February 16, 2011, http://www.nytimes.com/2011/02/17/opinion/ 17gailcollins.html.

231 *by far the highest* M. S. Kearney and P. B. Levine, "Why Is the Teen Birth Rate in the

United States So High and Why Does It Matter?," *Journal of Economic Perspectives* 26, no. 2 (2012): 141–166, http://www.ncbi.nlm.nih.gov/pubmed/22792555.

231   *highest in the so-called Bible Belt states*   "Presidential Election Results," *NBC News*, http://elections.nbcnews.com/ns/politics/2012/all/president, in conjunction with Martin et al., "Births: Final Data for 2013."

231   *And she is more than forty times*   United Nations Population Division, World Population Prospects, "Adolescent Fertility Rate (Births per 1,000 Women Ages 15–19)," World Bank, accessed July 16, 2015, http://data.worldbank.org/indicator/SP.ADO.TFRT.

231   *"Abstinence works"*   Rick Perry, "Gov. Rick Perry on Abstinence," video, *Texas Tribune*, October 15, 2010, accessed February 1, 2016, https://www.youtube.com/watch?v=SWlbN2b1PGg.

231   *Bristol Palin made roughly $1 million*   Nathan Francis, "Bristol Palin Made Close to $1 Million Pushing Abstinence-Only Policies, Now Is Pregnant with Child No. 2," *Inquisitr*, June 25, 2015, http://www.inquisitr.com/2203203/bristol-palin-made-close-to-1-million-pushing-abstinence-only-policies-now-is-pregnant-with-child-no-2.

231   *"But not everybody"*   Webber, interview.

233   *"For what a man"*   Bacon, *Novum Organum*.

236   *The foundational work in this field*   Charles G. Lord et al., "Biased Assimilation and Attitude Polarization: The Effects of Prior Theories on Subsequently Considered Evidence," *Journal of Personality and Social Psychology* 37, no. 11 (1979): 2098–2109, https://www.unc.edu/~fbaum/teaching/articles/jpsp-1979-Lord-Ross-Lepper.pdf.

236   *"But people are not"*   Eugenie Scott, interview by author, August 25, 2009.

237   *In 2014, Duke University researchers*   Troy H. Campbell and Aaron C. Kay, "Solution Aversion: On the Relation between Ideology and Motivated Disbelief," *Journal of Personality and Social Psychology* 107, no. 5 (2014): 809-824.

237   *a study showing that people use*   Harris et al., "The Neural Correlates of Religious and Nonreligious Belief."

237   *exploring how the brain responds*   Uffe Schjødt et al., "The Power of Charisma: Perceived Charisma Inhibits the Frontal Executive Network of Believers in Intercessory Prayer," *Social Cognitive and Affective Neuroscience* 6, no. 1 (2011): 119–127.

238   *call just world belief*   Melvin Lerner, *The Belief in a Just World: A Fundamental Delusion* (New York: Plenum Press, 1980).

238   *Not surprisingly, South Africans*   Adrian Furnham, "Just World Beliefs in an Unjust Society: A Cross-Cultural Comparison," *European Journal of Social Psychology* 15 (1985): 363–366, http://onlinelibrary.wiley.com/doi/10.1002/ejsp.2420150310/abstract.

238   *Canadians also have high levels*   M. J. MacLean and S. M. Chown, "Just World Beliefs and Attitudes toward Helping Elderly People: A Comparison of British and Canadian University Students," *International Journal of Aging and Human Development* 26, no. 4 (1988): 249–60, http://www.ncbi.nlm.nih.gov/pubmed/3170015.

238   *people get what they deserve*   Adrian Furnham, "Belief in a Just World: Research Progress over the Past Decade," *Personality and Individual Differences* 34, no. 5 (2002): 795–817.

238   *have a much stronger belief*   Roland Bénabou and Jean Tirole, "Belief in a Just World and

Redistributive Politics," *Quarterly Journal of Economics* 121, no. 2 (2006): 699–746, http://qje.oxfordjournals.org/content/121/2/699.abstract.

239   *the tendency to blame the victim*   Ibid.

239   *writer Nancy Raine describes how*   Nancy Venable Raine, *After Silence: Rape & My Journey Back* (New York: Crown, 1998), 91.

239   *more often be Protestant*   Bénabou and Tirole, "Belief in a Just World."

240   *a study showing that the just world belief*   Matthew Feinberg and Robb Willer, "Apocalypse Soon? Dire Messages Reduce Belief in Global Warming by Contradicting Just-World Beliefs," *Psychological Science* 22, no. 1 (2011): 34–38, http://www.climateaccess.org/resource/apocalypse-soon-dire-messages-reduce-belief-global-warming-contradicting-just-world-beliefs.

240   *"I think the evidence"*   Robb Willer, interview by author, December 16, 2010.

240   *"Conservatives are on average"*   Ibid.

240   *"What if you tried"*   Ibid.

241   *"We will be beseeching"*   Michele Bachmann, interview by Jan Markell, *Prophetic Views Behind The News*, radio broadcast, KKMS, 980-AM Twin Cities Christian Talk Radio, March 20, 2004.

242   *"What really surprised us"*   Willer, interview.

242   *"I think people"*   Quoted in *Flock of Dodos: The Evolution-Intelligent Design Circus*, motion picture, directed by Randy Olson, 2006.

243   *The more educated Republicans were*   "A Deeper Partisan Divide over Global Warming," Pew Research Center, May 28, 2008, http://people-press.org/report/417/a-deeper-partisan-divide-over-global-warming.

243   *he found that people*   Geoffrey D. Munro, "The Scientific Impotence Excuse: Discounting Belief-Threatening Scientific Abstracts," *Journal of Applied Social Psychology* 40, no. 3 (March 2010): 579–600.

244   *"When a person holds"*   Ibid.

244   *why the children of scientists*   Joseph Berger, *The Young Scientists: America's Future and the Winning of the Westinghouse* (New York: Perseus, 1993).

244   *Fully 70 percent*   Stuart Anderson, "New Research Finds 70 Percent of the Nation's Top High School Science Students Are the Children of Immigrants," news release, National Foundation for American Policy, May 23, 2011.

244   *"Our parents brought us up"*   David Kenneth Tang-Quan, quoted in Stuart Anderson, "The Impact of the Children of Immigrants on Scientific Achievement in America," policy brief, National Foundation for American Policy, May 2011, accessed February 1, 2016, http://www.nfap.com/pdf/Children_of_Immigrants_in_Science_and_Math_NFAP_Policy_Brief_May_2011.pdf.

245   *"Science education has been identified"*   Francis Eberle, quoted in "New Survey Finds Parents Need Help Encouraging Their Kids in Science," news release, National Science Teachers Association, May 10, 2010, http://www.nsta.org/about/pressroom.aspx?id=57403.

246   *denounced many science theories*   Danian Hu, "The Reception of Relativity in China," *Isis* 98 (2007): 539–557.

246 *Acupuncture had been banned* John J. Bonica, "Therapeutic Acupuncture in the People's Republic of China: Implications for American Medicine," *Journal of the American Medical Association* 228 (1974): 336, http://babel.hathitrust.org/cgi/pt?id=mdp.39015081270988 ;view=1up;seq=340.

246 *training peasant "barefoot doctors"* M. T. Jenkins, in *United States–China Science Cooperation: Hearings before the Subcommittee on Science, Research, and Technology of the Committee on Science and Technology, U.S. House of Representatives, Ninety-sixth Congress, First Session* (Washington, DC: US Government Printing Office 1979), 239, accessed July 29, 2015, http://babel.hathitrust.org/cgi/pt?id=mdp.39015081270988;view=1up;seq=1.

246 *"lecture on carcinogenesis"* Joseph Needham, "Science Reborn in China: Rise and Fall of the Anti-Intellectual 'Gang,'" *Nature* 274, no. 5674 (August 31, 1978): 832–834, http://www.nature.com/nature/journal/v274/n5674/abs/274832a0.html.

246 *political identity of the messenger* Campbell, *Solution Aversion.*

246 *"Tomorrow morning at 7:46 a.m."* Brian Williams, "An American Milestone," audiovisual footage, *NBC Nightly News*, October 16, 2006.

246 *"Every 11 seconds"* Mark Strassmann, "Countdown to 300 Million," audiovisual footage, *CBS News*, October 12, 2006, http://www.cbsnews.com/videos/countdown-to-300-million/.

247 *"Nobody knows the precise second"* Ed Pilkington, "300 Million and Counting . . . US Reaches Population Milestone," *Guardian*, October 13, 2006, http://www.guardian.co.uk/world/2006/oct/13/usa.topstories3.

247 *"We get what we celebrate"* Dean Kamen, interview by author, August 28, 2009.

248 *"The theory of relativity is"* "Counterexamples to Relativity," *Conservapedia*, March 5, 2011, http://www.conservapedia.com/Counterexamples_to_Relativity.

249 *"way of knowing"* Matthew Nisbet, interview by author, April 22, 2010.

250 *"Conflict lies at the heart of all stories"* Robert McKee, interview by author, October 7, 2010.

250 *a 2010* New York Times *article* Leslie Kaufman, "Darwin Foes Add Warming to Targets," *New York Times*, March 3, 2010, http://www.nytimes.com/2010/03/04/science/earth/04climate.html.

250 *a show on the antivaccine movement* Jon Palfreman et al., "The Vaccine War," *Frontline*, April 27, 2010, http://www.pbs.org/wgbh/frontline/film/vaccines/.

251 *"I think that people"* Eugenie Scott, interview.

252 *Pharyngula had some 2.5 million views* P. Z. Myers, "Ken Ham Brags about His Websites," *Pharyngula* (blog), March 22, 2011, http://scienceblogs.com/pharyngula/2011/03/22/ken-ham-brags-about-his-websit/.

252 *rivaling traffic on the websites* Hsiang Iris Chyi and Seth C. Lewis, "Use of Online Newspaper Sites Lags behind Print Editions," *Newspaper Research Journal* 30, no. 4 (2009): 38–52, https://www.academia.edu/194197/Use_of_Online_Newspaper_Sites_Lags_Behind_Print_Editions.

252 *"Chris Mooney calls what I do"* P. Z. Myers, interview by author, August 24, 2009.

252 *"It's not the fish"* Simon Levin, interview by author, November 21, 2010.

253 *"The same type of identity formation"* Nisbet, interview.

253 *"Saying that there's not"* Myers, interview.

254 *"[U]nfortunately, traditional reward systems"* Alan I. Leshner, "We Need to Reward Those Who Nurture a Diversity of Ideas in Science," *Chronicle of Higher Education,* March 6, 2011, http://chronicle.com/article/We-Need-to-Reward-Those-Who/126591/.

255 *"I can't tell you"* Scott Westphal, interview by author, April 14, 2010.

255 *"If we're teaching creationism"* Scott, interview.

Chapter 10. The Industrial War on Science

257 *"The conscious and intelligent manipulation"* Edward Bernays, *Propaganda* (New York: Ig Publishing, 1928).

257 *one hundred times faster* Gerardo Ceballos et al., "Accelerated Modern Human–Induced Species Losses: Entering the Sixth Mass Extinction," *Science Advances* 1, no. 5 (2015), http://advances.sciencemag.org/content/1/5/e1400253.full.

258 *"not propaganda as the Germans"* George Creel, *Rebel at Large: Recollections of Fifty Crowded Years* (New York: G. P. Putnam's Sons, 1947).

260 *the first class in public relations* Wendell Potter, *Deadly Spin: An Insurance Company Insider Speaks out on How Corporate PR Is Killing Health Care and Deceiving Americans* (New York: Bloomsbury, 2010).

260 *"Karl von Wiegand, foreign correspondent"* Edward Bernays, *Biography of an Idea: Memoirs of Public Relations Counsel* (New York: Simon and Schuster, 1965).

261 *"Hill called me in"* Ibid.

261 *"Some women regard cigarettes"* A. A. Brill, quoted in Bernays, *Biography of an Idea.*

262 *"Women!"* Ruth Hale, quoted in Allan Brandt, *The Cigarette Century: The Rise, Fall, and Deadly Persistence of the Product That Defined America* (New York: Basic Books, 2007).

262 *"Our parade of ten young women"* Bernays, *Biography of an Idea.*

262 *the Nazi government ran* Robert N. Proctor, "The Anti-Tobacco Campaign of the Nazis: A Little Known Aspect of Public Health in Germany, 1933–45," *British Medical Journal* 313, no. 1450 (1996), http://www.bmj.com/content/313/7070/1450.full.

263 *"On that December morning"* Naomi Oreskes and Erik M. Conway, *Merchants of Doubt: How a Handful of Scientists Obscured the Truth on Issues from Tobacco Smoke to Global Warming* (New York: Bloomsbury, 2010).

264 *and was given the title* Bruce Harrison's LinkedIn profile, accessed June 30, 2015, https://www.linkedin.com/in/envirocomm.

264 *"Along the way, they pioneered"* John C. Stauber, "Going Green: How to Communicate Your Company's Environmental Commitment," book review, *Earth First! Journal,* 1994, accessed February 1, 2016, http://www.eco-action.org/dod/no5/goinggreen.htm.

265 *"ignorance or biases"* William J. Darby, "Silence, Miss Carson," *Chemical & Engineering News* 40 (October 1, 1962): 62–63.

265 *doubled its PR budget* Frank Graham, *Since Silent Spring* (Boston: Houghton Mifflin, 1970).

265 *Some five thousand* "The Desolate Year," *Monsanto Magazine,* October 1962, 4-9, accessed February 1, 2016, http://iseethics.files.wordpress.com/2011/12/monsanto-magazine-1962-the-desolate-yeart.pdf.

265 *"greenwashing"*   E. Bruce Harrison, quoted in Clarke L. Caywood, *The Handbook of Strategic Public Relations and Integrated Marketing Communications*, 2nd ed. (New York: McGraw-Hill Education, 2011).

265 *"Rachel Carson's legacy"*   Paul R. Ehrlich, "Commentary: As Silent Spring's 50th Anniversary Nears, What Would Rachel Carson Be Saying Now?," *Environmental Health News*, June 25, 2012, http://www.environmentalhealthnews.org/ehs/news/2012/commentary-paul -ehrlich-on-rachel-carson.

266 *"In fact [. . .] the EPA today"*   Jack Lewis, "The Birth of the EPA," *EPA Journal*, November 1985, http://www2.epa.gov/aboutepa/birth-epa.

268 *In a 2015 Pew Research Center poll*   Cary Funk and Lee Rainie, "Americans, Politics and Science Issues," Pew Research Center, July 1, 2015, http://www.pewinternet.org/2015/07/ 01/americans-politics-and-science-issues/.

269 *Our understanding that increasing*   For a more complete bibliography of the scientific literature on global warming, see Spencer R. Weart, "The Discovery of Global Warming," American Institute of Physics, February 2016, https://www.aip.org/history/climate/ bibdate.htm.

269 *Fourier calculated that the planet*   Joseph Fourier, "Remarques Générales sur les Températures du Globe Terrestre et des Espaces Planétaires," *Annales de Chimie et de Physique* 27 (1824): 136–67, http://fourier1824.geologist-1011.mobi/.

269 *Irish physicist John Tyndall discovered*   John Tyndall, "The Bakerian Lecture: On the Absorption and Radiation of Heat by Gases and Vapours, and on the Physical Connexion of Radiation, Absorption, and Conduction," *Philosophical Transactions of the Royal Society of London* 151 (January 1, 1861): 1–36, http://rstl.royalsocietypublishing.org/content/151/1 .full.pdf.

269 *who estimated that a doubling*   Svante Arrhenius, "On the Influence of Carbonic Acid in the Air upon the Temperature of the Ground," *London, Edinburgh and Dublin Philosophical Magazine and Journal of Science* 41, no. 251 (1896): 237–276, http://nsdl .org/archives/onramp/classic_articles/issue1_global_warming/n4.Arrhenius1896.pdf.

269 *concluded that atmospheric CO_2 had increased*   G. S. Callendar, "The Artificial Production of Carbon Dioxide and Its Influence on Temperature," *Quarterly Journal of the Royal Meteorological Society* 64, no. 275 (April 1938): 223–40, http://onlinelibrary.wiley.com/ doi/10.1002/qj.49706427503/abstract;jsessionid=B113A5DD0B6EB336173963E9098DEF6E .f04t01.

269 *showed that the ocean saturation*   Gilbert N. Plass, "Effect of Carbon Dioxide Variations on Climate," *American Journal of Physics* 24 (1956): 376–87, http://scitation.aip.org/ content/aapt/journal/ajp/24/5/10.1119/1.1934233.

270 *showing that the oceans*   Roger Revelle and Hans E. Suess, "Carbon Dioxide Exchange between Atmosphere and Ocean and the Question of an Increase of Atmospheric CO_2 During the Past Decades," *Tellus* 9, no. 1 (February 1957): 18–27, http://onlinelibrary.wiley .com/doi/10.1111/j.2153-3490.1957.tb01849.x/abstract.

270 *Keeling's first measurements in 1958*   Charles D. Keeling, "The Concentration and Isotopic

Abundances of Carbon Dioxide in the Atmosphere," *Tellus* 12, no. 2 (1960): 200–203, http://onlinelibrary.wiley.com/doi/10.1111/j.2153-3490.1960.tb01300.x/epdf.

271   *briefed President Lyndon Johnson*   President's Science Advisory Committee, *Restoring the Quality of Our Environment: Report of the Environmental Pollution Panel* (Washington, DC: White House, November 1965), accessed February 1, 2016, http://dge.stanford.edu/labs/ caldeiralab/Caldeira%20downloads/PSAC,%201965,%20Restoring%20the%20Quality%20 of%20Our%20Environment.pdf.

272   *"Some countries would benefit"*   James F. Black, quoted in Neela Banerjee et al., "Exxon's Own Research Confirmed Fossil Fuels' Role in Global Warming Decades Ago," *InsideClimate News*, September 16, 2015, http://insideclimatenews.org/news/15092015/ Exxons-own-research-confirmed-fossil-fuels-role-in-global-warming.

272   *"Exxon responded swiftly"*   Banerjee et al., "Exxon's Own Research."

272   *"Faith in technologies"*   E. E. David, "Inventing the Future: Energy and the CO_2 'Greenhouse' Effect," in *Climate Processes and Climate Sensitivity*, ed. James E. Hansen and Taro Takahashi (Washington, DC: American Geophysical Union, 1984), http://sites.agu.org/publications/ files/2015/09/ch1.pdf.

273   *in the world's oil market*   Daniel Yergin, *The Prize: The Epic Quest for Oil, Money & Power* (New York: Simon & Schuster, 2012).

274   *"The first five months"*   James Hansen, quoted in Philip Shabecoff, "Global Warming Has Begun, Expert Tells Senate," *New York Times*, June 24, 1988, http://www.nytimes.com/1988/ -06/24/us/global-warming-has-begun-expert-tells-senate.html.

274   *"emphasize the uncertainty"*   Quoted in Katie Jennings, Dino Grandoni, and Susanne Rust, "How Exxon Went from Leader to Skeptic on Climate Change Research," *Los Angeles Times*, October 23, 2015, http://graphics.latimes.com/exxon-research/.

275   *"based on well-established scientific fact"*   Leonard S. Bernstein, "Predicting Future Climate Change: A Primer," Global Climate Coalition, 1995, accessed February 2, 2016, via *New York Times*, http://documents.nytimes.com/global-climate-coalition-aiam-climate-change -primer. This document was obtained via a 2007 lawsuit by the auto industry against the states of Vermont and California, challenging efforts by those states to restrict vehicle emissions of carbon dioxide. American auto companies fought turning over such documents during the discovery process, but the Association of International Automobile Manufacturers, representing import cars companies, provided the material.

275   *In a 1998 Ohio State University poll*   Jon A. Krosnick, Penny S. Visser, and Allyson L. Holbrook, "American Opinion on Global Warming," *Resources for the Future*, no. 133 (Fall 1998), http://www.rff.org/files/sharepoint/WorkImages/Download/RFF-Resources-133 -usopinion.pdf.

275   *"Let's agree there's a lot"*   Lee R. Raymond, "Energy—Key to Growth and a Better Environment for Asia-Pacific Nations," remarks to World Petroleum Conference, Beijing, People's Republic of China, October 13, 1997, in author's possession.

277   *the API commissioned*   John H. Cushman Jr., "Industrial Group Plans to Battle Climate Treaty," *New York Times*, April 26, 1998, http://www.nytimes.com/1998/04/26/us/

industrial-group-plans-to-battle-climate-treaty.html. A copy of the eight-page document is available at https://www.documentcloud.org/documents/784572-api-global-climate-science-communications-plan.html.

278 *"reposition global warming"* "Deception Dossier #5: Coal's 'Information Council on the Environment' Sham," *The Climate Deception Dossiers* (Cambridge, MA: Union of Concerned Scientists, 2015), http://www.ucsusa.org/sites/default/files/attach/2015/07/The-Climate-Deception-Dossiers.pdf.

282 *"have the courage to do nothing"* Catherine Upin, "Climate of Doubt," *Frontline*, October 23, 2012, http://www.pbs.org/wgbh/pages/frontline/climate-of-doubt/.

282 *"Hard-core environmentalist activists"* "About," Energy & Environment Legal Institute, accessed November 8, 2015, http://eelegal.org/?page_id=1657.

282 *"Environmental scepticism denies"* Peter J. Jacques et al., "The Organization of Denial: Conservative Think Tanks and Environmental Scepticism," *Environmental Politics* 17, no. 3 (2008): 349–85, http://www.tandfonline.com/doi/abs/10.1080/09644010802055576.

283 *joined the board of directors* Heartland Institute, IRS Form 990, 2005, 15–16, accessed November 8, 2015, http://www.guidestar.org/FinDocuments/2005/363/309/2005-363309812-0295fbb2-9.pdf.

283 *highlighting their attempts to manufacture uncertainties* Tim Lambert, "Tell the Truth about Heartland, Get Fired," *Deltoid* (blog), April 16, 2008, accessed November 7, 2015, http://scienceblogs.com/deltoid/2008/04/16/tell-the-truth-about-heartland/.

283 *Donors Trust now provides* Robert J. Brulle, "Institutionalizing Delay: Foundation Funding and the Creation of U.S. Climate Change Counter-Movement Organizations," *Climatic Change* 122, no. 4 (February 2014): 681–94, http://drexel.edu/~/media/Files/now/pdfs/Institutionalizing%20Delay%20-%20Climatic%20Change.ashx?la=en.

284 *cofounded by Charles and David Koch* Peter Overby, "Who's Raising Money for Tea Party Movement?," National Public Radio, February 9, 2010, http://www.npr.org/templates/story/story.php?storyId=123859296.

284 *originally the Charles Koch Foundation* "Restated Articles of Incorporation," Cato Institute, April 30, 1994, accessed November 8, 2015, http://dbapress.com/wp-content/uploads/2011/03/Koch-Cato-restated-articles-of-incorporation-from-Charles-koch-Foundation-1994.pdf.

284 *declined beginning in 1990* Peter J. Jacques, Riley E. Dunlap, and Mark Freeman, "The Organization of Denial: Conservative Think Tanks and Environmental Skepticism," *Environmental Politics* 17, no. 3 (2008): 349–385, http://www.tandfonline.com/doi/pdf/10.1080/09644010802055576.

284 *"We have seen this all before"* George Miller, Congressional Record Vol. 144, No. 48. P. H2323-H2327 (April 27, 1998), accessed July 9, 2015, http://www.gpo.gov/fdsys/pkg/CREC-1998-04-27/html/CREC-1998-04-27-pt1-PgH2323.htm.

286 *Climate change is real* Committee on the Science of Climate Change, National Research Council, *Climate Change Science: An Analysis of Some Key Questions* (Washington, DC: National Academies Press, 2001), http://www.nap.edu/catalog/10139/climate-change-science-an-analysis-of-some-key-questions.

287   *a major speech*   George W. Bush, "President Bush Discusses Global Climate Change,"
      press release, June 11, 2001, http://georgewbush-whitehouse.archives.gov/news/releases/
      2001/06/20010611-2.html.

287   *in the last 650,000 years*   Ralph J. Cicerone, "Climate Change: Evidence and Future
      Projections," statement before the Oversight and Investigations Subcommittee, Committee
      on Energy and Commerce, US House of Representatives, July 27, 2006, http://www
      .nationalacademies.org/ocga/109session2/testimonies/ocga_150983.

287   *increased warming may lead*   Committee on Radiative Forcing Effects on Climate, Climate
      Research Committee, National Research Council, *Radiative Forcing of Climate Change:
      Expanding the Concept and Addressing Uncertainties* (Washington, DC: National Academies
      Press, 2005), http://www.nap.edu/catalog/11175/radiative-forcing-of-climate-change
      -expanding-the-concept-and-addressing.

288   *reproduction rates to decline*   Jim Morrison, "The Incredible Shrinking Polar Bears,"
      National Wildlife, February 1, 2004, https://www.nwf.org/News-and-Magazines/National
      -Wildlife/Animals/Archives/2004/The-Incredible-Shrinking-Polar-Bears.aspx.

288   *Center for Biological Diversity petitioned*   Kassie Siegel and Brendan Cummings, "Petition
      to List the Polar Bear (*Ursus maritimus*) as a Threatened Species under the Endangered
      Species Act," Center for Biological Diversity, February 16, 2005, accessed February 1, 2016,
      http://www.biologicaldiversity.org/species/mammals/polar_bear/pdfs/15976_7338.pdf.

288   *due to global warming*   Kassie Siegel, "Conservation Group Petitions United States
      Government to List the Polar Bear as a Threatened Species under the Endangered Species
      Act," press release, Center for Biological Diversity, February 16, 2005, http://www
      .biologicaldiversity.org/news/press_releases/polarbear2-16-05.html.

288   *"We wanted to force"*   Kassie Siegel, interview by author, June 13, 2011.

288   *some six hundred thousand public comments*   Michael Shnayerson, "The Edge of
      Extinction," *Vanity Fair*, May 2008, http://www.vanityfair.com/politics/features/2008/05/
      polarbear200805.

288   *to prepare nine detailed studies*   Ian Stirling et al., *Polar Bear Population Status in the
      Northern Beaufort Sea* (Reston, VA: US Geological Survey, 2007), accessed February 1, 2016,
      http://www.usgs.gov/newsroom/special/polar_bears.

288   *"not supported by the evidence"*   David R. Legates, quoted in H. Sterling Burnett, "Are
      Polar Bears Dying?," *Environment and Climate News*, May 1, 2006, http://www.heartland
      .org/policybot/results/18971/Are_Polar_Bears_Dying.html.

288   *"mission is to seek"*   David R. Legates, *Climate Science: Climate Change and Its Impacts*
      (Dallas, TX: National Center for Policy Analysis, 2006), accessed February 1, 2016, http://
      www.ncpa.org/pub/st285?pg=7.

288   *Legates was listed*   Burnett, "Are Polar Bears Dying?"

288   *and the Koch foundations*   Brendan DeMelle, "Disinformation Database: David Legates,"
      DeSmogBlog, accessed February 1, 2016, http://www.desmogblog.com/node/2830.

289   *arguing that the polar-bear population*   Mitchell Taylor, "Last Stand of Our Wild Polar
      Bears," *Toronto Star*, May 1, 2006, http://meteo.lcd.lu/globalwarming/Taylor/last_stand
      _of_our_wild_polar_bears.html.

289 *overall polar-bear populations had exploded* H. Sterling Burnett, "ESA Listing Not Needed for Polar Bears," *Environment and Climate News*, March 1, 2007, http://news .heartland.org/newspaper-article/2007/03/01/esa-listing-not-needed-polar-bears.

289 *The actual estimate* "Polar Bear Politics," *Wall Street Journal*, January 3, 2007, http:// www.wsj.com/articles/SB116778985966865527.

289 *polar bears were recovering* Ian Stirling, "Polar Bears and Seals in the Eastern Beaufort Sea and Amundsen Gulf: A Synthesis of Population Trends and Ecological Relationships over Three Decades," *Arctic* 55, suppl. 1 (2002): 59–76, http://arctic.journalhosting .ucalgary.ca/arctic/index.php/arctic/article/view/735.

289 *The government of Nunavut had just increased* Government of Nunavut, "Minister Accepts Decisions of the Nunavut Wildlife Management Board on Polar Bear Management," press release, January 7, 2005, http://www.gov.nu.ca/Nunavut/English/news/2005/jan/jan7.pdf.

289 *"concerned that listing"* Mitchell Taylor, quoted in Jane George, "Global Warming Won't Hurt Polar Bears, GN Says," *Nunatsiaq News*, May 26, 2006, http://www.nunatsiaqonline .ca/archives/60526/news/climate/60526_01.html.

289 *brings about $2 million* John Thompson, "Polar Bear Die-Off Unlikely: GN Official," *Nunatsiaq News*, September 14, 2007, http://www.nunatsiaqonline.ca/archives/2007/709/ 70914/news/nunavut/70914_498.html.

289 *a paper in an obscure science journal* Willie Soon et al., "Reconstructing Climate and Environmental Changes of the Past 1000 Years: A Reappraisal," *Energy and Environment* 14, no. 2–3 (2003): 233–296, http://ruby.fgcu.edu/courses/twimberley/EnviroPhilo/Climatic.pdf.

289 *Any claim that the decline* M. G. Dyck et al., "Polar Bears of Western Hudson Bay and Climate Change: Are Warming Spring Air Temperatures the 'Ultimate' Survival Control Factor?," *Ecological Complexity* 4, no. 3 (2007): 73–84, http://ruby.fgcu.edu/courses/ twimberley/EnviroPhilo/HudsonBay.pdf.

289 *"a case, or several cases"* Peter Shaw, *The Philosophical Works of Francis Bacon, Baron of Verulam, Viscount St. Albans, and Lord High-Chancellor of England: Methodized, and Made English, from the Originals*, vol. 2 (London: J. J. and P. Knapton et al., 1733).

290 *other discredited, industry-funded* Irene Sanchez, "Warming Study Draws Fire: Harvard Scientists Accused of Politicizing Research," *Harvard Crimson*, September 12, 2003, http:// www.thecrimson.com/article/2003/9/12/warming-study-draws-fire-a-study.

290 *ethical rules about disclosure* "Guide for Authors: Conflict of Interest," Elsevier, accessed February 1, 2016, http://www.elsevier.com/wps/find/journaldescription.cws_home/701873/ authorinstructions#7000.

290 *He received more than $1.2 million* "Deception Dossier #1: Dr. Wei-Hock Soon's Smithsonian Contracts," *The Climate Deception Dossiers* (Cambridge, MA: Union of Concerned Scientists, 2015), http://www.ucsusa.org/sites/default/files/attach/2015/07/The-Climate -Deception-Dossiers.pdf.

291 *just one of many such exchanges* Kert Davies, "Willie Soon Harvard Smithsonian Documents Reveal Southern Company Scandal," Climate Investigations Center, February 21, 2015, accessed July 16, 2015, http://www.climateinvestigations.org/willie-soon-harvard -smithsonian-documents-reveal-southern-company-scandal.

291   *His many obfuscations*   Brendan DeMelle, "Disinformation Database: An Extensive
Database of Individuals Involved in the Global Warming Denial Industry," *DeSmogBlog*,
accessed February 1, 2016, http://www.desmogblog.com/global-warming-denier-database.
For a broader sampling of denialist publications, see Edward T. Wimberley, "An Archive
of Resources on Global Warming," Florida Gulf Coast University, http://ruby.fgcu.edu/
courses/twimberley/EnviroPhilo/Critical1.html.

291   *"churnalism"*   Waseem Zakir, quoted in Karin Wahl-Jorgensen and Thomas Hanitzsch,
eds., *The Handbook of Journalism Studies* (New York: Taylor and Francis, 2009).

291   *AM talk-radio hosts picked up*   Rush Limbaugh, "Global Warming Hoax: Polar Bears Are
Just Fine!," *Rush Limbaugh Show*, March 8, 2007, http://www.rushlimbaugh.com/home/
daily/site_030807/content/01125108.LogIn.html.

291   *the American Enterprise Institute*   Kenneth P. Green, "Is the Polar Bear Endangered, or Just
Conveniently Charismatic?," *Environmental Policy Outlook*, no. 2 (May 2008), American
Enterprise Institute for Public Policy Research, accessed February 1, 2016, https://www.aei
.org/publication/is-the-polar-bear-endangered-or-just-conveniently-charismatic/.

291   *the Heritage Foundation*   Conn Carroll, "Morning Bell: Blame Canada," *Daily Signal*,
May 1, 2008, http://dailysignal.com//2008/05/01/morning-bell-blame-canada. See also
Ben Lieberman, "Do Polar Bears belong on the Endangered Species List? No: Bears Are
Thriving; Greens Tread on Thin Ice," *Seattle Times*, February 21, 2008, http://www
.seattletimes.com/opinion/do-polar-bears-belong-on-the-endangered-species-list-no
-bears-are-thriving-greens-tread-on-thin-ice/.

291   *the George C. Marshall Institute*   "Climate Change," George C. Marshall Institute, accessed
February 1, 2016, http://marshall.org/climate-change/.

292   *"high-pointing"*   Clarke L. Caywood, *The Handbook of Strategic Public Relations and
Integrated Marketing Communications*, 2nd Ed. (New York: McGraw-Hill Education, 2011).

292   *slanted press materials*   "Polar Bear Politics."

293   *these letters were from*   Kevin Grandia, "Five More Forged Letters Uncovered from
Bonner & Associates' Work for DC Coal Lobby—Read Them Here," *Huffington Post*,
September 18, 2009, http://www.huffingtonpost.com/kevin-grandia/five-more-forged
-letters_b_262496.html.

293   *began with a major speech*   "Inhofe Speech on Polar Bears and Global Warming," US Senate
Committee on Environment and Public Works, January 5, 2007, accessed February 1, 2016,
http://www.epw.senate.gov/public/index.cfm/2007/1/post-f339c09a-802a-23ad-4202
-611ef8047a6b.

294   *five energy-industry organizations*   The American Petroleum Institute, the National
Association of Manufacturers, the US Chamber of Commerce, the National Mining
Association, and the American Iron and Steel Institute.

294   *joining Palin's effort*   Kari Lydersen, "Oil Group Joins Alaska in Suing to Overturn Polar
Bear Protection," *Washington Post*, August 31, 2008, http://www.washingtonpost.com/
wp-dyn/content/article/2008/08/30/AR2008083001538.html.

294   *The documentation accompanying*   Sarah Palin, "State Comments on Proposed FWS
Polar Bear Rule," Alaska Department of Fish and Game, accessed February 1, 2016,

http://www.adfg.alaska.gov/static/species/specialstatus/pdfs/polarbear_2007_soa
_comments_4_9.pdf.

294 *"While the legal standards"* Dirk Kempthorne, "Secretary Kempthorne Announces
Decision to Protect Polar Bears under Endangered Species Act: Rule Will Allow
Continuation of Vital Energy Production in Alaska," Office of the Secretary, US
Department of the Interior, May 14, 2008, http://www.doi.gov/archive/news/08_News
_Releases/080514a.html.

295 *testified before Congress* On Thin Ice: The Future of the Polar Bear, Hearing before the
Select Committee on Energy Independence and Global Warming, House of Representatives,
One Hundred Tenth Congress, Second Session (Washington, DC: US Government
Printing Office, 2010), https://www.gpo.gov/fdsys/pkg/CHRG-110hhrg58416/html/
CHRG-110hhrg58416.htm.

295 *gave nearly $25 million* Greenpeace, *Koch Industries Secretly Funding the Climate Denial
Machine* (Washington, DC: Greenpeace USA, March 2010), http://graphics8.nytimes.com/
images/blogs/greeninc/koch.pdf.

295 *American Petroleum Institute* Karoli Kuns, "Koch Industries Denies Funding Tea Parties,
but Official Filings Say Otherwise," Crooks and Liars, April 18, 2010, http://crooksandliars
.com/karoli/koch-industries-denies-funding-freedomworks.

295 *more than $2 billion* Joe Romm, "Dirty Money: Oil Companies and Special Interests
Spend Millions to Oppose Climate Legislation," *ClimateProgress*, September 27, 2010,
http://thinkprogress.org/climate/2010/09/27/206784/dirty-money-oil-companies-special
-interest-polluters-spend-millions-to-kill-climate-bil/.

295 *an estimated $73 million more* Rebecca Lefton and Noreen Nielsen, "Interactive: Big
Polluters' Big Ad Spending," Center for American Progress Action Fund, October 27,
2010, https://www.americanprogressaction.org/issues/green/news/2010/10/27/8530/
interactive-big-polluters-big-ad-spending/.

295 *gave $48 million overall* Greenpeace, *Koch Industries Secretly Funding.*

295 *published by the George C. Marshall Institute* Willie Soon and Sallie Baliunas, *Lessons
and Limits of Climate History: Was the 20th Century Climate Unusual?* (Washington, DC:
George C. Marshall Institute, 2003), accessed February 1, 2016, http://marshall.org/
wp-content/uploads/2013/08/Soon-and-Baliunas-Lessons-Limits-of-Climate-History.pdf.

296 *the 2001 report issued by* Working Group I, Intergovernmental Panel on Climate Change,
"Summary for Policymakers," in *Climate Change 2001: The Scientific Basis* (Cambridge,
UK: Cambridge University Press, 2001), accessed February 1, 2016, http://www.grida.no/
climate/ipcc_tar/wg1/pdf/WG1_TAR-FRONT.pdf.

296 *The findings by Mann* Michael E. Mann et al., "Global-Scale Temperature Patterns and
Climate Forcing over the Past Six Centuries," *Nature* 392 (April 23, 1998): 779–787, http://
www.nature.com/nature/journal/v392/n6678/abs/392779a0.html.

297 *backed the call for a science debate* To view videos of these appointees calling for a
science debate, see ScienceDebate.org's Youtube channel, https://www.youtube.com/
playlist?list=FLsb0dssXRG8sTQNqCmkkmvA.

297 *"timid"* Jon Stewart, quoted in CNN Wire Staff, "Obama Visits Jon Stewart and 'The Daily

Show' in D.C.,'' CNN, October 28, 2010, http://www.cnn.com/2010/POLITICS/10/28/
obama.daily.show/index.html.

297   *But Americans for Prosperity*   "Pledge Signers Sweep into Office," press release, No Climate
Tax, Americans for Prosperity, November 3, 2010, http://site.americansforprosperity.org/
noclimatetax/2010/11/03/pledge-signers-sweep-into-office/.

298   *annual revenues of nearly $135 billion*   "American's Largest Private Companies 2014,"
Forbes, November 5, 2014, http://www.forbes.com/sites/andreamurphy/2014/11/05/
americas-largest-private-companies-2014/. In 2015, after the oil glut, Cargill took the
number-one rank, and Koch Industries was listed as number two.

298   *AFP was one of the lead organizations*   Jane Mayer, "Covert Operations," *New Yorker*,
August 30, 2010.

298   *"We certainly did radio ads"*   "Tim Phillips: The Case against Climate Legislation,"
Frontline, October 23, 2012, http://www.pbs.org/wgbh/pages/frontline/environment/
climate-of-doubt/tim-phillips-the-case-against-climate-legislation/.

299   *did not represent a winning message*   Suzanne Goldenberg, "Revealed: The Day Obama
Chose a Strategy of Silence on Climate Change," *Guardian*, November 1, 2012, http://www
.theguardian.com/environment/2012/nov/01/obama-strategy-silence-climate-change.

299   *All of these men*   "CFACT Board of Advisors," accessed July 7, 2015, https://www.cfact
.org/about/cfact-board-of-advisors/.

299   *"a new civil war"*   "Lord Monckton Tells Obama Global Warming Is Bull Shit," audiovisual
footage, Brain Stream Media, April 15, 2010, accessed February 1, 2016, http://www
.youtube.com/watch?v=gJdRwZG5ssA.

300   *"There's nothing like a loss"*   John Kerry, quoted in Catherine Upin, "Climate of Doubt,"
Frontline, October 23, 2012, http://www.pbs.org/wgbh/pages/frontline/climate-of-doubt/.

301   *he no longer had the votes*   Gail Russell Chaddock, "Harry Reid: Senate Will Abandon
Cap-and-Trade Energy Reform," *Christian Science Monitor*, July 22, 2010, http://www
.csmonitor.com/USA/Politics/2010/0722/Harry-Reid-Senate-will-abandon-cap-and-trade
-energy-reform.

301   *climate-skeptic blogs the* Air Vent   Jeff Id, "Leaked FOIA Files 62 mb of Gold," *Air Vent*
(blog), November 19, 2009, http://noconsensus.wordpress.com/2009/11/19/leaked-foia
-files-62-mb-of-gold.

301   *and* Watts Up with That?   Anthony Watts, "Breaking News Story: CRU Has Apparently
Been Hacked—Hundreds of Files Released," *Watts Up with That?* (blog), November 19,
2009, http://wattsupwiththat.com/2009/11/19/breaking-news-story-hadley-cru-has
-apparently-been-hacked-hundreds-of-files-released.

301   *as well as the blog* RealClimate   "The CRU Hack," *RealClimate* (blog), November 20, 2009,
http://www.realclimate.org/index.php/archives/2009/11/the-cru-hack/#more-1853.

303   *"Phil Jones has gone on record"*   Michael Mann, interview by author, April 27, 2010.

303   *was subsequently independently verified*   Committee on Surface Temperature
Reconstructions for the Last 2,000 Years, National Research Council, *Surface Temperature
Reconstructions for the Last 2,000 Years* (Washington, DC: National Academies Press,
2006), http://www.nap.edu/catalog/11676.html.

303 *Briffa and his colleagues explained*   K. R. Briffa et al., "Influence of Volcanic Eruptions on Northern Hemisphere Summer Temperature over the Past 600 Years," *Nature* 393, no. 6684 (June 4, 1998): 450–455, http://www.nature.com/nature/journal/v393/n6684/abs/393450a0.html.

306 *"Take a look at the press release"*   John Bohannon, "I Fooled Millions into Thinking Chocolate Helps Weight Loss. Here's How," *io9*, May 27, 2015, http://io9.com/i-fooled-millions-into-thinking-chocolate-helps-weight-1707251800.

307 *"More people are starting"*   Rush Limbaugh, "UN Climate Change Plan Fits with Obama's Anti-Capitalism Scheme," *Rush Limbaugh Show*, March 27, 2009, http://www.rushlimbaugh.com/daily/2009/03/27/un_climate_change_plan_fits_with_obama_s_anti_capitalism_scheme.

307 *"snake-oil science"*   Sarah Palin's Facebook page, December 3, 2009, accessed February 1, 2016, https://www.facebook.com/notes/sarah-palin/mr-president-boycott-copenhagen-investigate-your-climate-change-experts/188540473434?__fns&hash=Ac3cv7OFKQcY5bke.

307 *"Too many lazy journalists"*   Mann, interview.

308 *eighteen years earlier*   "Deception Dossier #5: Coal's 'Information Council on the Environment' Sham," *The Climate Deception Dossiers* (Cambridge, MA: Union of Concerned Scientists, 2015), http://www.ucsusa.org/sites/default/files/attach/2015/07/Climate-Deception-Dossier-5_ICE.pdf.

308 *"The people who have been preaching"*   Rush Limbaugh, "Three Trees Said to Prove Warming!," *Rush Limbaugh Show*, November 24, 2009, http://www.rushlimbaugh.com/daily/2009/11/24/3_trees_said_to_prove_warming.

308 *"one of the best-funded"*   Mann, interview.

309 *"In the last few weeks"*   David Archer et al., "An Open Letter to Congress from U.S. Scientists on Climate Change and Recently Stolen Emails," Union of Concerned Scientists, December 4, 2009, http://www.ucsusa.org/assets/documents/global_warming/scientists-statement-on.pdf.

309 *"doomsday scare tactics"*   Palin, "Mr. President: Boycott Copenhagen."

310 *"It's not about science literacy"*   Matthew Nisbet, interview by author, April 22, 2010.

311 *"Given the controversy"*   Bill Sammon, quoted in Ben Dimiero, "FOXLEAKS: Fox Boss Ordered Staff to Cast Doubt on Climate Science," Media Matters for America, December 15, 2010, http://mediamatters.org/blog/2010/12/15/foxleaks-fox-boss-ordered-staff-to-cast-doubt-o/174317.

311 *the* New York Times   Clark Hoyt, "Stolen E-Mail, Stoking the Climate Debate," *New York Times*, December 5, 2009, http://www.nytimes.com/2009/12/06/opinion/06pubed.html?scp=54&sq=climategate&st=nyt.

311 *Politifact.com*   Catharine Richert, "Inhofe Claims That E-Mails 'Debunk' Science behind Climate Change," Politifact.com, December 11, 2009, http://politifact.com/truth-o-meter/statements/2009/dec/11/james-inhofe/inhofe-claims-cru-e-mails-debunk-science-behind-cl.

311 *FactCheck.org*   Jess Henig, "'Climategate': Hacked E-Mails Show Climate Scientists in a Bad Light but Don't Change Scientific Consensus on Global Warming," FactCheck.org, December 10, 2009, http://www.factcheck.org/2009/12/climategate.

311 *the Associated Press*   Seth Borenstein et al., "Review: E-Mails Show Pettiness, Not Fraud," Associated Press, December 12, 2009, http://www.msnbc.msn.com/id/34392959/ns/us _news-environment.

311 *McClatchy Newspapers*   "'Climategate' Is a Lesson in the Politics of Science," McClatchy Newspapers, December 15, 2009, http://www.mcclatchydc.com/2009/12/15/v-print/ 80663/commentary-climategate-is-a-lesson.html.

311 *"caused grave damage"*   Jim Sensenbrenner, "Sensenbrenner Urges IPCC to Exclude Climategate Scientists," Select Committee on Energy Independence and Global Warming, December 7, 2009, accessed February 1, 2016, http://republicans.globalwarming .sensenbrenner.house.gov/press/PRArticle.aspx?NewsID=2749.

312 *"exposes a highly politicized scientific circle"*   Sarah Palin, "Sarah Palin on the Politicization of the Copenhagen Climate Conference," *Washington Post*, December 9, 2009, http:// www.washingtonpost.com/wp-dyn/content/article/2009/12/08/AR2009120803402.html.

312 *"Glaciers in the Himalayas"*   Working Group II, Intergovernmental Panel on Climate Change, "The Himalayan Glaciers," in *Climate Change 2007: Impacts, Adaptation and Vulnerability* (New York: Cambridge University Press, 2007), accessed February 1, 2016, http://www.ipcc.ch/publications_and_data/ar4/wg2/en/ch10s10-6-2.html.

312 *The IPCC classed the statement*   Working Group II, Intergovernmental Panel on Climate Change, "Introduction to the Working Group II Fourth Assessment Report," in *Climate Change 2007: Impacts, Adaptation and Vulnerability* (New York: Cambridge University Press, 2007), accessed February 1, 2016, http://www.ipcc.ch/pdf/assessment-report/ar4/ wg2/ar4-wg2-intro.pdf.

313 *published by science journalist Fred Pearce*   Fred Pearce, "Flooded Out," *New Scientist*, no. 2189 (June 5, 1999).

313 *"the extrapolar glaciation of the earth"*   V. M. Koylyakov, ed., *Variations of Snow and Ice in the Past and at Present on a Global and Regional Scale* (Paris, France: International Hydrological Programme, UNESCO, 1996), http://unesdoc.unesco.org/images/0010/ 001065/106523e.pdf.

313 *"alarmist"*   Jairam Ramesh, quoted in Pallava Bagla, "Himalayan Glaciers Melting Deadline 'a Mistake,'" *BBC News*, December 5, 2009, http://news.bbc.co.uk/2/hi/south_ asia/8387737.stm.

315 *"The CRU controversy"*   US Senate Committee on Environment and Public Works Minority Staff, *"Consensus" Exposed: The CRU Controversy* (Washington, DC: US Senate, February 2010), accessed February 1, 2016, http://www.inhofe.senate.gov/download/?id=ce35055e -8922-417f-b416-800183ab7272&download=1.

316 *"People said I should go"*   Phil Jones, quoted in Aislinn Laing, "'Climategate' Professor Phil Jones 'Considered Suicide over Email Scandal,'" *Daily Telegraph*, February 7, 2010, http://www.telegraph.co.uk/earth/environment/climatechange/7180154/Climategate -Professor-Phil-Jones-considered-suicide-over-email-scandal.html.

316 *"You communistic dupe"*   Quoted in David Fogarty, "Climate Debate Gets Ugly as World Moves to Curb CO_2," Reuters, April 26, 2010, www.reuters.com/article/2010/04/26/us -climate-abuse-2-feature-idUSTRE63P00A20100426.

316 *subpoenaed the papers and e-mails* Courtenay Stuart, "Oh, Mann: Cuccinelli Targets UVA Papers in Climategate Salvo," *Hook*, April 29, 2010, accessed February 1, 2016, http://www.readthehook.com/67811/oh-mann-cuccinelli-targets-uva-papers-climategate-salvo.

316 *"knowing inconsistencies"* Ken Cuccinelli, quoted in Rosalind S. Helderman, "State Attorney General Demands Ex-Professor's Files from University of Virginia," *Washington Post*, May 4, 2010, http://www.washingtonpost.com/wp-dyn/content/article/2010/05/03/AR2010050304139.html.

316 *"The revelations of Climate-gate"* Ken Cuccinelli, "Updated Statement Regarding University of Virginia CID and Investigation," press release, Commonwealth of Virginia, Office of the Attorney General, May 19, 2010, accessed February 1, 2016, https://www.highbeam.com/doc/1G1-226751851.html.

316 *"It's totally unacceptable"* Francesca Grifo, interview by author, May 7, 2010.

316 *the book* Science as a Contact Sport   Stephen Schneider, *Science as a Contact Sport* (Washington, DC: National Geographic Society, 2009).

317 *"One e-mail I got"* Katharine Hayhoe, quoted in Katherine Bagley and Naveena Sadasiva, "Climate Denial's Ugly Side: Hate Mail to Scientists," *InsideClimate News*, December 11, 2015, http://insideclimatenews.org/news/11122015/climate-change-global-warming-denial-ugly-side-scientists-hate-mail-hayhoe-mann.

317 *from a departing Hummer* Seth Shulman, "Climate Fingerprinter," Union of Concerned Scientists, accessed November 8, 2015, http://www.ucsusa.org/global_warming/science_and_impacts/science/climate-scientist-benjamin-santer.html#.Vj_Mf66rQUE.

318 *"Certainly I'm concerned"* Will Steffen, quoted in Marian Wilkinson and Deb Masters, "The Carbon War," *Four Corners*, Australian Broadcasting Corporation, September 20, 2011, http://www.abc.net.au/4corners/stories/2011/09/15/3318364.htm.

318 *which specializes in this sort of harassment* "About," Energy & Environment Legal Institute.

318 *It was first registered* Kaye Fissinger, "Who Benefited from Outside Influence?," *Free Range Longmont*, February 12, 2010, accessed November 8, 2015, http://www.freerangelongmont.com/2010/02/12/who-benefited-from-outside-influence/.

318 *who pleaded guilty that same year* Andrew Tilghman, "Jury Returns Conviction in Case of Business Tied to Bob Schaffer," *Talking Points Memo*, May 29, 2008, http://talkingpointsmemo.com/muckraker/img-src-http-www-talkingpointsmemo-com-images-schaffer1-jpg-vspace-5-hspace-5-align-left-jury-returns-conviction-in-case-of-business-tied-to-bob-schaffer.

318 *In 2010, WTP changed its name* American Tradition Institute, IRS Form 990, 2010, accessed November 8, 2015, http://www.southernstudies.org/sites/default/files/ATI_990_2010_final.pdf.

318 *"settled facts"* Board on Atmospheric Sciences and Climate, National Research Council, *Advancing the Science of Climate Change* (Washington, DC: National Academies Press, 2010), 22, http://www.nap.edu/read/12782/chapter/1.

318 *"E&E Legal's Board of Directors"* "About," Energy & Environment Legal Institute.

318 *targeted similar standards* Shawn Lawrence Otto, "Climate Scientist Wins a Round for

America," *Huffington Post*, November 1, 2011, http://www.huffingtonpost.com/shawn
-lawrence-otto/climate-scientist-wins-a-_b_1070426.html.

318 *"so that it effectively becomes"*   Rich Porter and John Droz, "National PR Campaign
Proposal," January 3, 2012, accessed February 1, 2016, https://assets.documentcloud.org/
documents/355257/national-pr-campaign-proposal.txt.

319 *E&E broke state campaign laws*   Mike Dennison, "Ruling Says Western Tradition
Partnership Broke State Campaign Law," *Missoulian*, October 22, 2010, http://missoulian
.com/news/local/article_9b902fd2-dd81-11df-9486-001cc4c002e0.html.

319 *corruption and money laundering*   Daniel Person, "State Suggests WTP Involved in
Corruption, Money Laundering," *Bozeman Daily Chronicle*, October 22, 2010, http://www
.bozemandailychronicle.com/news/article_493daff8-dd76-11df-99b2-001cc4c03286.html.

319 *"have created a hostile environment"*   Earl Lane, "AAAS Board: Attacks on Climate
Researchers Inhibit Free Exchange of Scientific Ideas," American Association for the
Advancement of Science, June 29, 2011, accessed November 8, 2015, http://www.aaas.org/
news/aaas-board-attacks-climate-researchers-inhibit-free-exchange-scientific-ideas.

319 *"Scientists should be able"*   Grifo, interview by author.

319 *a letter to the UVA president*   American Association of University Professors to Teresa
A. Sullivan, April 14, 2011, Union of Concerned Scientists, accessed November 8, 2015,
http://www.ucsusa.org/sites/default/files/legacy/assets/documents/scientific_integrity/
Letter-to-UVA-President-Teresa-Sullivan-regarding-academic-freedom.pdf.

320 *to reject cap-and-trade*   Max Pappas, "The Election Mandate: 67 Signers of the Contract
from America Won Their House and Senate Races," FreedomWorks.org, November 3,
2010, reprinted on *Conservative Refocus* (blog), accessed February 1, 2016, http://
conservativerefocus.com/blogs/blog5.php/2010/11/03/the-election-mandate-67-signers
-of-the-contract-from-america-won-their-house-and-senate-races.

320 *the "No Climate Tax" pledge*   "Pledge Signers Sweep into Office."

320 *"I am vindicated"*   James Inhofe, quoted in Jim Snyder and Kim Chipman, "Global
Warming Skeptics Ascend in Congress," *Bloomberg Businessweek*, November 24, 2010,
http://www.businessweek.com/magazine/content/10_49/b4206033143446.htm.

320 *Forty-six of the freshmen*   Shawn Lawrence Otto, "American Denialism," ShawnOtto
.com, May 22, 2011, http://web.archive.org/web/20110619063618/http://shawnotto.com/
blog20110522.html.

320 *"Today's global warming doomsayers"*   Bill Huizenga, "On the Issues," Huizenga for
Congress, accessed February 1, 2016, http://web.archive.org/web/20101104213929/http://
huizengaforcongress.com/on-the-issues/.

320 *"George, the idea that"*   John Boehner, quoted in Kate Galbraith, "Boehner: Calling Carbon
Dioxide Dangerous Is 'Almost Comical,'" *New York Times*, April 21, 2009, http://green.blogs
.nytimes.com/2009/04/21/boehner-calling-carbon-dioxide-dangerous-is-almost-comical.

321 *"I call on my fellow Republicans"*   Sherwood Boehlert, "Science the GOP Can't Wish
Away," *Washington Post*, November 19, 2010, http://www.washingtonpost.com/wp-dyn/
content/article/2010/11/18/AR2010111806072.html.

321 *full of compromises* James Valvo and Carl Oberg, *Exposing the Special Interests behind Waxman-Markey* (Arlington, VA: Americans for Prosperity, 2009).

322 *"There needs to be some reeducation"* Douglas Holtz-Eakin, interview by author, November 22, 2010.

322 *has saved an estimated $70 billion annually* *Cap and Trade: Acid Rain Program Results* (Washington, DC: Environmental Protection Agency, 2003), accessed February 1, 2016, http://www.epa.gov/capandtrade/documents/ctresults.pdf.

324 *"Misinformation was presented as fact"* "Into Ignorance," *Nature* 471, no. 7338 (March 17, 2011): 265–266, http://www.nature.com/nature/journal/v471/n7338/full/471265b.html.

324 *a former oil distributor* "Biography," Congressman Ed Whitfield, accessed February 1, 2016, http://whitfield.house.gov/about/biography.

325 *"will continue to be a leader"* "Energy," Congressman Ed Whitfield, accessed February 1, 2016, http://whitfield.house.gov/issues/energy.

325 *legislatures in three US states* Debora MacKenzie, "Battle over Climate Science Spreads to US Schoolrooms," *New Scientist* 205, no. 2751 (March 10, 2010), https://www.newscientist.com/article/mg20527514-100-battle-over-climate-science-spreads-to-us-schoolrooms/.

325 *"The South Dakota Legislature urges"* South Dakota Legislative Assembly, 85th Session, House Concurrent Resolution No. 1009, March 2, 2009, http://legis.sd.gov/docs/legsession/2010/Bills/hcr1009p.htm.

326 *"No large stands of rice"* John Moyle, "Wild Rice in Minnesota," *Journal of Wildlife Management* 8, no. 3 (1944): 177–184.

326 *that arbitrarily increased* House of Representatives, State of Minnesota, 87th Legislative Session, House File 1010, May 18, 2011.

326 *the only permitted numbers* Alon Harish, "New Law in North Carolina Bans Latest Scientific Predictions of Sea-Level Rise," *ABC News*, August 2, 2012, http://abcnews.go.com/US/north-carolina-bans-latest-science-rising-sea-level/story?id=16913782.

326 *they were ordered to refrain* Tristram Korten, "In Florida, Officials Ban Term 'Climate Change,'" Florida Center for Investigative Reporting, March 8, 2015, http://fcir.org/2015/03/08/in-florida-officials-ban-term-climate-change/.

326 *"It's not a part of our sole mission"* Matt Adamczyk, quoted in Eric Roston, "For Some Wisconsin State Workers, 'Climate Change' Isn't Something You Can Talk About," *Bloomberg Business*, April 8, 2015, http://www.bloomberg.com/news/articles/2015-04-08/for-some-wisconsin-state-workers-climate-change-isn-t-something-you-can-talk-about.

327 *"Doomsday predictions [...] can no longer be met"* Pope Francis, *Encyclical Letter Laudato Si' of the Holy Father on Care for our Common Home* (Vatican City: Holy See, 2015), accessed July 8, 2015, http://w2.vatican.va/content/francesco/en/encyclicals/documents/papa-francesco_20150524_enciclica-laudato-si.html.

327 *two-year degree in chemistry* Thomas Reese, "Does Pope Francis Have a Master's Degree in Chemistry?," *National Catholic Reporter*, June 3, 2015, http://ncronline.org/blogs/ncr-today/does-pope-francis-have-masters-degree-chemistry.

327 *"Oh, everyone's going to ride"* James Inhofe, quoted in Suzanne Goldenberg, "Republicans'

Leading Climate Denier Tells the Pope to Butt Out of Climate Debate," *Guardian*, June 11, 2015, http://www.theguardian.com/environment/2015/jun/11/james-inhofe-republican-climate-denier-pope-francis.

329 *analysis of contributions*   Claire Moser and Matt Lee-Ashley, "The Fossil-Fuel Industry Spent Big to Set the Anti-Environment Agenda of the Next Congress," Center for American Progress, December 22, 2014, accessed July 13, 2015, https://www.americanprogress.org/issues/green/news/2014/12/22/103667/the-fossil-fuel-industry-spent-big-to-set-the-anti-environment-agenda-of-the-next-congress/.

329 *118 anti-climate front groups*   Brulle, "Institutionalizing Delay."

329 *In 1990, Democrats in the US Senate*   Author's analysis of League of Conservation Voters scorecards published for 1990 and 2014.

329 *"The church has gotten it wrong"*   Rick Santorum, quoted in Dom Giordano, "Rick Santorum on Pope Francis' Letter on Climate Change: 'Leave the Science to the Scientist,'" *CBS News*, June 1, 2015, http://philadelphia.cbslocal.com/2015/06/01/rick-santorum-on-pope-francis-letter-on-climate-change-leave-the-science-to-the-scientist/.

330 *"a hoax" and "junk science"*   Rick Santorum, quoted in Darren Samuelsohn, "Rick Santorum: Climate Change is 'Junk Science,'" *Politico*, June 9, 2011, http://www.politico.com/news/stories/0611/56599.html.

330 *"ought to be about making"*   Jeb Bush, quoted in Steve Holland and Amanda Becker, "Jeb Bush Thinks Pope Francis Should Avoid Climate Change: Stick to 'Making Us Better as People,'" Reuters, June 6, 2015, http://www.rawstory.com/2015/06/jeb-bush-thinks-pope-francis-should-avoid-climate-change-stick-to-making-us-better-as-people/.

330 *"The last fifteen years"*   Ted Cruz, quoted in Dana Bush and Deirdre Walsh, "Cruz to CNN: Global Warming Not Supported by Data," CNN, February 20, 2014, http://politicalticker.blogs.cnn.com/2014/02/20/cruz-to-cnn-global-warming-not-supported-by-data/.

330 *Cruz would also repeal*   Ted Cruz, "Sen. Cruz and Rep. Bridenstine Introduce American Energy Renaissance Act," Senator Ted Cruz official website, March 18, 2015, accessed July 13, 2015, http://www.cruz.senate.gov/?p=press_release&id=2267.

330 *"has a negative impact"*   American Energy Renaissance Act of 2014, S. 2170, 113th Cong. (2014), https://www.gpo.gov/fdsys/pkg/BILLS-113s2170is/html/BILLS-113s2170is.htm.

330 *"with the notion that some"*   Marco Rubio, quoted in Annie Rose-Strasser, "Marco Rubio Goes Full-On Climate Denier," *ClimateProgress*, May 11, 2014, accessed July 13, 2015, http://thinkprogress.org/climate/2014/05/11/3436606/marco-rubio-goes-full-on-climate-denier/.

330 *Americans for Prosperity's "no climate tax" pledge*   Americans for Prosperity website, accessed July 13, 2015, http://americansforprosperity.org/noclimatetax/pledge-takers/.

330 *"we may be warming"*   Ben Carson, quoted in John McCormick, "Ben Carson Not Convinced on Global Warming," *Bloomberg Politics*, November 26, 2014, http://www.bloomberg.com/politics/articles/2014-11-26/ben-carson-not-convinced-on-global-warming.

331 *"not sure anybody exactly"*   Rand Paul, quoted in IOP Staff, "Sen. Rand Paul at the IOP," University of Chicago Institute of Politics, April 23, 2014, accessed July 13, 2015, http://politics.uchicago.edu/news/entry/watch-sen-rand-paul-at-the-iop.

331 *"I'm not a believer in global warming"* Donald Trump, quoted in Hugh Hewitt, "Donald Trump: I Am Not a Believer in Global Warming," *Hugh Hewitt Show*, September 21, 2015, accessed October 29, 2015, https://www.youtube.com/watch?v=rfrpSzEPx7I.

331 *"In retrospect, I was over-enthusiastic"* Peter Gwynne, "Climate Change Mea Non Culpa," *Slate*, December 7, 2014, http://www.slate.com/articles/technology/future_tense/2014/12/_1975_newsweek_article_on_global_cooling_how_climate_change_deniers_use.html.

331 *but Kasich didn't think* Ben Geman, "Ohio Gov. Kasich Concerned by Climate Change, but Won't 'Apologize' for Coal," *The Hill*, May 2, 2012, http://thehill.com/policy/energy-environment/225073-kasich-touts-climate-belief-but-wont-apologize-for-coal.

331 *"I think there will be"* Lindsey Graham, quoted in Humberto Sanchez, "Will Republicans Need a New Message on Climate Change in 2016?," *WGDB* (blog), Roll Call, November 18, 2014, http://blogs.rollcall.com/wgdb/graham-urges-gop-to-take-on-climate-change-ahead-of-2016/.

332 *Some Republicans are trying a new approach* Mark Reynolds, "Give GOP a Chance to Embrace Climate Change Solutions," *Wausau Daily Herald*, July 13, 2015, http://www.wausaudailyherald.com/story/opinion/columnists/2015/07/12/give-gop-chance-embrace-climate-change-solutions/29979517/.

332 *plans to spend $889 million* Timothy Cama, "Green Billionaire's '16 Gameplan? Shame GOP on Climate Change," *The Hill*, April 6, 2015, http://thehill.com/policy/energy-environment/238026-steyer-readies-new-2016-attacks.

332 *encouraging them to follow up on an idea* Jagadish Shukla et al., "Letter to President Obama, Attorney General Lynch, and OSTP Director Holdren," September 1, 2015, accessed February 1, 2016, http://web.archive.org/web/20150920110942/http:/www.iges.org/letter/LetterPresidentAG.pdf.

333 *and Hillary Clinton* Katherine Bagley, "Hillary Clinton Joins Call for Justice Dept. to Investigate Exxon," *InsideClimate News*, October 29, 2015, http://insideclimatenews.org/news/29102015/hillary-clinton-joins-call-justice-department-investigate-exxon-RICO.

333 *"terrorism, international drug trafficking"* "About Us," Europol, accessed October 30, 2015, https://www.europol.europa.eu/content/page/about-us.

333 *saying the issue warrants further review* Margaret Harding McGill, "Full 2nd Circ. Won't Rehear EU RICO Suit Against RJ Reynolds," Law360, April 13, 2015, http://www.law360.com/articles/642254/full-2nd-circ-won-t-rehear-eu-rico-suit-against-rj-reynolds.

334 *his office had issued a subpoena* Justin Gillis and Clifford Krauss, "Exxon Mobil Investigated for Possible Climate Change Lies by New York Attorney General," *New York Times*, November 5, 2015, http://www.nytimes.com/2015/11/06/science/exxon-mobil-under-investigation-in-new-york-over-climate-statements.html.

336 *Several independent investigations* "What Do the 'Climategate' CRU Emails Tell Us?" Skeptical Science, accessed February 1, 2016, https://www.skepticalscience.com/Climategate-CRU-emails-hacked.htm.

337 *"fly lovers and people haters"* Trofim Lysenko, quoted in Oren S. Harman, "Cyril Dean Darlington: The Man Who 'Invented' the Chromosome," *Nature Reviews Genetics* 6 (January 2005): 79–85, http://www.nature.com/nrg/journal/v6/n1/box/nrg1506_BX3.html.

337 *"caste priests of ivory-tower bourgeois pseudoscience"* Trofim Lysenko, quoted in David Joravsky, *The Lysenko Affair* (Chicago: University of Chicago Press, 1970), 242.

337 *Mao was concerned that scientists* Laurence Schneider, *Biology and Revolution in Twentieth-Century China* (Lanham, MD: Rowman & Littlefield, 2005), 179.

337 *and often executed* William Harms, "China's Great Leap Forward," *University of Chicago Chronicle* 15, no. 13 (1996), http://chronicle.uchicago.edu/960314/china.shtml.

Chapter 11. Freedom and the Commons

341 *"The only way to have real success"* Richard P. Feynman, *"What Do You Care What Other People Think?": Further Adventures of a Curious Character* (New York: W. W. Norton, 2011).

342 *the American Psychological Association's collusion* James Risen, "Outside Psychologists Shielded U.S. Torture Program, Report Finds," *New York Times*, July 10, 2015, http://www.nytimes.com/2015/07/11/us/psychologists-shielded-us-torture-program-report-finds.html.

343 *"upended three decades"* Helene Cooper and Robert F. Worth, "In Arab Spring, Obama Finds a Sharp Test," *New York Times*, September 24, 2012, http://www.nytimes.com/2012/09/25/us/politics/arab-spring-proves-a-harsh-test-for-obamas-diplomatic-skill.html.

344 *Tunisia adopted a constitution* "The Constitution of Tunisia," accessed October 15, 2015, http://confinder.richmond.edu/admin/docs/Tunisiaconstitution.pdf.

344 *"watershed moment"* Ayman Mohyeldin, "Beji Caid Essebsi: Tunisia's Unifying Leader," *Time*, April 16, 2015, http://time.com/3823201/beji-caid-essebsi-2015-time-100/.

344 *"Unprecedented levels of global connectedness"* Martin E. Dempsey, *Quadrennial Defense Review 2014* (Washington, DC: US Department of Defense, March 4, 2014), http://archive.defense.gov/pubs/2014_Quadrennial_Defense_Review.pdf.

345 *"the tragedy of the commons"* Garrett Hardin, "The Tragedy of the Commons," *Science* 162, no. 3859 (December 13, 1968): 1243–1248, www.sciencemag.org/content/162/3859/1243.full.

346 *the only "rational" economic conclusion* Ayn Rand, *The Virtue of Selfishness* (New York: Signet, 1964).

347 *writing the Declaration of Independence* Hume, "An Enquiry Concerning Human Understanding."

348 *"In the Gulf I see this"* McNutt, interview by author.

352 *"widely regarded as sufficiently important"* Milton Friedman, *Capitalism and Freedom* (Chicago: University of Chicago Press, 1962), 34.

353 *"By creating new knowledge"* Jane Lubchenco, interview by author, March 28, 2010.

353 *"I see a real split"* Levin, interview by author.

354 *I'm responsible for a burden* The EPA estimates that, in 2007, discarded electronics totaled about 2.25 million US tons. By 2010, it was about 3.3 million. Placing that on a very rough growth curve going back to 1980 gives a figure of about 27 million US tons over that time. At 2,000 pounds in a ton and a population of about 300 million people, each American is responsible for roughly 180 pounds of electronic waste. In China, e-waste in 2010 totaled about 2.5 million tons—fast approaching US levels—and that country has more than four times the population.

354 *generated in the rest of the world* "Hazardous E-Waste Surging in Developing Countries," ScienceDaily, February 23, 2010, www.sciencedaily.com/releases/2010/02/100222081911 .htm.

354 *"The Nation behaves well"* Teddy Roosevelt, quoted in Hugh Rawson and Margaret Miner, *The Oxford Dictionary of American Quotations* (New York: Oxford University Press, 2006), 223.

356 *embarked on a quest for metrics* Robert Costanza et al., "The Value of the World's Ecosystem Services and Natural Capital," *Nature* 387, no. 6630 (May 15, 1997): 253–260, http://www.nature.com/nature/journal/v387/n6630/full/387253a0.html.

356 *The economics behind this valuation* Wade Roush, "Putting a Price Tag on Nature's Bounty," *Science* 276, no. 5315 (May 16, 1997): 1029, http://science.sciencemag.org/ content/276/5315/1029.

356 *"Within a decade"* Stuart Pimm, interview by author, September 7, 2010.

357 *"We deny . . . that Earth's climate system"* Cornwall Alliance, "An Evangelical Declaration on Global Warming," May 1, 2009, http://www.cornwallalliance.org/2009/05/01/ evangelical-declaration-on-global-warming/.

357 *"Many are concerned that liberty"* Cornwall Alliance, "The Cornwall Declaration on Environmental Stewardship," May 1, 2000, http://www.cornwallalliance.org/docs/the -cornwall-declaration-on-environmental-stewardship.pdf.

358 *"The wheels of government"* Robert May, interview by author, June 1, 2011.

358 *"It speaks to a very deep division"* Pimm, interview.

359 *Fewer than 30 percent* Natalie Avon, "Why More Americans Don't Travel Abroad," CNN, February 4, 2011, http://www.cnn.com/2011/TRAVEL/02/04/americans.travel .domestically/.

359 *"you get maybe eight people"* Jonathan Moreno, interview by author, August 23, 2010.

359 *"The idea that we can go"* Pimm, interview.

359 *costly storms from global warming* Evan Mills et al., "Availability and Affordability of Insurance under Climate Change: A Growing Challenge for the United States," *Journal of Insurance Regulation* 25, no. 2 (Winter 2006): 109–150, http://evanmills.lbl.gov/pubs/pdf/ jir-zb-25-02.pdf.

359 *estimates that the United States* Richard Feely, interview by author, November 4, 2010.

359 *about $17.7 trillion* Table 1.1.5: Gross Domestic Product, Bureau of Economic Analysis, http://www.bea.gov/national/txt/dpga.txt.

360 *particularly in coastal areas* Mills, "Availability and Affordability of Insurance."

360 *global natural disasters had risen* Ernst Rauch, in *2014 Natural Catastrophe Year in Review*, Munich Re, January 7, 2015, accessed July 12, 2015, http://www.munichre.com/ us/property-casualty/events/webinar/2015-01-natcatreview/index.html.

360 *"both too much precipitation"* Carl Hedde, in *2014 Natural Catastrophe Year in Review*, Munich Re, January 7, 2015, accessed July 12, 2015, http://www.munichre.com/us/ property-casualty/events/webinar/2015-01-natcatreview/index.html.

361 *"this month has been the warmest"* Peter Höppe, *2014 Half-Year Natural Catastrophe*

Review, Munich Re, July 9, 2014, accessed July 12, 2015, http://www.munichre.com/us/property-casualty/events/webinar/2014-07-natcatreview/index.html.

361   *expected to reach 9.6 billion*   "World Population Projected to Reach 9.6 Billion by 2050," United Nations Department of Economic and Social Affairs, June 13, 2013, http://www.un.org/en/development/desa/news/population/un-report-world-population-projected-to-reach-9-6-billion-by-2050.html.

361   *now lives in coastal regions*   "UN Atlas of the Oceans," accessed July 12, 2015, http://www.oceansatlas.org/.

361   *sea levels could rise by 6 meters*   A. Dutton et al., "Sea-Level Rise due to Polar Ice-Sheet Mass Loss During Past Warm Periods," *Science* 349, no. 6244 (July 10, 2015), http://www.sciencemag.org/content/349/6244/aaa4019.

362   *"The best definition"*   John Hubble, quoted in Christianson, *Edwin Hubble*, 23.

362   *the best way to handle it*   Robert Costanza et al., "The Perfect Spill: Solutions for Averting the Next Deepwater Horizon," *Solutions*, June 16, 2010, www.thesolutionsjournal.com/node/629.

363   *an idea that has been developing*   Ibid.

364   *has now shown a return on investment*   Andrew Balmford et al., "Economic Reasons for Conserving Wild Nature," *Science* 297, no. 5583 (August 9, 2002): 950–953, http://science.sciencemag.org/content/297/5583/950.

Chapter 12. Battle Plans

369   *"Scientists would see no reason why"*   Snow, *The Two Cultures*.

371   *The summer of 2014 was the hottest*   "Global Land and Ocean Temperature Anomalies, June–August 1880–2015," US National Oceanographic and Atmospheric Administration, accessed October 12, 2015, http://www.ncdc.noaa.gov/cag/time-series/global/globe/land_ocean/3/8/1880-2015.

371   *"It's a testament to how powerful"*   Ricken Patel, quoted in Lisa W. Foderaro, "Taking a Call for Climate Change to the Streets," *New York Times*, September 21, 2014, http://www.nytimes.com/2014/09/22/nyregion/new-york-city-climate-change-march.html.

371   *What made the march so successful*   "Social Movement Expansion at the People's Climate March," University of Maryland Program for Society and the Environment, October 12, 2015, accessed October 12, 2015, http://www.cse.umd.edu/blog/social-movement-expansion-at-the-peoples-climate-march.

372   *by far the hottest June*   "Global Land and Ocean Temperature Anomalies, June–August 1880–2015."

376   *American Indian tribes are using*   Audrea Lim, "How First Nations in Canada Are Winning the Fight against Big Oil," *Nation*, September 10, 2014, http://www.thenation.com/article/how-first-nations-canada-are-winning-fight-against-big-oil/.

383   *"Public sentiment [. . .] is everything"*   Abraham Lincoln, in *The Complete Lincoln-Douglas Debates of 1858*, ed. Paul M. Angle (Chicago: University of Chicago Press, 1991), 128.

384   *Most Americans (87 percent)*   Shawn Lawrence Otto and Mary Woolley, *Americans' Views*

on Science and the Candidates (Research!America and ScienceDebate.org., September 2015), http://sciencedebate.org/goods/Poll2015.pdf.

386 *Karl Popper frequently warned against*   Karl R. Popper, *The Myth of the Framework: In Defence of Science and Rationality*, rev. ed. (New York: Routledge, 1996).

388 *abolished its national science advisor position*   Hannah Hoag, "Canada Abolishes Its National Science Adviser," *Nature* 451, no. 7178 (January 30, 2008), http://www.nature.com/news/2008/080130/full/451505a.html.

388 *"We're making two recommendations"*   Katie Gibbs, interview by author, July 28, 2015.

389 *"A science advisor is not just a scientist"*   Peter Gluckman, interview by author, July 19, 2015.

389 *same as that of Texas*   Alan Johnson, *Striking a Better Balance: A State of the Nation Report from the Salvation Army* (Manukau, New Zealand: Salvation Army Social Policy and Parlimentary Unit, February 2014), 27, http://www.salvationarmy.org.nz/sites/default/files/uploads/20140211SONStriking%20a%20Better%20Balance%20-%20Final%20Web.pdf.

392 *"It's fairly arrogant"*   Peg Chemberlin, interview by author, November 23, 2010.

392 *"Science and religion"*   Pope Francis, *Encyclical Letter Laudato Si'*.

392 *"Science and technology are wonderful products"*   Pope John Paul II, "Address to Scientists and Representatives of the United Nations," University of Hiroshima, February 25, 1981.

392 *"expresses the inner tension"*   Benedict XVI, *Encyclical Letter Caritas in Veritate* (Vatican City: Holy See, 2009), accessed July 14, 2015, http://w2.vatican.va/content/benedict-xvi/en/encyclicals/documents/hf_ben-xvi_enc_20090629_caritas-in-veritate.pdf.

394 *"Harshness, mockery, and intolerance"*   Westphal, interview.

397 *Chinese parents also are more likely*   Huabin Chen, "Parents' Attitudes and Expectations Regarding Science Education: Comparisons among American, Chinese-American, and Chinese Families," *Adolescence* 36, no. 142 (2001), http://www.csub.edu/~rhewett/english100/chen.pdf.

397 *has been borne out by other studies*   Dave Breitenstein, "Asian Students Carry High Expectations for Success," *Fort Myers News-Press*, August 4, 2013, http://www.usatoday.com/story/news/nation/2013/08/04/asian-students-carry-high-expectations-for-success/2615483/.

397 *parental involvement in kids' science homework*   Liyanage Devangi H. Perera et al., "Parents' Attitudes towards Science and their Children's Science Achievement," discussion paper, Monash University Department of Economics, February 2014, accessed February 1, 2016, http://www.buseco.monash.edu.au/eco/research/papers/2014/0214parespererabomhofflee.pdf.

399 *initially made by UK science advisor Sir David King*   Pallab Ghosh, "UK Science Head Backs Ethics Code," *BBC News*, September 12, 2007, http://news.bbc.co.uk/2/hi/science/nature/6990868.stm.

402 *antismoking laws around the world*   Danny Hakim, "U.S. Chamber of Commerce Works Globally to Fight Antismoking Measures," *New York Times*, June 30, 2015, http://www.nytimes.com/2015/07/01/business/international/us-chamber-works-globally-to-fight-antismoking-measures.html.

402 *quit its membership*   Phil Whaba, "CVS Health Leaving U.S. Chamber of Commerce over

Smoking Spat," *Fortune*, July 8, 2015, http://fortune.com/2015/07/08/cvs-chamber -cigarettes/.

403 *"The mission of the Department of State"* National Research Council, *Diplomacy for the 21st Century*.

405 *once described changing the culture* Jane Lubchenco, comment on a panel at the CapSci Conference organized by the author, National Science Foundation, March 28, 2010.

409 *"The most surprising thing I found"* Mary Lynk, interview by author, July 26, 2015.

411 *97 percent consensus* Earth Science Communications Team, "Consensus: 97% of Climate Scientists Agree," National Aeronautics and Space Administration, accessed July 29, 2015, http://climate.nasa.gov/scientific-consensus/.

411 *man-made climate change is happening* John Cook et al., "Quantifying the Consensus on Anthropogenic Global Warming in the Scientific Literature," *Environmental Research Letters* 8, no. 2 (May 15, 2013), http://iopscience.iop.org/1748-9326/8/2/024024/article.

411 *debunk climate-denier myths* John Cook and Stephan Lewandowsky, *The Debunking Handbook* (St. Lucia, Australia: University of Queensland, January 23, 2012), http://www .skepticalscience.com/docs/Debunking_Handbook.pdf.

411 *believing conspiracy theories* John Cook and Stephan Lewandowsky, "Recursive Fury: Facts and Misrepresentations," Skeptical Science, March 21, 2013, http://www.skepticalscience .com/Recursive-Fury-Facts-misrepresentations.html.

411 *deniers on the defensive* Joseph Bast and Roy Spencer, "The Myth of the Climate Change '97%,'" *Wall Street Journal*, May 26, 2014, http://www.wsj.com/articles/SB1000142405270 2303480304579578462813553136.

Chapter 13. Truth and Beauty

413 *"Believe in yourself"* Eleanor Roosevelt, quoted in *It Seems to Me: Selected Letters of Eleanor Roosevelt*, ed. Leonard C. Schlup and Donald W. Whisenhunt, (Lexington, KY: University Press of Kentucky, 2001).

413 *but he ran anyway* David Sanders, interview by author, December 14, 2010.

416 *a common argument of climate-science deniers* Christopher Monckton, *Greenhouse Warming? What Greenhouse Warming?* (Washington, DC: Science and Public Policy Institute, September 2007), http://scienceandpublicpolicy.org/images/stories/papers/ monckton/whatgreenhouse/moncktongreenhousewarming.pdf.

417 *outside of Pasadena, California* Christianson, *Edwin Hubble*.

417 *inspired Noyes to write* Alfred Noyes, *The Torch Bearers: Watchers of the Sky* (New York: Frederick A. Stokes, 1922).

417 *"fantasy of a utopian laboratory"* Robert R. Wilson, *Starting Fermilab* (Batavia, IL: Fermi National Accelerator Laboratory, 1992), http://history.fnal.gov/GoldenBooks/ gb_wilson2.html.

418 *"Is there anything here"* John Pastore and Robert Wilson, quoted in Mike Perricone, "Some Words of Wisdom," *Fermi News* 23, no. 2 (January 28, 2000), www.fnal.gov/pub/ ferminews/ferminews00-01-28/p3.html.

419   *"did not have to be cheap"*   Wilson, *Starting Fermilab.*

419   *an eloquent criticism*   Stanley Fish, "The Value of Higher Education Made Literal," *New York Times*, December 13, 2010, http://opinionator.blogs.nytimes.com/2010/12/13/the -value-of-higher-education-made-literal.

419   *"Our proposals put students"*   John Browne et al., *Securing a Sustainable Future for Higher Education: An Independent Review of Higher Education Funding and Student Finance* (October 12, 2010), https://www.gov.uk/government/uploads/system/uploads/ attachment_data/file/422565/bis-10-1208-securing-sustainable-higher-education -browne-report.pdf.

421   *safeguarding intellectual and academic freedom*   Jennifer Washburn, "Academic Freedom and the Corporate University," *Academe* 97, no. 1, (January–February 2011), http://www .aaup.org/article/academic-freedom-and-corporate-university.

421   *"Surgeons today have seen"*   Henry Buchwald, interview with the author, October 2015.

422   *"Science is no longer"*   Sanders, interview.

424   *"And that's about all I have"*   Ronald Reagan, "Farewell Address to the Nation," Miller Center of Public Affairs, January 11, 1989, http://millercenter.org/scripps/archive/speeches/ detail/3418.

424   *"My colleagues think"*   Sanders, interview.

424   *"I think that the American ethos"*   McKee, interview.

425   *the sixth mass extinction*   Ceballos, "Accelerated Modern Human–Induced Species Losses."

425   *the death of the oceans*   Alex Renton, "The Disaster We've Wrought on the World's Oceans May Be Irrevocable," *Newsweek*, July 2, 2014, http://www.newsweek.com/2014/07/11/ disaster-weve-wrought-worlds-oceans-may-be-irrevocable-256962.html.

425   *the end of antibiotics*   Tom Frieden, "The End of Antibiotics. Can We Come Back from the Brink?," US Centers for Disease Control and Prevention, May 5, 2014, http://blogs.cdc .gov/cdcdirector/2014/05/05/the-end-of-antibiotics-can-we-come-back-from-the-brink/.

425   *the robosourcing of jobs*   For one example of how this is beginning to occur, see Grant, "Driverless Trucks Could Mean 'Game Over' for Thousands of Jobs."

425   *the deployment of killer robots*   "Autonomous Weapons."

INDEX

Erika Ludwig

SHAWN OTTO is an award-winning science advocate, writer, teacher, and speaker. He is the cofounder of ScienceDebate.org and the producer of the US Presidential Science Debates, for which he received the IEEE-USA's National Distinguished Public Service Award. He has advised science debate efforts in several countries. He is also a novelist and filmmaker. His novel, *Sins of Our Fathers*, was a finalist for the Los Angeles Times Book Prize, and his film *House of Sand and Fog* was nominated for three Academy Awards. He lives in Minnesota with his wife, Rebecca Otto, in a wind-powered, green, solar home he designed and built with his own hands. The couple have one son, Jacob.

Interior design by Connie Kuhnz
Typeset in Warnock Pro
by Bookmobile Design and Digital Publisher Services